Ecological Genetics and Air Pollution

George E. Taylor, Jr.   Louis F. Pitelka
Michael T. Clegg
Editors

Ecological Genetics and Air Pollution

With 34 Illustrations

Springer-Verlag
New York Berlin Heidelberg London
Paris Tokyo Hong Kong Barcelona

George E. Taylor, Jr.
Desert Research Institute and
Department of Environmental and
Resource Sciences
University of Nevada, Reno
Reno, Nevada 89506-0220, USA

Louis F. Pitelka
Ecological Studies Program
Electric Power Research Institute
3412 Hillview Avenue
Palo Alto, California 94303, USA

Michael T. Clegg
Department of Botany and Plant Science
University of California-Riverside
Riverside, California 92521, USA

Library of Congress Cataloging-in-Publication Data
Ecological genetics and air pollution/George E. Taylor, Jr., Louis F. Pitelka, Michael T.
Clegg, editors.
    p.   cm.
  Papers from a symposium held in Phoenix, Ariz., Apr. 1989.
  Includes bibliographical references and index.
  ISBN 0-387-97414-8 (alk. paper)
  1. Plants, Effect of air pollution on—Genetic aspects—Congresses.   2. Plant ecological
genetics—Congresses. I. Taylor, George E.   II. Pitelka, Louis F.   III. Clegg, Michael T.,
1941– .
QK751.E27   1991
581.5'222—dc20   90-10209

Printed on acid-free paper.

Typeset by Best-set Typesetter Ltd., Chai Wan, Hong Kong.
Printed and bound by Edwards Brothers, Inc., Ann Arbor, Michigan.
Printed in the United States of America.

9 8 7 6 5 4 3 2 1

ISBN 0-387-97414-8 Springer-Verlag New York Berlin Heidelberg
ISBN 3-540-97414-8 Springer-Verlag Berlin Heidelberg New York

Preface

The theme of this book is the role of air pollution in governing the genetic structure and evolution of plant species. The book is the product of a symposium convened in Phoenix, Arizona, in April 1989. Its goal is to explore applied and basic issues for which air pollution is recognized as an existing or potential forcing function underlying the distribution of genotypes in both natural and managed ecosystems. This objective is addressed by presenting an initial suite of background chapters (Chapters 1–3), a progression of disciplinary topics (Chapters 4–10), and a final chapter (Chapter 11) on the application of knowledge in both the applied and basic sciences. The disciplinary contributions encompass atmospheric chemistry, genetics of inheritance, population-level processes and methodologies, plant physiological ecology, and molecular genetics.

One of the most significant challenges to the success of the symposium and book was to integrate the diversity of disciplines, ranging from molecular genetics to plant ecology to atmospheric chemistry. To foster interaction among disciplines, the symposium and now this book have the unusual feature of two types of contributions. The first and more traditional are the 11 commissioned chapters, the objective of which is to present authoritative, critical analyses of the subject. The ten commentaries constitute the second type of contribution. Their objective is to further explore and challenge aspects of the subject, seizing upon the dissimilar perspectives offered by the array of disciplines in attendance at the symposium. After the symposium's discussions, the commentators penned their contributions to reflect the discussions, concentrating less on their personal analysis and more on interdisciplinary perspectives. Commentaries were developed for Chapters 4–11, which constitute the technical contributions of the volume.

The editors express their appreciation to several individuals and organizations for their support. Norma F. Caldwell of Oak Ridge National Laboratory handled the organizational aspects of the symposium in Phoenix. Whereas the role of authors is acknowledged by their contributions to the book, a number of other invited individuals participated in the discussions,

v

and their contributions are anonymously reflected in the commentaries. These individuals were Patricia M. Irving (National Acid Precipitation Assessment Program), Hector Flores (Pennsylvania State University), Ron Robberecht (University of Idaho), Ann M. Bartuska (U.S. Forest Service) and Don H. DeHayes (University of Vermont). Each of the contributions, authoritative chapters and commentaries, was peer reviewed, and the manuscripts benefited significantly from the review process. The editorial and production staff at Springer-Verlag were instrumental in keeping the volume on schedule. The symposium was supported by the Ecological Studies Program of the Electric Power Research Institute.

George E. Taylor, Jr.
Louis F. Pitelka
Michael T. Clegg

Contents

Contributors

CHRISTIAN P. ANDERSEN
U.S. Environmental Protection Agency, Environmental Research Laboratory, Corvallis, Oregon 97330, USA

MIKE R. ASHMORE
Department of Pure and Applied Biology, Imperial College, Silwood Park, Ascot, Berkshire SL5 7PY, UK

SPENCER C.H. BARRETT
Department of Botany, University of Toronto, Toronto, M5S 1A1, Canada

J. NIGEL B. BELL
Department of Pure and Applied Biology, Imperial College, Silwood Park, Ascot, Berkshire SL5 7PY, UK

ANTHONY D. BRADSHAW
Department of Botany, University of Liverpool, Liverpool L69 3BX, UK

ELIZABETH J. BUSH
Department of Botany, University of Toronto, Toronto, M5S 1A1, Canada

MICHAEL T. CLEGG
Department of Botany and Plant Science, University of California-Riverside, Riverside, California, 92521, USA

JAMES S. COLEMAN
Department of Organismic and Evolutionary Biology, Biological Laboratories, Harvard University, Cambridge, Massachusetts 02138, USA

CHRIS A. CULLIS
Biology Department, Case Western Reserve University, Cleveland, Ohio 44106, USA

JAMES R. EHLERINGER
Department of Biology, University of Utah, Salt Lake City, Utah 84112, USA

CHRISTOPHER GILLESPIE
Department of Plant Pathology and Physiology, Virginia Polytechnic Institute and State University, Blacksburg, Virginia 24067, USA

SUSAN KALISZ
Kellogg Biological Station, Michigan State University, Hickory Corners, Michigan 49060, USA

DAVID F. KARNOSKY
School of Forestry, Michigan Technological University, Houghton, Michigan 49931, USA

PATRICIA A. LAYTON
Scott Paper Company, Scott Plaza, Philadelphia, Pennsylvania 19113, USA

ALAN A. LUCIER
National Council of the Paper Industry for Air and Stream Improvement, New York, New York 10016, USA

MARK R. MACNAIR
Department of Biological Sciences, University of Exeter, Exeter EX4 4PS, UK

SAMUEL B. MCLAUGHLIN
Environmental Sciences Division, Oak Ridge National Laboratory, Oak Ridge, Tennessee 37831-6038, USA

T. MCNEILLY
Department of Environmental and Evolutionary Biology, University of Liverpool, Liverpool L69 3BX, UK

HAROLD A. MOONEY
Department of Biological Sciences, Stanford University, Stanford, California 93405, USA

RONALD T. NAGAO
Botany Department, University of Georgia, Athens, Georgia 30602, USA

RICHARD J. NORBY
Environmental Sciences Division, Oak Ridge National Laboratory, Oak Ridge, Tennessee 37831-6038, USA

DAVID J. PARSONS
National Park Service, Three Rivers, California 93271, USA

EVA J. PELL
211 Buckhout Laboratory, Pennsylvania State University, University Park, Pennsylvania 16802, USA

LOUIS F. PITELKA
Ecological Studies Program, Electric Power Research Institute, Palo Alto, California 94303, USA

H. JAMES PRICE
Department of Soils and Crop Science, Texas A&M University, College Station, Texas 77843, USA

T. MIKE ROBERTS
Institute of Terrestrial Ecology, Monks Woods Experiment Station, Abbots Ripton, Huntingdon PE17 2LS, UK

MIKEAL L. ROOSE
Department of Botany and Plant Science, University of California-Riverside, Riverside, California 92521, USA

FLORIAN SCHOLZ
Institut für Forstgenetik und Forstpflanzenzüchtung der Bundesforschungsanstalt für Forst und Holzwirtschaft, 2070 Großhansclorfz, Federal Republic of Germany

J. JONATHAN SHAW
Department of Biology, Ithaca College, Ithaca, New York 14850, USA

BOYD R. STRAIN
Department of Botany, Duke University, Durham, North Carolina 27706, USA

DUNCAN R. TALBOT
Department of Molecular and Cell Biology, University of Connecticut, Storrs, Connecticut 06269-3125, USA

GEORGE E. TAYLOR, JR.
Desert Research Institute and Department of Environmental and Resource Sciences, University of Nevada, Reno, Reno, Nevada 89506-0220, USA

DAVID T. TINGEY
U.S. Environmental Protection Agency, Environmental Research Laboratory, Corvallis, Oregon 97330, USA

STEPHEN J. TONSOR
Kellogg Biological Station, Michigan State University, Hickory Corners, Michigan 49060, USA

G.B. WILSON
Department of Pure and Applied Biology, Imperial College, Silwood Park, Ascot, Berkshire SL5 7PY, United Kingdom

WILLIAM E. WINNER
Department of General Science, Oregon State University, Corvallis, Oregon 97331-6505, USA

1
Introduction

GEORGE E. TAYLOR, JR., LOUIS F. PITELKA, and
MICHAEL T. CLEGG

The chemical and physical properties of the atmosphere are changing as a consequence of anthropogenic activities. Notable examples are increases in tropospheric ozone (O_3), carbon dioxide (CO_2), oxides of nitrogen (NO_x) and sulfur (SO_x), ammonia (NH_3), particles, heavy metals, ultraviolet radiation in the 290 to 320 nm range (UV-B radiation), multiple ions in wet deposition, and organic vapors. Some of these are likely to decline in importance over the next several decades (e.g., NO_2, SO_2), whereas others will remain near current levels (e.g., heavy metals) or increase (e.g., UV-B, O_3, and CO_2). From an ecologic perspective, there is concern that many of these changes influence terrestrial landscapes through effects on plant growth and development or modification of ecosystem processes. These anthropogenic stresses supplement and potentially interact with the suite of naturally occurring factors that govern the productivity of terrestrial ecosystems.

These atmospheric modifications can be unusual in their intensity and duration and, in some cases, chemically unique in their mode of action. Whereas few if any of these stresses are entirely novel in terrestrial landscapes, many now occur at higher levels and over broader geographic areas than ever before. Furthermore, they have become more important and pervasive stresses over a relatively short period of time from an evolutionary perspective. Thus, they may represent a novel evolutionary challenge for sensitive plants in natural and managed ecosystems. Concerns about these changes are reflected in the variety of national and international research programs designed to understand how these changes, singly or in combination, are affecting the earth's biotic resources.

One of the outstanding features of the response of terrestrial vegetation to many of the anthropogenic changes in the atmosphere is the variable phenotypic responses at both the interspecific and intraspecific level. This feature is manifest following both acute and chronic exposure regimes and is characteristic of response parameters ranging from visual leaf necrosis to a mulititude of physiological and biochemical traits. Through carefully

1

controlled studies, it has been repeatedly documented that a major component underlying the breadth of phenotypic expression is the genotype. Such genetic variation is common to a variety of life forms (e.g., woody perennials, herbaceous annuals), occurs in both domesticated and native species, and exists in response to all of the principal atmospheric modifications that have been demonstrated to influence the physiology, growth, and productivity of terrestrial plants.

Given the existence of an intrinsic pool of heritable variation within plant species, it is hypothesized that anthropogenic changes in the atmosphere are acting differentially within populations as sensitive individuals are less able to grow and compete for limited resources. As a result, the population's genetic structure is shifting progressively over time as sensitive individuals are lost from the population or contribute less to the gene pool of subsequent generations. Whereas it is commonplace to assume that the genetic structure of most plant populations reflects environmental stresses of natural origin (Clausen et al. 1940), it is apparent that the genetic structure may also be evolving as a consequence of anthropogenic factors (Bradshaw 1952; Bishop and Cook 1981; Pitelka 1988; Scholz et al. 1988). Thus, genetic heterogeneity in space and time of sensitive plant species will be characterized by "pockets" or clines of genetic variation reflecting changes in the physical or chemical properties of the atmosphere. These anthropogenic modifications are of recent origin and potentially present a novel "experimental stage" on which natural selection can interact with genetic variation.

The objective of this book is to explore the issue of ecological genetics, terrestrial vegetation, and air pollution stress and to identify important areas for future research. The volume addresses issues ranging from basic (e.g., evolutionary biology of plant populations) to applied (e.g., biotechnological implications of genetic variation in pollution response).

Issues

A number of important issues and considerations bear on the topic of ecological genetics and air pollution stress. Each is discussed below, and collectively they serve as a platform for recognizing the value of investigating microevolutionary responses to air pollutants.

Regional and Global Distribution of the Stress

Most studies of ecotypic differentiation in response to anthropogenic stresses have focused on intense selection pressure operating over small spatial scales (e.g., heavy metal tolerance on mine spoils) so that changes in genetic structure are restricted to scales equal to or less than meters (Antonovics et al. 1971; McNair 1981). In contrast, many changes in atmospheric chemistry due to anthropogenic activities are not highly

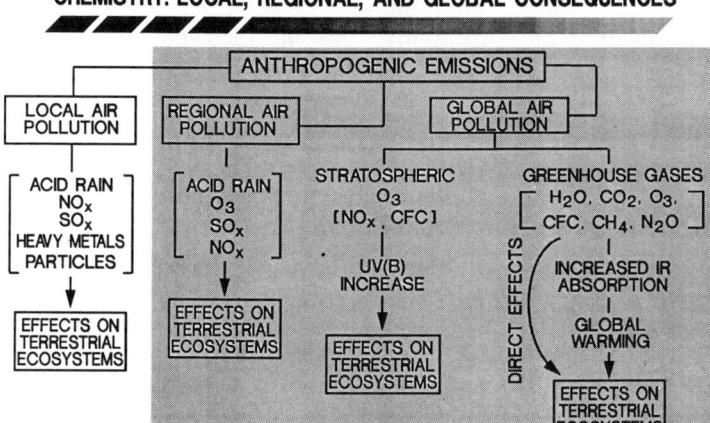

FIGURE 1. Linkages among changes in atmospheric chemistry, ranging from effects on local and regional air sheds to impacts on global tropospheric and stratospheric processes. NO_x = oxides of nitrogen; SO_x = oxides of sulfur; O_3 = ozone; CFC = chlorofluorocarbons; H_2O = water vapor; CO_2 = carbon dioxide; CH_4 = methane; N_2O = nutrous oxide; IR = infrared radiation. (Adapted from A.E. Green, personal communication.)

localized but rather are distributed in a gradient fashion from a point source (e.g., SO_2, heavy metals, NH_3), uniformly across regions (e.g., O_3, acidic precipitation), or over the entire globe (e.g., CO_2, UVB). Many of these regional and global modifications are also linked mechanistically with anticipated changes in global climate (Fig. 1). Thus, microevolutionary responses involving an increase or descrease in the frequency of alleles conferring resistance to a pollutant may extend beyond a single population and potentially to the entire distribution of a species. Genetic responses of plants to pollutants were once primarily of academic interest, but they now must be considered as an important component of the response of terrestrial vegetation to human activities.

Exposure Dynamics and Measures

Many natural selective agents that affect the growth, distribution, and abundance of plants have been characterized by ecologists in terms of their spatial and temporal patterns and their modes of action. This is not the case for most of the anthropogenic modifications of the atmosphere. For instance, it is only in recent years that the spatial and temporal patterns of phenomena such as O_3 and acidic deposition have been documented, particularly for rural areas while those of others (e.g., UVB) are poorly known. Among the changes in atmospheric chemistry, there are tremendous differences in both spatial and temporal patterns of exposure. At one extreme, atmospheric CO_2 is increasing at a relatively slow and constant rate and is invariant over large spatial scales. Heavy metals concentrations in soils can vary dramatically over short distances but at any particular location are essentially constant over time. Conversely, air pollutants such as O_3, acidic deposition, and UVB exhibit tremendous temporal variation, where episodes of high exposure levels can be followed by prolonged periods of intermediate to low concentrations (e.g., Hogsett et al. 1988). For some pollutants, and certainly for many parts of the world, scientists are only now documenting the exposures to natural vegetation. An even more critical problem, in contrast to the situation with some natural stresses, is that biologists do not fully understand the critical components of the exposure regime. Is a low chronic level of exposure more or less harmful or effective as a selection pressure than occasional acute exposures? Without a better understanding of the mechanisms by which the pollutants alter plant metabolism, it will remain difficult to identify relevant measures of exposure.

Toxic Versus Beneficial Changes in Atmospheric Chemistry

Whereas a variety of anthropogenic changes in the atmosphere are recognized by the community of atmospheric chemists, only a subset are relevant to the physiology and growth of terrestrial vegetation. Of this subset, some compounds elicit toxic effects causing an inhibition of physiological function; this subset includes tropospheric O_3, UV-B, and heavy metals. Another class of compounds enhances physiological function and is recognized as being "beneficial," supplying an element that exists in the environment below optimum levels for growth and reproductive success. The two most notable examples of this class are CO_2 and the multiple forms of wet and dry deposited nitrogen (NO_2, nitric acid vapor, NH_3, and nitrate in cloudwater and rainfall). Although these "beneficial" compounds may ameliorate an otherwise growth limiting condition, it is important to recognize that they constitute an environmental stress in an ecological context, altering an individual's relationship with its environment. Moreover, as the concentration approaches and exceds an optimum level for growth, further increases to supraoptimal levels may lead to toxicity, as documented for nitrogen and sulfur. Consequently, both classes of anthropogenic changes in the atmosphere are important in terms of microevolutionary responses and ecological effects.

Indirect Versus Direct Effects and the Concept of Physiological Costs

Whereas direct effects of air pollutants on terrestrial vegetation are well documented in intensively managed ecosystems (Guderian 1985; Winner et al. 1985; Heck et al. 1988), indirect effects are thought to be far more important for vegetation in natural ecosystems (Johnson and Taylor 1990). This latter category of effects is one in which pollutant-induced modifications of physiological processes influence the plant's ability to either (1) compete for limited resources (e.g., nutrients, water, light) or (2) withstand the mix of abiotic and biotic (e.g., pathogens, competition) factors that determine fitness. Notable examples are increases in foliar water use efficiency due to stomatal closure, enhanced herbivory or pathogenic attacks, changes in winter hardiness, and alteration in nutrient use efficiency. A shift in the genetic structure of a population as a consequence of a specific air pollutant stress may be manifest in changes in physiology that are significant in terms of how individuals interact in an ecological context. Thus, selection for resistance to air pollution stress may have "physiological costs" to the population that determine its ecological amplitude and fitness (Roose et al. 1982). It is important to note that the potential for significant fitness costs of pollution resistance is a reason why the evolution of resistance can be viewed as undesirable.

Multiple Interacting Stresses of Anthropogenic Origin

It is common to focus on a single stress as the sole factor governing an evolutionary response. In the context of air pollution, it is increasingly apparent that most natural ecosystems are not impacted by a single air pollutant but rather experience concurrent interacting stresses from other changes of anthropogenic origin. For example, because one of the primary anthropogenic precursors of tropospheric O_3 is NO_2, areas experiencing high levels of O_3 tend to have high levels of nitrogen input through wet or dry deposition. Similarly, many existing and anticipated changes in atmospheric properties associated with global climate change (i.e., UVB, O_3, drought, temperature) are characterized as increasing the general oxidative nature of the environment through the production of free radicals. Thus, while plant populations may be experiencing dissimilar chemical or physical stresses of atmospheric origin, the mode of action of the stresses may be similar. The potential for multiple stress interactions is important because unlike other well-known instances of ecotypic differentiation, the intensity of each individual stress may be low but, in combination with other stresses, capable of exerting significant selection pressure on alleles that determine sensitivity.

Methodological Approaches and Experimental Designs

Unlike the situation where microevolution occurs in response to sharp gradients in natural or anthropogenic stresses, the methodological problems of demonstrating microevolutionary responses to chronic low levels of multiple pollutants are difficult. Local or regional variation in the distribution of the stress, temporal variation in intensity, and the prominent role of the environment in masking the expression of the genotype all place constraints on the ability of traditional experimental designs (e.g., reciprocal transplants) to isolate genetic differences among individuals or populations. Novel experimental approaches are needed, including the use of herbarium materials, seed collections, and unique biomarker techniques identifying biochemical changes that are more tightly coupled to genic expression than to growth and reproduction.

Preadaptation

The concept of preadaptation is relevant in understanding the physiological mechanisms underlying differential sensitivity among members of a population, the speed of evolutionary change, and the likelihood of response to a novel chemical stress in the environment. It is clear that many stresses of anthropogenic origin represent a change in the intensity or frequency of an already existing natural factor (e.g., CO_2 concentration, soil acidity, and O_3). In such cases a preadaptation appears likely. However, even when a pollutant superficially represents a qualitatively new stress to a species or population, the mode of action may not be unique (e.g., generation of free radicals) such that the potential for preadaptation to play a role is present. The conventional wisdom in evolutionary biology is that preadaptation is not necessarily correlated with a species' capacity to evolve resistance, although clearly it is a component underlying a population's ecological amplitude (Antonovics 1975).

Evolution as a Component of a Species' Ecological Amplitude

Ecological amplitude is commonly described as being invariant over time and best characterized by a population's physiological characteristics governing plasticity. For stresses that are progressively increasing in intensity and duration and for which interactions are common, a change in ecological amplitude (mean, mode, and limits) may arise through evolution. The fact that ecological amplitude can respond to anthropogenic changes in atmospheric chemistry is important to recognize as one of the strategies available to plants, particularly given the regional/global distribution of air pollution stresses and the episodic intensity of many exposure regimes. With longer time frames over which stresses operate, it is hypothesized that shifts in genetic structure due to evolution are likely to increase in importance, whereas the role of physiological accommodation in sensitive species will diminish.

Limits to Microevolution

Species vary in their capacity to evolve resistance to environmental stresses of anthropogenic origin; they differ in the asymptotic limits to which resistance can evolve. These limits are not fully resolved genetically or physiologically but are clearly species specific and stress dependent and determined by such factors as mode of inheritance, reproductive biology, physiological costs, mechanism of stress effects, and intensity of selection pressure. Exploring the limits to microevolution is important in predicting how much microevolution can shift the ecological amplitude of a species in light of the relatively rapid rise in the prominence of anthropogenic stresses. Intrinsic limits on the capacity to evolve resistance may mean that some sensitive species will be unable to "compensate" for modifications in the atmospheric or global climate change through microevolution.

Opportunities for Research

The study of how changes in atmospheric chemistry are influencing the genetic structure of plant populations in natural and intensively managed ecosystems offers a number of unusual and important opportunities for research beyond the most obvious ones of simply documenting the occurrence of the phenomenon. The variety of pollutants, their diverse spatial and temporal distributions, their relatively recent and sudden increases in concentration, and their diverse modes of action present a wide range of natural experiments on which to base research programs. A variety of fundamental questions regarding the mechanisms and limits governing evolutionary change in plant populations can be addressed. Similarly, there is a lack of understanding of the longer term ecological and evolutionary consequences of the evolution of pollution resistance. Knowledge in this area will contribute to basic understanding and to the applied problem of how to deal with pollutants and manage ecosystems affected by pollution stress.

Future Aspects of Air Pollution Stress

Whereas several changes in atmospheric chemistry are recognized as already being ecologically relevant in some regions (e.g., tropospheric O_3, nitrogen deposition, heavy metals, and CO_2), additional changes in atmospheric chemistry are anticipated (Caldwell 1979; Guderian 1985; White 1989). Notable examples are further increases in levels of tropospheric O_3, CO_2, UVB, some heavy metals (e.g., mercury), and volatile organics. Many of these are linked mechanistically with some of the anticipated changes in global climate (Fig. 1). Hence, natural vegetation is facing increases in both the array and quantity of atmospheric pollutants as well as the prospect of climate change.

Application of Knowledge

Increasing attention in the environmental sciences is being placed on emission source reduction and site remediation to combat existing or anticipated negative impacts on biotic resources. Prominent examples of recent or proposed targets of source reductions are emissions contributing to acidic precipitation (sulfur and nitrogen compounds) and UVB (chloro-fluorocarbons). However, changes in atmospheric chemistry will continue to influence the physiology and growth of vegetation in some areas, and remediation techniques must be considered. One avenue is through genetic engineering of resistance in domesticated species that are sensitive to air pollution stress. The efficacy of this option is predicated on identification of genetic variation at the molecular level amenable to engineering, the evaluation of environmental stress interactions that are most relevant, and the physiological costs of engineering that are most important to the performance of the species in an ecological context. Moreover, the application of knowledge is relevant to natural ecosystems in order to understand the capacity of a species' ecological amplitude to change. This issue is particularly important to the concept of biosphere reserves where ecosystems are maintained in perpetuity as reservoirs of genetic variation experiencing minimal impact from anthropogenic inputs. Subtle, long-term shifts in genetic structure due to regional or global changes in atmospheric chemistry may compromise this objective.

Summary

The fact that air pollution and other changes in atmospheric properties are likely to result in microevolutionary changes in terrestrial plant populations represents a problem in applied ecology and population biology that requires immediate attention. In addition, the unusual aspects of air pollution as a complex of environmental stresses eliciting microevolutionary responses in terrestrial vegetation present a number of opportunities to explore a range of basic issues. The breadth of relevant concerns requires that the effort be interdisciplinary in nature, establishing bridges between the disciplines of plant physiology, molecular genetics, population genetics, population biology, pollution ecology, and atmospheric chemistry. While such interactions have immediate implications for understanding how plant species repond over time to air pollution stress, the ramifications of this effort extend into the issues of global climate change, biotechnology, breeding for resistance, and basic questions regarding the biology of plant populations.

References

Antonovics J (1975) Predicting evolutionary response of natural populations to increased UV radiation. In: Impacts of Climatic Change on the Biosphere. Part 1. Ultraviolet Radiation Effects. DOT-TST-75-55. Department of Transportation, Climatic Impact Assessment Program, Washington, D.C.

Antonovics J, Bradshaw AD, Turner R (1971) Heavy metal tolerance in plants. Advances Ecological Research 7:1–85

Bishop JA, Cook LM (eds) (1981) Genetic Consequences of Man Made Change. Academic Press, New York

Bradshaw AD (1952) Populations of *Agrostis tenuis* resistant to lead and zinc poisoning. Nature 169: 1098–1099

Caldwell M (1979) Plant life and ultraviolet radiation: some perspective in the history of the earth's UV climate. Bioscience 29:520–525

Clausen J, Keck DD, Hiesy WM (1940) Experimental studies on the nature of species. 1. Effect of varied environments on western North American plants. Carnegie Institute of Washington Publication 520:1–442

Guderian RG (1985) Air Pollution by Photochemical Oxidants. Springer-Verlag, New York

Heck WW, Tingey DT, Taylor OC (eds) (1988) Assessment of Crop Loss from Air Pollutants. Elsevier Applied Science, New York

Hogsett WE, Tingey DT, Lee EH (1988) Ozone exposure indices: concepts for development and evaluation of their use. In Heck WW, Tingey DT, Taylor OC (eds) Assessment of Crop Loss from Air Pollutants. Elsevier Applied Science, New York, pp 107–138

Johnson DW, Taylor GE Jr (1990) Role of air pollution in forest decline in eastern North America. Water Air Soil Pollution 48:21–43

McNair MR (1981) Tolerance of higher plants to toxic materials. In: Bishop JA, Cook LM (eds) Genetic Consequences of Man Made Change. Academic Press, New York, pp 177–214

Pitelka, LF (1988) Evolutionary responses of plants to anthropogenic pollutants. Trends Ecology Evolution 39:233–236

Roose ML, Bradshaw AD, Roberts TM (1982) Evolution of resistance to gaseous air pollutants. In: Unsworth MH, Ormrod, DP (eds) Effects of Gaseous Air Pollution on Agriculture and Horticulture. Butterworth, New York, pp 379–409

Scholz F, Gregorius H-R, Rudin D (eds) (1988) Genetic Effects of Air Pollutants in Forest Tree Populations. Springer-Verlag, New York

White JC (ed) (1989) Global Climate Change Linkages. Acid Rain, Air Quality, and Stratospheric Ozone. Elsevier Science, New York

Winner WE, Mooney HA, Goldstein RA (eds) (1985) Sulfur Dioxide and Vegetation. Physiology, Ecology, and Policy Issues. Stanford University Press, Stanford

2
Evolution in Relation to Environmental Stress

ANTHONY D. BRADSHAW and T. MCNEILLY

Introduction

To understand what evolutionary changes might occur in plant species affected by air pollutants, a useful place to start is in other situations where environmental stresses occur. This could be anywhere, since stress—the occurrence of a factor, or factors, reducing growth or reproduction— occurs everywhere in nature. Stress factors can be anything from extremes of temperature to competition, as well as more subtle stresses.

Over the years since Turesson (1922) first demonstrated the occurrence of evolutionary differentiation within plant species, a great deal of work has been done on evolution in plants. It is necessary therefore to be rather selective, using situations where anthropogenic factors are involved and choosing those examples which are valuable in:

1. Indicating the nature of the evolutionary mechanisms which might operate in conditions of air pollution;
2. Indicating the characteristics of plant species that would foster this evolution;
3. Suggesting the evolutionary changes to be expected as a consequence.

Basic Mechanisms

Despite all the work and arguments, no other mechanism is known for evolution than that proposed by Darwin—the action of selection on inherited variation. Since any stress factor reduces performance, it follows that a stress factor can exert selection on any variability that is present in the population. The effect the stress factor will have will depend on its severity. If it can reduce performance by 50% and therefore exert a coefficient of selection of 0.5, it can have quite remarkable effects on gene frequency (Fig. 1). A 50% reduction in growth is not extreme in many stress situations; indeed, reductions in growth of 90% are not uncommon in polluted situations (Jain and Bradshaw 1966). Animals can usually

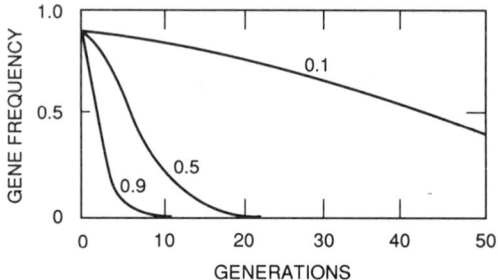

FIGURE 1. Changes in frequency of resistance controlled by a gene of intermediate dominance where the fitness of sensitive genotypes is either 10, 50, or 90% of the resistant (i.e., selection pressures of 0.9, 0.5, or 0.1).

escape stress conditions by running away; plants by contrast have to stay put and suffer them—and so be subject to strong selection (Bradshaw 1972). Yet to judge by textbooks, until recently many people have assumed that coefficients of selection were unlikely to be more than about 0.1 to 0.2. This makes a considerable difference to speeds of evolution, although change is still observable over a few generations (Fig. 1).

It is often easy to underestimate the potential power of selection in a given situation. Absence of any visible change in the size of a population does not mean that substantial death and replacement is not occurring. A 4-year census of the individuals of the grass *Agrostis canina* on a copper mine showed that although the population remained more or less constant in size, the numbers of deaths and replacements were equal to the total population size (Fig. 2) (Farrow et al. 1981). The only way, therefore, to understand the potential power of selection, is to make careful demographic studies combined with critical assessments of the relative performance of contrasting genetic material. We will return to this point later.

Of course the other requirement, rarely considered, is the presence of appropriate variability, that is, variation which can provide relief to the problems caused by the stress factor. This will be considered after some examples of evolution under stress have been examined.

What Evolution Can Achieve

There are plenty of examples of what evolution can achieve, given time, within a single species. An old example, which still has considerable lessons to offer, is the differentiation found in *Achillea millefolium* in California by Clausen et al. (1948). The morphological differences they reported are considerable, but what was outstanding is the degree of

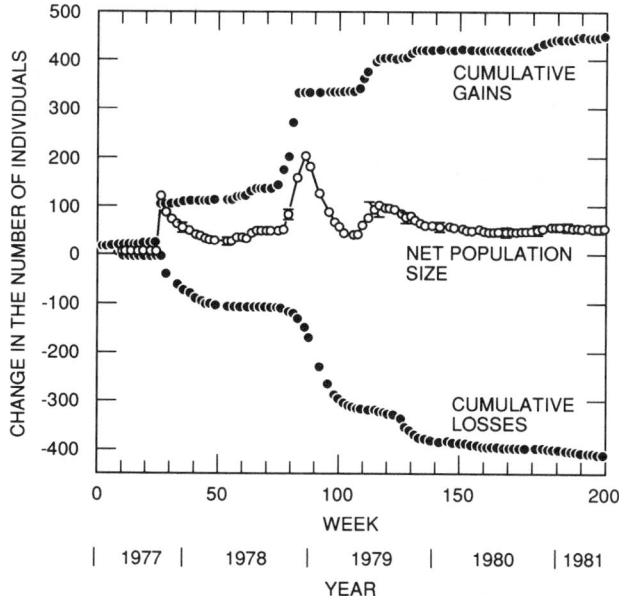

FIGURE 2. The dynamics of a *Agrostis canina* population on a copper mine over 5 years in terms of the number of individuals present, establishing or dying (Farrow et al. 1981).

physiological distinctiveness of the populations demonstrated by reciprocal transplants. Even in the absence of the effects of competition, populations from different parts of the range of the species were not able to survive in each other's habitats.

But this differentiation could have developed over a long period of time, at least since the last glaciation, if not longer. A very interesting, different example is, therefore, the situation found in the populations of *Anthoxanthum odoratum* occurring in the Park Grass fertilizer/lime experiment at Rothamsted in the United Kingdom. This experiment was originally set up in 1852; in 1903 a liming treatment was superimposed. Since then the populations of *A. odoratum* have become substantially different in both morphology and physiology. The differences in adaptation to soil calcium are closely related to the pHs of the plots from which the populations originated (Fig. 3) (Snaydon 1970). Reciprocal transplant experiments, in which natural competition was retained, showed that, in any plot average fitness of aliens was 50% of that of natives (Davies and Snaydon 1976).

In many ways, the evolution of metal resistance provides the most relevant evidence, since it is in relation to a new, manmade stress. It is not necessary to consider this example in detail since there are a number of

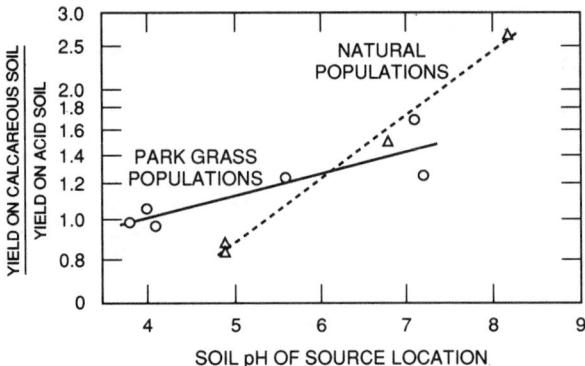

FIGURE 3. The relative adaptation to calcareous soil of populations of *Anthoxan-thum odoratum* taken from contrasting pH plots of the Park Grass Experiment, Rothamsted (Snaydon 1970).

recent reviews (McNair 1981, Baker 1987). The first point of note is that the degree of adaptation is remarkable; resistant plants are able to grow in conditions that are completely lethal to plants from normal habitats. This is by virtue of a metal-specific cellular mechanism which does not prevent uptake of the metal but renders it innocuous within the plant (hence the use of the term "metal tolerance"). The second point is that although the resistance is found in populations on metal-contaminated areas that have been in existence for a very long time, the same resistance can evolve in just a few years when a population becomes newly exposed to metal pollution (Wu et al. 1975). Selection can rapidly screen out the few resistant individuals that are found in normal populations and build up a resistant population (Walley et al. 1974).

When the evolution of metal resistance was first demonstrated (Prat 1934), little attention was paid to it, either because people did not believe it or perhaps because they thought it was a rather special case of evolution that had little relevance to other situations. Yet numerous cases are known of similarly rapid evolution in plants, particularly where crop cultivars have been moved for seed production from their normal location. *Anthoxan-thum odoratum* in the Park Grass experiment is, of course, another good example.

But the most illuminating recent discovery is that of the widespread evolution of herbicide resistance in agricultural weeds, reviewed by LeBaron and Gressel (1982). Although the evolution may be slowed down by irregular application of herbicide to individual populations, and sections of a population may escape exposure to herbicide by seed dormancy, evolution can be very rapid. Indeed, herbicide resistance has become a serious practical problem in the space of 15 years.

Variation

All of these examples suggest that evolution in stress conditions can be expected to be be swift and considerable. But a superficial observation of stressed environments suggests that because only a few species seem to occur in them the evolution of resistance must be limited to particular species. That these are species that have evolved resistance can be confirmed by experiment. What then has happened to the other species? Evolution must have failed to occur. But why?

Evolution depends on the occurrence of (i) selection and (ii) heritable variation. In a particular stress situation, the stress must have affected all the species originally present. It will also, over time, have affected all those species that arrived as immigrants—by seeds or other propagules. Therefore, all these species will have experienced selection. In which case it follows that failure to evolve can only be due to lack of the appropriate variation. What evidence is there for this?

Most people will presume that, except in pure lines arising by inbreeding or apomixis, genetic variability is always present in populations. Credence to this view has been given by the demonstration of very considerable amounts of variation at the level of isoenzymes. But it is too often forgotten that any sort of variation is not enough. It has to be appropriate variation, i.e., variation that will increase fitness in relation to the stress.

TABLE 1. Examples of the different sources used by plant breeders in their search for useful genes.

Sources	Adaptations
From original gene pools	
Potato	Blight resistance within *S. tuberosum*
Alfalfa	Spotted aphid resistance
Sugar beet	Sugar content
Rye	Reduced height
From other gene pools— other cultivars	
Barley	Yellow dwarf resistance from Abyssinian cultivars
Wheat	Dwarfing genes from Japanese cultivars
Grapes	Root aphis resistance from American material
Cotton	Blackarm resistance from African cultivars
From other gene pools— other species	
Oats	Mildew resistance from *A. ludoviciana*
Bread wheat	Stem rust resistance from *T. dicoccum*
Bread wheat	Eye spot resistance from *Aegilops ventricosa*
Rice	Grassy stunt resistance from *O. nivara*
Delphinium	Red flower colour from *D. cardinale*
Potato	Blight resistance from *S. demissum*

From Bradshaw 1984.

There is, in fact, much to show that this can be the stumbling block to evolution. There is no a priori reason why such variation should be present. It is usually assumed that mutation is a random occurrence; there is plenty of evidence for this. But this randomness is in occurrence and not in direction—certain mutations occur, others do not. As a result particular variation is to be found in some species and not in others. This is well borne out by the efforts which have to be expended by plant breeders to find particular important characteristics. The genes for these may be available within the population, within other populations, or only available in other species (Table 1). Of course, they may not be available at all; but this is usually not reported.

This situation is well shown in the evolution of metal resistance. Over the world as a whole, only a certain number of species have evolved resistance. From geographical considerations, those that have done it have probably done it several times. This can be seen if we consider a single site, such as the area around a copper smelter at Prescot near Liverpool. Before the smelter began to operate around 1900, there would have been a large number of plant species in the area, which would have been exposed to the pollution when it began. However, after 80 years only five species are to be found in the polluted zone (Table 2) (Bradshaw 1984). These are species which have been found to evolve resistance, quite separately, in other contaminated situations as well.

It can readily be shown that only certain species possess the necessary, appropriate, heritable variability in their normal populations. A range of species, some of which do and some of which do not evolve resistance, can be screened for such variability using the sensitive method of sowing several thousand seeds in metal solution in floating polythene beads (Table

TABLE 2. Species occurring in the contaminated grassland adjacent to a copper refinery near Liverpool, compared with those in similar but uncontaminated grasslands further away

Copper in soil (ppm)	Species found	Similar species
>2000	*Agrostis stolonifera*	*Festuca rubra*
Adjacent to refinery	*A. capillaris*	*Agropyron repens*
		Holcus lanatus
<500	*Ranunculus repens*	*Achillea millefolium*
Away from refinery	*R. bulbosus*	*Hypochaeris radicata*
	Cerastium vulgatum	*Leontodon autumnale*
	Trifolium repens	*Luzula campestris*
	T. pratense	*Lolium perenne*
	Taraxacum officinale	*Poa annua*
	Rumex obtusifolius	*P. pratensis*
	Prunella vulgaris	*P. trivialis*
	Plantago lanceolata	*Dactylis glomerata*
	Bellis perennis	*Cynosurus cristatus*
		Hordeum murinum

TABLE 3. The occurrence of copper-tolerant individuals in normal populations of various grass species in relation to the possession by these species of copper-tolerant populations in copper mine areas.

Species	Occurrence of tolerant individuals (%)	Presence of species on mines	Tolerance of collected adult plants
Holcus lanatus	0.16	+	+
Agrostis capillaris	0.13	+	+
Festuca ovina	0.07	−	−
Dactylis glomerata	0.05	+	+
Deschampsia flexuosa	0.03	+	+
Anthoxanthum odoratum	0.02	−	−
Festuca rubra	0.01	+	+
Lolium perenne	0.005	−	−
Poa pratensis	0.0	−	−
Poa trivialis	0.0	−	−
Phleum pratense	0.0	−	−
Cynosurus cristatus	0.0	−	−
Alopecurus pratensis	0.0	−	−
Bromus mollis	0.0	−	−
Arrhenatherum elatius	0.0	−	−

From Ingram in Bradshaw 1984.

3) (Ingram in Bradshaw 1984). Evolution of resistance is only found in those species that possess the appropriate variation.

It would be understandable if there was some phylogenetic pattern to the occurrence of this variation. For metal resistance there are some indications that it is found more commonly in some families, such as the Caryophyllaceae, and certainly in some genera, such as *Agrostis*, *Festuca*, and *Silene*. But an early extensive survey (Antonovics et al. 1971) failed to detect any broader phylogenetic picture. Resistance to each pollutant stress must, of course, be considered separately. For air pollution no such systematic survey has been carried out, but the evidence gathered in the survey by Roose et al. (1982) does not suggest any clearer phylogenetic pattern than for metal resistance.

It must be remembered that, in many cases where variability is present, it will have limits depending on the number of genes involved, since these can be rapidly fixed by selection. The population runs out of additive variation and an evolutionary plateau is reached. Even if suitable mutations occurred, mutation is such a rare event that it can do little to relieve this. This will explain why even in those species which evolve metal resistance, there are levels of metal to which they cannot adapt. As a result there are many bare areas on even those mine sites where metal-resistant populations of several species abound.

If resistance is controlled by several genes, it is possible that maximum levels of resistance are attained only after selection followed by recombina-

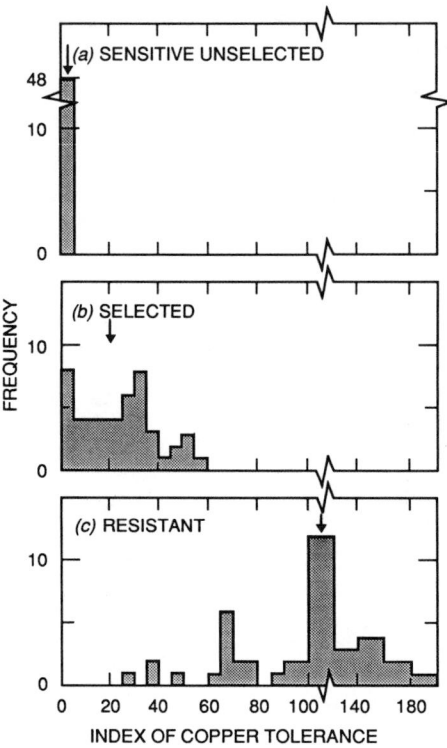

FIGURE 4. Distributions of the indices of copper resistance (tolerance) (at $0.15\,\mu g\,Cu\,cm^{-3}$) of *Agrostis capillaris* seedlings from (*a*) unselected normal population, (*b*) selected normal population and (*c*) established mine population (*arrows* indicate means).

tion of the different genes involved, to give trangressive segregation, followed by further selection. This can be seen in several experiments involving the selection of metal resistance out of normal populations. Full resistance is not achieved in the first cycle of selection (Fig. 4) (Ingram 1988). In situations where the character is controlled by only a single pair of alleles, this will not occur.

If the relevant variability is present at low frequency in the unselected populations, then another factor can come into play, that of chance occurrence or absence as a result of normal sampling variance. If a particular genotype is present at low frequency in a population, its appearance in successive samples of the population will be given by the Poisson distribution. Selection acts on individual generations and genotypes. Therefore, from an evolutionary point of view, in any single generation being selected, the significant parameter is the number of occasions on which that genotype is not found in the sample. This is affected both by the frequency of the genotype in the base population and the sample size.

All this is very familiar, often discussed as genetic drift. But the actual chances of a given sample not containing the genotype in question are very interesting, since, although they are obviously affected by the mean

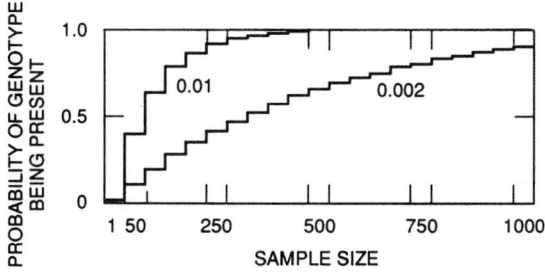

FIGURE 5. The probabilities of a given genotype occurring in a base population at a frequency of either 0.002 or 0.01 being present in random samples of different size taken from those populations.

frequency of the genotype, they are higher than might be expected (Fig. 5). If, for instance, the mean frequency of the genotype is 2 in 1000, it will be completely absent in over half of samples composed of 300 individuals. If a population has recently gone through a demographic bottleneck, during which its numbers were reduced drastically (perhaps to about 10 to 50 individuals for a single generation), the effects of this on the presence or, more importantly, the absence of the relevant adaptive variation are very great indeed. This is the so-called founder effect.

Important variability may, therefore, not be present in all populations of a species even when it is present in some. This is again well known to plant breeders and dominates strategies used in germ plasm sampling (Frankel and Hawkes 1975). It is also found in the occurrence of metal resistance in normal populations of a species such as *Agrostis capillaris*

TABLE 4. The percentage* occurrence of individuals resistant to different metals in different populations of *Agrostis capillaris*.

| | Tolerance to | | | | |
Population	Cd	Cu	Pb	Ni	Zn
1 Cullingworth	0	0	0	0	0
2 Denholme	0	0	0	0	1
3 Merllyn Gwyn	1	0	5	0	3
4 Ruthin	0	8	0	0	1
5 Dwrnudion	4	0	0	0	0
6 Highland	0	23	0	4	4
7 NZ Browntop	11	3	3	6	0
8 Parys	10	97	44	38	60
9 Goginan	10	70	78	40	93

From Symeonidis et al. 1985.
*Obtained by screening the 80 most tolerant individuals out of 15,000 original seedlings, showing more than 50% tolerance to different metals.

(Table 4) (Symeonidis et al. 1985), and seems to be the cause of different patterns of evolution of zinc resistance under different electricity pylons (Al-Hiyaly et al. 1988). If the required variation occurs differently in populations, we should not necessarily expect the same patterns of evolution between them, even if the selective situation is the same.

There is, finally, the possibility that, although variability giving resistance to the stress factor is present, it is not completely appropriate. It could have such negative effects on normal fitness characters that its overall effect is negative. This could be due either to pleiotropic physiological effects or to tight linkage with other, adverse, genes. Although there is the possibility that the latter can broken by recombination, the former are permanent. In this case the only possibility is that there are other genes present with mitigating effects which could be selected. In some species, but not all, metal resistance has been shown to have adverse effects on fitness in normal soils. It is, perhaps, due to the development of a metal requirement, as in *Anthoxanthum*, *Armeria*, and other species (reviewed by MacNair 1981). However, in these cases the resistance must have been of overriding importance, because it has been selected.

Selection

It was assumed in the previous section that selection was not limiting. Is this a fair assumption? Anything that depresses growth is able to cause selection. A reduction in performance of only 10% due to a stress factor is difficult to demonstrate experimentally. Yet if genotypes differing in this order of performance are present in the population, significant changes in the genetic composition of the populations can be predicted (Fig. 1). Differences in performance do not have to be manifest as differences in mortality. They can be manifest as difference in growth, since this will cause the genotypes to alter the total areas they cover. Such differences will presumably be manifest subsequently in reproductive success. But the latter itself can be the only manifestation of differences in performance. In the selection for metal resistance, it can be readily demonstrated that selection operates through both differential mortality and differential growth (Wu et al. 1975).

The actual process of change can be followed in some situations. Figure 6 shows the changes in the occurrence of copper resistance found in populations of different ages of *Agrostis stolonifera* at the copper refinery at Prescot (Wu et al. 1975). This resistance shows continuous variation. Because of the size of the samples taken (30 plants) in the control, unselected populations, no resistant individuals were found. Such variability is, however, present, at a frequency of about 5 resistant, or partially resistant individuals per 1000. In the population exposed to

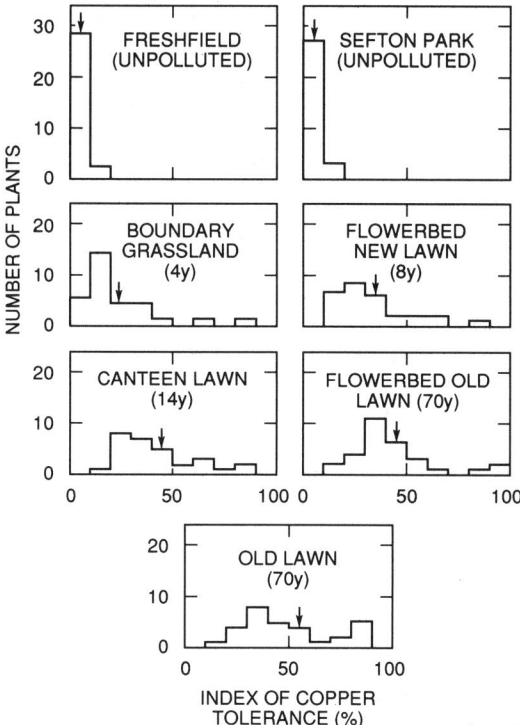

FIGURE 6. The distribution of copper resistance (tolerance) in *Agrostis stolonifera* populations of different ages around a copper refinery near Liverpool (*arrows* indicate means) (Wu et al. 1975).

copper pollution for only 4 years, a marked change is already apparent; there is a considerable reduction in the proportion of sensitive individuals and resistant individuals have reached a sufficient frequency to be picked up in the small sample. In the 8-year-old population, all the sensitive individuals have been eliminated. The 14-year-old population shows this process continuing, but now the slightly resistant individuals are disappearing.

In the two 70-year-old populations, this process has gone further, only there is an unexplained lack of build-up in frequency of the very resistant individuals, although these are clearly present. This may be due to an associated lack of fitness in other respects, particularly in competitive ability, which will be of major importance in a lawn. Certainly, the oldest populations retain an interestingly high level of variability.

This progression can be simulated very simply in the laboratory using a soil mixture series of increasing toxicity (Gartside and McNeilly 1974). This showed that the more toxic the soil the greater the resistance of the

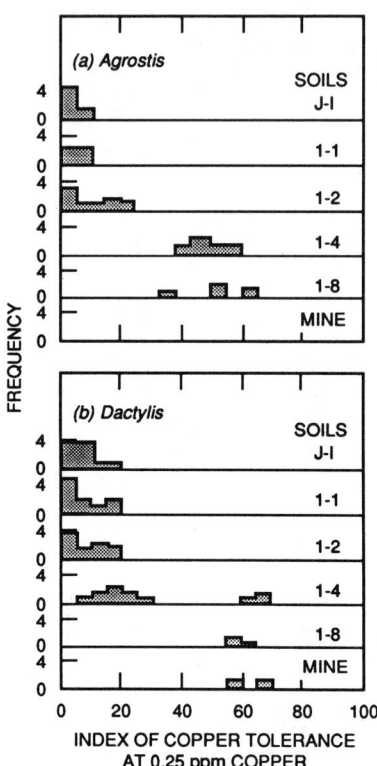

FIGURE 7. The indices of copper resistance (tolerance) of seedlngs of (*a*) *Agrostis capillaris* and (*b*) *Dactylis glomerata* surviving on soils of different copper contamination (made by mixing John Innes potting compost and copper mine soil in different proportions).

surviving individuals (Fig. 7). It is also clear from these data that some individuals with lower resistance can survive alongside more resistant neighbors.

Despite the slight complexity in this example, the general pattern of evolutionary change is clear—the progressive elimination of the most sensitive individuals and an increase in the frequency of resistant ones as soil contamination due to copper has built up. All this can occur over a very short period of time, even under conditions where reproduction by seed is limited.

It is worthwhile generalizing from this example and the related evidence to suggest that, where resistance shows continuous variation, population change takes place in three stages:

Stage 1. Elimination of the most sensitive genotypes
Stage 2. Elimination of all genotypes except the most resistant
Stage 3. Interbreeding of the survivors to give even more resistant genotypes which are then further selected.

Progression through these stages will be dependent on both the time and level of selection (i.e., intensity of stress) and the nature of the species' reproductive processes. In many situations progress may be no further than stage 1. In the present example, stage 2 has clearly not been reached, although in other situations, e.g., on old copper mines (McNeilly and Bradshaw 1968), it has been. Stage 3 will normally be difficult to separate from stage 2, but the possibility always exists that a population can produce, by genetic recombination, more resistant genotypes.

Stage 3 will be most readily achieved in species reproducing mainly by seed and least readily in species reproducing vegetatively or by apomixis. Mutation could also play a part in stage 3, but since it normally occurs at very low frequency, its contribution to short-term response to selection can be disregarded.

A further selective factor which must be considered is the presence or absence of competition. Startling differences in performance—survival, growth and reproduction—of different populations can be found in many species by means of reciprocal transplant experiments, even when done with spaced plants in the absence of competition, e.g., in *Achillea millefolium* (Clausen et al. 1948). But the real magnitude of the differences in performance (and therefore potential selection pressures) may only become manifest in the presence of full competition, e.g., in rice (Jennings and de Jesus 1968) and in *Anthoxanthum* (Davies and Snaydon 1976). This is also true for metal resistance (Cook et al. 1972). Since competition is normal in most situations, including those where environmental stress is operating, one should be cautious of making judgements on the presence or level of selection operating in a given situation if competitive effects are not included in the assessment. The same can be true for abiotic factors such as moisture or nutrient shortage.

Nevertheless, it must not be presumed that selection is always operating powerfully in apparent stress situations. It may, for instance, be affecting some species, but not others because of other features of their general biology. For instance certain weeds of escape herbicide effects because of their seasonal growth pattern, either late germination (annuals) or late emergence (perennials) (Putwain et al. 1982).

There is also the possibility that the stress factor fluctuates in intensity. In these situations it must be remembered that evolution does not operate in the manner of a rachet. Selection can go backwards or forwards. Its outcome depends on the overall effects of the stress on fitness over time. Any changes occurring under the influence of a stress factor can be nullified if, in the absence of that stress, the resistant form is less fit than the normal one. The best evidence for this is in crop plant breeding, especially in outbreeding species where it is not possible to fix variability in a pure line. Breeding for higher levels of carbohydrate content in *Lolium perenne*, for instance, is limited because there is a correlated reduction in seed yield, so that during seed multiplication, when selection cannot be maintained, the improved material tends to change back to the unimproved form because of the negative selection exercised by seed yield (Hayward and Abdullah 1985).

Reproductive Biology

Evolutionary change can only take place in a population if some individuals increase in size and/or frequency at the expense of others. In other words the normal processes of growth and reproduction are involved. In the long term these should not cause any absolute limitations to change. But in the short term, in the sort of conditions created by pollution, they are very likely to affect the speed of change.

Stage 1 in the process of change involves the disappearance of more sensitive individuals, but the general characteristics, i.e., the structure of the whole population, are not transformed. It is likely that, as these individuals disappear, those remaining can (in a population of perennials) expand with time to take up the space made available. This process is well understood and described by plant population biologists (e.g., Kays and Harper 1974). The process of replacement can be so immediate in rapidly tillering species such as grasses that no obvious signs of selection are apparent. But in slow growing perennials the space occupied by the dying individuals may not be taken up by neighbors immediately. In this case the process of selection may be obvious, as in populations of eastern white pine (*Pinus strobus*) (Dochinger and Seliskar 1970).

Stage 2 involves the replacement of the original population by the most resistant genotypes. In this case the whole population structure is revolutionized. The initial frequency of the most resistant genotypes is very important, but the critical process is the rate at which these genotypes can grow and reproduce, to colonize the space made available by the death of the rest. This will depend first on the reproductive biology of the species involved. The process is likely to be faster in annuals and biennials because of their seed production, and in colonizing or "r"-selected rather than "k"-selected species. But it will also depend on the general ecological conditions. If they are poor, perhaps due to the pollution, then growth and reproduction may be severely limited. As a result the polluted population may go through a phase when its numbers are depleted and it consists of a few widely scattered individuals. Only later, after a period of one or many years, does the population recover.

Stage 3, in which genetic recombination occurs to produce even more resistant individuals on which selection can act, can obviously only take place if normal sexual reproduction occurs with outcrossing. It will therefore be most likely in outbreeding species, although the small amount of outcrossing that occurs in most inbreeding species can generate enough heterozygosity to allow some recombination. It will be slowed considerably in perennial species if these take many years before first flowering. It will be prevented completely if the species is apomictic.

The populations of *Agrostis stolonifera* at Prescot show these possibilities clearly. This is a self-incompatible outbreeding perennial grass. Where the process of selection is only beginning, and also in the oldest populations, plant cover is continuous and it is difficult to see selection in progress, although in the young populations some genotypes reveal their lack of resistance by a certain amount of chlorosis. But in populations around the refinery which have been exposed to rather severe copper pollution for 5 to 10 years, a very different situation can be found. There is a wide scattering of individuals in otherwise barren ground from which the original continuous cover of sensitive individuals has disappeared (Fig. 8). In these situations the differences in relative resistance of individuals already referred to can be seen. The rate of recovery is slowed by the practice of mowing, which restricts reproduction by seed, but is speeded up by the application of fertilizer, which increases individual growth rate. The contrasting situations in three of these populations are given in Table 5 (Bradshaw 1975). Since the populations are not as resistant as the much older populations on abandoned mine workings, it is likely that insufficient time has elapsed for recombination and selection to produce full resistance even though the species is a good outbreeder.

In natural situations, where present selection pressures are likely to have been in existence for a long time, any evolutionary adjustments are likely

FIGURE 8. Patchiness due to the few resistant individuals surviving in a copper contaminated lawn sown 8 years previously with normal *Agrostis capillaris*.

TABLE 5. The probable ecological and evolutionary history of three populations of *Agrostis stolonifera* in the face of increasing copper pollution in sites adjacent to a copper refinery near Liverpool.

	Boundary grassland	Flowerbed new lawn	Canteen lawn
Original inoculum and environment	Existing rough grassland containing scattered *A. stolonifera* on unpolluted soil	Repeated seed sowings on very contaminated soil	Normal turf containing scattered *A. stolonifera* laid over contaminated soil
Age of population (in years)	4 (from beginning of contamination)	7	15
Nature of metal contamination	None until new refinery built, then building up rapidly	Preexisting in very contaminated soil	Preexisting in underlying soil and building up by aerial fallout
Levels of metal contamination (ppm)	Nil → 1900 in 4 years	4800 all the time	Nil → 2600 in 14 years
Frequency of individuals with tolerances >40% in original population	4/1000 0.01/m^2	4/1000 6/m^2	4/1000 0.1/m^2
Rate of vegetative spread of selected individuals	200 mm/year	50 mm/year	100 mm/year
Opportunity for seed production	Considerable	Low	Low
End product	Closed mixed sward with *A. stolonifera* frequency 40%	Sward with 50% bare ground, composed of separate clones about 100 mm diameter	Continuous sward composed of individual clones about 400 mm in diameter
Mean copper tolerance[a]	21%	32%	42%
Other species	Several, including grasses and herbs	No other species	Only *Agrostis tenuis*
Conclusions	Population still mixed with other species but tolerance already evolving	Population with fairly well developed tolerance, but not yet enough time elapsed for sward to close	Population with well-developed tolerance, forming closed sward, but fullest tolerance possible not yet achieved

From Bradshaw 1975.
[a] Fully tolerant mine population would be 70%, pasture population 5%.

to have been completed. By contrast, however, in new situations caused by pollution, the stress may have occurred so recently that not only may the evolutionary change not be complete, but also the populations themselves may not have reached ecological equilibrium with the new conditions. Since these two processes are interlinked, such populations may be in an interesting dynamic, transitory state. This is most likely in slow-growing perennials, rather than in vigorous, rapidly reproducing annuals.

Implications for Air Pollution

It is the purpose of this book to look closely at what is happening with regard to air pollution. The emphasis must be on *is happening* because, except in a few special natural situations such as close to volcano vents, air pollution is a relatively new phenomenon. In some cases, for example, NO_x, it is a very recent stress; in other cases, such as SO_2, it is older, at least in cities, and may be declining. There are plenty of examples of related evolution having taken place. But from the evidence that is reviewed here it seems there are a number of important points. Several of these have been raised previously (Roose et al. 1982), but they still need further investigation.

1. Since there are many different air pollutants with quite different effects, from an evolutionary point-of-view air pollution is not just one phenomenon but a complex of several. Each air pollutant must be considered as a separate evolutionary factor with its own potentially separate effects. It is possible that evolution of resistance to one air pollutant may give resistance to another, but not necessarily. No presumptions can be made, unless there are good grounds, for instance, where the resistance appears to be due to some change in cuticle or stomatal characteristics which confers resistance to air pollutants in general. This can only make for complications in work on air pollution, since it is commonplace for several different air pollutants to occur together.

2. There is no doubt that air pollutants can cause powerful damaging— and therefore potentially selective—effects. However, often as a result of recent control measures, much air pollution has rather weak effects, sufficiently weak to be difficult to demonstrate (Roberts 1984). In such cases potential selective effects may be difficult or impossible to demonstrate and may indeed not be present (Colvill et al., 1985). What is relevant here is how low a selection pressure can cause evolutionary differentiation in theory, over what time scale, and how far can we translate this to match real situations with their multitude of other variables.

3. If resistance has an associated cost, the point below which resistance is not selected for may be higher than expected, as suggests by Pitelka

(1988). The situation will be further complicated if the level of pollution fluctuates hourly, daily, or seasonally. This is again commonplace with air pollutants. There is a need to obtain more critical evidence for a variety of situations about the levels of selection.

4. Selection is ineffectual unless appropriate variation is present. There is good evidence for genetic variation within populations of a number of species in sensitivity to various air pollutants. Its genetic basis has been clearly demonstrated (Taylor 1978). But it needs further investigation. Is it present only in some species but not others? What is its range in original populations, after selection, and after recombination? How does this relate to the presence or absence of different species in areas suffering from air pollution? Is it always genetically based?

5. Since resistance has costs, what is the full range of performance of resistant genotypes in both normal and polluted environments?

6. How does this relate to field situations where air pollution is operating? What stages of the evolutionary process can be found? Do situations exist, particularly in long-lived perennial species, where equilibrium has not been reached?

7. From all this, can sensible predictions be made about final equilibrium states? Can many situations be envisaged where evolution of resistance might be a valuable and important way in which wild and cultivated plants survive in areas suffering air pollution which we cannot eliminate or obviate? Do the examples of crop cultivars in which unconscious selection for resistance seems to have occurred, for instance, tobacco (Menser and Hodges 1972), onions (Gabelman in Ryder 1973), and alfalfa (Howell et al. 1971) hold exciting possibilities? The possibility that the maintenance of yields by the evolution of resistance needs to be tested.

8. Unlike other forms of environmental contamination air pollution is not tidy in that it sprawls across geographical areas. It is also rather difficult to work with unless good experimental facilities are available. Yet, it represents an effectively new and rigorous evolutionary challenge to both plants and animals. It is, therefore, like other forms of pollution, certainly rewarding to study.

Summary

There are plenty of theoretical analyses of evolutionary mechanisms. The critical question is how evolution in relation to environmental stress occurs in practice. The basic mechanism must be that of selection acting on inherited variation.

Environmental stress has clearly caused major evolutionary changes within species. Some of these could have occurred over many millenia, but there are now several examples of evolutionary changes occurring in only a few years. These are mainly in relation to anthropogenic factors such as metal pollution and herbicides.

For these changes to occur, the first prerequisite is the occurrence of appropriate variation. Despite the assumptions of many people, evidence from plant breeding and the evolution of metal resistance suggests that such variability is not always available, therefore producing a severe initial constraint to evolution in some species and populations. A further constraint then occurs when the supply of variability initially available becomes exhausted by the selection process.

Selection, by contrast, is usually not likely to be a constraint. Selection coefficients produced by anthropogenic factors can be very high. Three stages in the progress of such selection acting on a population are recognized: (1) elimination of the most sensitive genotypes; (2) elimination of all genotypes except the most resistant; (3) interbreeding of these survivors to give even more resistant genotypes which are then further selected. But during such changes the population will only maintain itself if the survivors can grow and reproduce rapidly enough to replace those which have been eliminated. How effectively this will occur must depend on the reproductive biology of the species. There is good evidence that in some species and in some situations this provides a serious constraint.

Air pollution is, on the whole, a recent phenomenon. As a result we can expect to find evolution in progress and equilibria not yet reached. At the same time, due to developing control measures, air pollution effects are not longer always severe and are often varying. The coefficients of selection generated may, therefore, be low. Resistance may also be associated with loss of fitness in normal environments. Thus, although some excellent examples of evolution in relation to air pollution have been found, only further work will indicate how widespread such evolution is. This work will not only provide material by which the effects of air pollution can be combatted but also provide valuable evidence about the processes of evolution as they occur in practice.

References

Al-Hiyaly SA, McNeilly T, Bradshaw AD (1988) The effects of zinc contamination from electricity pylons—evolution in a replicated situation. New Phytologist 110:571–580

Antonovics J, Bradshaw AD, Turner RG (1971) Heavy metal tolerance in plants. Advances in Ecological Research 7:1–85

Baker AJ (1987) Metal tolerance. New Phytologist 106 (Suppl): 93–111

Bradshaw AD (1972) Some of the evolutionary consequences of being a plant. Evolutionary Biology 5:25–47

Bradshaw AD (1975) The evolution of heavy metal tolerance and its significance for vegetation establishment on metal contaminated sites. In: Hutchinson TC (ed) Heavy Metals in the Environment. Toronto University Press, Toronto, pp 599–622

Bradshaw AD (1984) The importance of evolutionary ideas in ecology—and vice versa. In: Shorrocks B (ed) Evolutionary Ecology. Blackwell, Oxford, pp 1–25

Clausen J, Keck DD, Hiesey WM (1948) Experimental studies on the nature of plant species. 3. Environmental responses of climatic races of *Achillea*. Carnegie Institute of Washington Publication 581. Carnegie Institute, Washington

Colvill KE, Horsman DC, Roose ML, Roberts TM, Bradshaw AD (1985). Field trials on the influence of air pollutants, and sulphur dioxide in particular, on the growth of ryegrass *Lolium perenne* L. Environmental Pollution (Series A) 39: 235–266

Cook SA, Lefebvre C, McNeilly T (1972) Competition between metal tolerant and normal plant populations on normal soil. Evolution 26:366–372

Davies MS, Snaydon RW (1976) Rapid population differentiation in a mosaic environment. 3. Measures of selection pressures. Heredity (London) 36:59–66

Dochinger LS, Seliskar CE (1970) Air pollution and the chlorotic dwarf disease of eastern white pine. Forest Science 16:46–55

Farrow S, McNeilly T, Putwain PD (1981) The dynamics of natural selection for tolerance in *Agrostis canina* L. subsp. montana Hartm. International Conference, Heavy Metals in the Environment, Amsterdam 1981. C.E.P. Consultants, Edinburgh, pp 289–295

Frankel OH, Hawkes JG (eds) (1975) Crop Genetic Resources for Today and Tomorrow. Cambridge University Press, Cambridge

Gartside DW, McNeilly T (1974) The potential for evolution of heavy metal tolerance in plants. 2. Copper tolerance in normal populations of different plant species. Heredity (London) 32:335–348

Hayward MD, Abdullah IB (1985). Selection and stability of synthetic varieties of *Lolium perenne*. 1. The selected character and its expression over generations of multiplication. Theoretical and Applied Genetics 70:48–51

Howell RK, Devine TE, Hanson CH (1971) Resistance of selected alfalfa strains to ozone. Crop Science 11:114–115

Ingram C (1988) The evolutionary basis of ecological amplitude of plant species. Ph.D. thesis, University of Liverpool

Jain SK, Bradshaw AD (1966) Evolutionary divergence among adjacent plant populations. 1. The evidence and its theoretical analysis. Heredity (London) 21:407–421

Jennings PR, de Jesus J (1968) Studies on competition in rice. 1. Competition in mixtures of varieties. Evolution 22:119–124

Kays K, Harper JL (1974) The regulation of plant and tiller density in a grass sward. Journal of Ecology 62:97–105

LeBaron HM, Gressel J (1982) Herbicide Resistance in Plants. Wiley, New York

McNair MR (1981) Tolerance of higher plants to toxic materials. In: Bishop JA, Cook LM (eds) Genetic Consequences of Man Made Change. Academic Press, London, pp 177–207

McNeilly T, Bradshaw AD (1968) Evolutionary processes in copper tolerant populations of *Agrostis tenuis*. Evolution 22:108–118

Menser HA, Hodges GH (1972) Oxidant injury to shade tobacco cultivars developed in Connecticut for weather fleck resistance. Agronomy Journal 64:189–192

Pitelka LF (1988) Evolutionary responses of plants to anthropogenic pollutants. Trends in Ecology and Evolution 3:233–237

Prat S (1934) Die Ehrblichkeit der Resistenz gegen Kupfer. Bering Deutsch Botanische Gesellschaft 102:65–67

Putwain PD, Scott KR, Holliday RJ (1982) The nature of resistance to triazine herbicides: case histories of phenology and population studies. In: Le Baron HM, Gressel J (eds) Herbicide Resistance in Plants. Wiley, New York, pp 99–116

Roberts TM (1984) Long-term effects of sulphur dioxide on crops: an analysis of dose-response relations. In: Beament J, Bradshaw AD, Chester PF, Holdgate MW, Sugden M, Thrush BA (eds) The Ecological Effects of Deposited Sulphur and Nitrogen Compounds. Royal Society, London, pp 41–58

Roose ML, Bradshaw AD, Roberts TM (1982) Evolution of resistance to gaseous air pollutants. In: Unsworth MH, Ormrod DP (eds) Effects of Gaseous Air Pollution in Agriculture and Horticulture. Butterworth, London, pp 379–409

Ryder EJ (1973) In: Naegele JA, (ed) Air Pollution Damage to Vegetation. Advances in Chemistry Series 122:75–82. American Chemical Society, Washington

Snaydon RW (1970) Rapid population differentiation in a mosaic environment. 1. The response of *Anthoxanthum odoratum* populations to soils. Evolution 24:257–269

Symeonidis L, McNeilly T, Bradshaw AD (1985) Interpopulation variation in tolerance to cadmium, copper, lead, nickel and zinc in nine populations of *Agrostis capillaris* L. New Phytologist 101:317–324

Taylor GE (1978) Genetic analysis of ecotypic differentiation of an annual plant species, *Geranium carolinianum* L., in response to sulfur dioxide. Botanical Gazette 139:362–368

Turesson G (1922) The genotypical response of the plant species to the habitat. Hereditas 3:211–350

Walley KA, Khan MSI, Bradshaw AD (1974) The potential for evolution of heavy metal tolerance in plants. 1. Copper and zinc tolerance in *Agrostis tenuis*. Heredity (London) 32:309–319

Wu L, Bradshaw AD, Thurman DA (1975) The potential for evolution of heavy metal tolerance in plants. 3. The rapid evolution of copper tolerance in *Agrostis stolonifera*. Heredity (London) 34:165–187

3
Ecological Genetics and Chemical Modifications of the Atmosphere

J. NIGEL B. BELL, MIKE R. ASHMORE, and G.B. WILSON

Introduction

The influence of atmospheric pollutants on vegetation has been studied since the middle of the 19th century and numerous cases have demonstrated deleterious effects, ranging from subtle growth reductions to the destruction of entire ecosystems. It is thus surprising that such a powerful environmental stress was not identified earlier as a potent force for the natural selection of air pollution resistance under field conditions. The first reference to such a possiblity is the observation by Dunn (1959) that *Lupinus* populations in Los Angeles were more resistant to the local photochemical smog than populations originating from cleaner locations. This observation remained largely unrecognized until the early 1970s, when the first systematic attempts were made to detect the evolution of resistance to SO_2 in herbaceous species in polluted areas. In this respect, Bradshaw and McNeilly (1981) have drawn a parallel with their own research on the evolution of resistance to heavy metals, noting the work of Prat (1934) on copper resistance in *Silene dioica* populations from copper mines in Czechoslovakia, that remained forgotten in the literature for some 20 years.

This lack of interest by botanists in the evolution of resistance to air pollution in plant species is also unexpected in view of the long history of studies by entomologists on industrial melanism in moths and ladybirds in coal-smoke polluted locations in the United Kingdom (Lees et al. 1973). In particular the classic study of Kettlewell (1955) demonstrated experimentally the selective advantage of the dark form of the peppered moth *Biston betularia* on smoke-blackened tree trunks via a reduction in predation, while later work showed a shift back towards the nonmelanic form accompanying improvements in air quality (Askew et al. 1971).

This chapter will review the published data on evolution of resistance to gaseous air pollutants arising from natural selection at polluted sites. This is aimed at producing a definitive overview of the evidence for such evolution occurring in response to the single pollutants SO_2, NO_2, NH_3,

O_3, and their mixtures, as well as to wet deposition. The evidence includes field observations, with subsequent experimental investigations to demonstrate the existence of resistance in the plant populations concerned. Consideration will be given to the mechanisms at a population level by which evolution has taken place, but not to the physiological processes underlying these. The information will be assessed critically with respect to the strength of the evidence and major gaps in current knowledge will be highlighted.

Criteria for Determination of Intraspecific Evolution of Resistance

There are numerous problems inherent in the unequivocal demonstration of evolution of resistance to air pollution by natural selection. In particular, unlike the case of heavy metal pollution in the soil, air pollutants do not display sharp boundaries in their distribution. Consequently there are difficulties in identifying clean air control sites for the collection of unpolluted populations, which are similar in all their characteristics with the exception of the ambient pollution regime. This inevitably raises the possibility that the results of comparisons between populations collected from widely separated sites may be confounded by differences in response to other environmental variables, particularly climate and soils. This is an even greater problem where commercially bred cultivars of a species have been used as the control site population, as the breeding process may have inadvertently selected for changes in pollutant sensitivity. In the case of O_3, where gradients of concentration only occur on a regional scale the difficulties in identifying suitable control populations present an even more intractable problem, while for CO_2 the global nature of the pollutant effectively precludes spatial studies on response to elevated concentrations.

There are two approaches which potentially afford a high level of certainty in the identification of the evolution of resistance. First, the same populations can be screened for changes in their level of resistance at intervals from the time of their establishment at a polluted location; ideally, a parallel experiment should be performed at a similar unpolluted location, using the same genetic material. The second approach is to investigate the level of resistance at different distances away from a point source of pollution, situated in an otherwise relatively unpolluted area. However, very few investigations have adopted either of these ideal approaches.

A further important criterion for the demonstration of resistance is that this character must be shown to be stable when the plants are transferred to an unpolluted environment and retained for a prolonged period of vegetative propagation. This provides some indication that there may be a genetic basis for the observations. A further useful character is the demonstration of heritability via retention of resistance in plants grown from seed collected from polluted sites. Although in a number of studies resistance has been shown in such seed grown plants, in only one case (Taylor 1978) has this also been examined in the parent material from which the seed was collected. However, the ultimate proof of genetic control of tolerance must be sought on the basis of biometrical genetic analysis from breeding experiments, using differentially tolerant material.

Sulfur Dioxide

Nonpoint Source Studies

The first study on evolution of resistance to SO_2 was initiated by observations at an experimental farm at Helmshore in the polluted Rossendale Valley, in the Pennine Hills of northwest England. Introduced bred grass cultivars showed poor growth and on occasion displayed symptoms characteristic of air pollution injury while adjacent permanent pastures containing wild populations of the same species remained apparently healthy (Bell and Mudd 1976). From these observations, it was postulated that natural selection for SO_2 resistance had taken place in the permanent pastures. This hypothesis was tested by growing wild clones of *Lolium perenne* selected from Helmshore alongside the S23 cultivar (cv.) in a fumigation experiment where the plants were grown for 9 weeks in outdoor chambers, ventilated with charcoal-filtered air or with air containing a mean of 130 ppb SO_2 (Bell and Clough 1973). At the end of this period the shoot dry weight of the S23 plants was reduced in the presence of SO_2 by 41%, while the Helmshore clones showed no difference between the treatments. The growth reduction in the S23 plants was accounted for by a decrease in numbers of tillers and living leaves, leaf area, and dry weights of living leaves and stubble, accompanied by an increase in senescence; in contrast, none of these parameters was affected in the Helmshore population. In addition, the S23 plants grown in SO_2 were obviously chlorotic, while the Helmshore clones retained a healthy, green appearance, reflecting the field observations.

These initial studies on a limited number of *L. perenne* clones prompted research on a population basis, using individually identified clones of *L. perenne* collected randomly from an indigenous pasture at Helmshore, and parkland sites with different levels of SO_2 pollution in and around the Merseyside conurbation (Horsman et al. 1978, 1979a). Experiments were carried out to assess the performance of the populations in response to an 8-week SO_2 fumigation. These confirmed the SO_2 resistance of the Helmshore population and also indicated that plants from polluted areas in Merseyside were unaffected by a fumigation which reduced the growth of

plants collected from adjacent clean locations. This investigation represented an important step forward in that differences in resistance were
related to the pollutant regime at sites in the same general area, thereby
minimizing the possible confounding effects of genetic adaptations to other
environmental factors such as climate. An analysis of the degree of
resistance in the individual clones of the Merseyside populations showed a
wide range of response to SO_2, but this was lower in the polluted site
plants, providing evidence of selection for resistance resulting in dimi-
nished genetic variation (Fig. 1).

Following these early studies with *L. perenne* a much broader-based
investigation was carried out by Ayazloo and Bell (1981). In this case five
grass species collected from two polluted sites in the north of England were

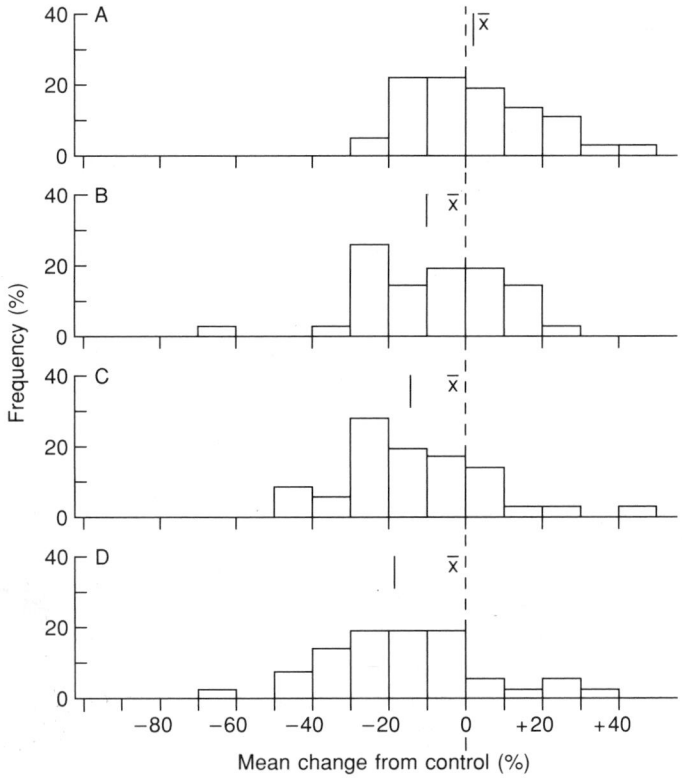

FIGURE 1. Differential response in total dry matter of four Merseyside ryegrass
populations to an 8-week exposure with 247 ppb SO_2. (A, B) Polluted site
populations; (C, D) clean site populations. Mean percent change from control is
the change in yield of each individual clone due to exposure to the SO_2-polluted air
expressed as a percentage of its growth in control air, $\bar{x}$ is arithmetic mean of
frequency histogram. (From Horsman et al. 1979a.)

screened for resistance to both acute and chronic SO_2 injury. The major site of investigation was again at Helmshore, where *Lolium perenne*, *Festuca rubra*, *Dactylis glomerata*, and *Holcus lanatus* plants were collected randomly from a permanent pasture or from adjacent wasteland. In addition a *Phleum bertolonii* population was sampled from a field which had been sown with the cv. S50 24 years earlier. The other location was Philips Park in the city of Manchester, from which randomly selected *Lolium perenne* plants were collected from a mown grass sward which had been sown with cv. S23 17 years previously.

Chronic fumigations were designed to produce a maximum growth reduction due to SO_2, but without producing any acute injury in the form of necroses, and were carried out in outdoor, climatically uncontrolled, chambers at mean concentrations ranging between 95 ppb and 183 ppb for 61 to 180 days, depending on the season. Response to SO_2 was assessed on the basis of a range of growth parameters at the end of the fumigation compared with the performance of the same genetic material grown in a similar chamber ventilated with clean charcoal-filtered air. Acute fumigations were performed in a controlled environment cabinet with 1900 to 3800 ppb SO_2 over 6 h, followed by assessment of the percentage of leaf length destroyed. The response of the polluted site populations was compared with that of the bred cultivars *Lolium perenne* cv. S23, *Festuca rubra* cv. Engina, *Dactylis glomerata* cv. S37, and *Phleum bertolonii* cv. S50, and in the case of *Holcus lanatus*, which is not an agricultural grass, a wild population collected from Ascot, a clean location in southern England.

Three out of the five species collected from Helmshore—*L. perenne*, *D. glomerata*, and *F. rubra*—showed a significant reduction in acute injury compared with their corresponding bred cultivars, and a similar pattern was seen in the case of the *L. perenne* population from Philips Park. In the case of the chronic fumigations, a large number of significant SO_2/population interactions occurred (Table 1). The Philips Park *L. perenne* population showed a smaller reduction in living leaf weight and a decreased degree of senescence (based on the dead leaf/living leaf ratio). *Festuca rubra* plants from Helmshore showed significant SO_2/population interactions for all characters measured, including the root/shoot ratio, as well as reduced senescence. *Lolium perenne* displayed a similar pattern, except for the absence of a significant interaction with respect to root weight and the root/shoot ratio. *Dactylis glomerata* plants from Helmshore were less adversely affected by SO_2 than cv. S37 with respect to all characters other than shoot weight. In the case of *Phleum bertolonii* significantly smaller reductions in SO_2 occurred in the Helmshore population than the cv. S50 plants for total and root weights and the root/shoot ratio. The exception was *Holcus lanatus*, where despite SO_2 causing a large suppression in growth and increased senescence, there was no clear indication of a differential response between the two populations.

TABLE 1. Percentage change in dry weight of SO₂ fumigated plants compared with clean air controls, showing significant gas/population interactions.

Species	Population	Living Leaf wt	Dead Leaf wt	Shoot wt	Root wt	Toal wt	Dead: Living Leaf wt	Root: Shoot wt × 100%
Lolium perenne	Philips Park	−27 ***	+31 ***	−14	−10	−13	+88 ***	+11
	cv. S23	−38	+79	−21	−10	−18	+217	+11
	Helmshore	−4 ***	+15 ***	−6 ***	−10 ***	−8 ***	+20 **	−2 *
Festuca rubra	cv. Engina	−24	+86	−28	−39	−33	+180	−14
	Helmshore	−18 ***	+38 ***	−9 ***	−35 ***	−16 ***	+71 **	−29
Lolium perenne	cv. S23	−26	+24	−18	−33	−23	+94	−20
	Helmshore	−23 *	+5 ***	−14 *	−29 ***	−21 *	+35 ***	−19 *
Dactylis glomerata	cv. S37	−44	+49	−21	−45	−33	+171	−30
	Helmshore	−22 ***	+4 **	−20 ***	−18 ***	−19 ***	+34 ***	+3
Holcus lanatus	Ascot	−24	+27	−21	−20	−21	+72	+1
	Helmshore	−26	+57	−15	−9 ***	−12 **	+104	+7 ***
Phleum bertolonii	cv. S50	−35	+57	−20	−38	−29	+142	−24

* $p < 0.05$; ** $p < 0.01$; *** $p < 0.001$.
Modified from Ayazloo and Bell, 1981.

An interesting feature of this study was that there was no indication in any of the populations of a relationship between the amounts of chronic injury and acute injury on the same genetic material. This absence of such a relationship has been reported elsewhere (Horsman et al. 1979b; Wilson and Bell 1985, 1986) and implies that the mechanisms of the two types of injury must be entirely different.

Further studies on SO_2 resistance have been carried out on populations of plants from areas with different pollution regimes in The Netherlands and East Germany. In the case of the work of Ernst et al. (1985), who studied seed-grown populations of *Silene cucubalus* the results are extremely difficult to interpret, with an SO_2 fumigation stimulating the growth of plants from a polluted site and reducing the root/shoot ratio in a clean site population. However, the main interest in this work is that it represents one of the few examples of a study into SO_2 resistance with respect to reproductive characters: a reduction was demonstrated in the weight of seeds produced per plant in the clean site population only. A further study by Dueck et al. (1987) on populations of *Agrostis capillaris* populations from the Netherlands failed to show any clear differential effect of SO_2 between populations from different locations.

Point Source Studies

It is surprising that relatively few studies on resistance have been carried out around point sources of SO_2, in view of their potential for this purpose. The first such study was reported by Taylor and Murdy (1975), who collected seeds of *Geranium carolinianum* from two locations within 700 m of a coal-fired power station in Georgia. Although SO_2 concentrations were not measured at these sites, they were high enough to induce acute injury on a number of weed species in the area, including *G. carolinianum* itself. Similar collections were made of seed from four unpolluted sites. Two sets of fumigations were performed on seedlings grown from seed collected in successive years. In both cases the polluted site populations showed a significantly smaller amount of injury than those from corresponding clean locations. All populations showed considerable heterogeneity in their response to the fumigations, but there was some evidence of this being reduced in the polluted site plants, supporting the observations of Horsman et al. (1979a) who suggested that this presented evidence of selection for tolerance reducing genetic variability. A similar study on an annual species was described by Murdy (1979), working with *Lepidium virginicum* from a heavily polluted area around a smelter at Copper Basin, Tennessee. Seedlings of nine populations collected from Copper Basin or a clean area were subjected to an acute 9-h SO_2 fumigation and the percentage of seed sterility estimated 1 week later by counting the number of aborted fruits on the raceme. Both sets of populations showed a small but similar degree of sterility when they had not been fumigated. However, SO_2 significantly increased sterility by 119% and 202% in the polluted and clean site populations, respectively, with a significant SO_2/population interaction.

These two studies are important in a number of respects. First, more than one population was sampled at both polluted and corresponding clean locations, thereby lending weight to the evidence for evolution of resistance. Second, the ambient SO_2 regimes at the polluted sites were very severe and could be expected to cause damage to sensitive species. Finally, and of particular interest, is the demonstration of resistance in an annual species where reproduction is only via seed production and germination. The work of Murdy (1979) indicates that resistance may have developed in the reproductive structures themselves, suggesting that a reduced potential for sexual reproduction is the key mechanism by which SO_2 has acted as a selective force in the field. These two studies remain, with one exception, the only investigations into annual species, with nearly all other work being concerned with perennial grasses where selection is likely to be primarily via intraspecific competition between individual genotypes in a sward.

Another location where the evolution of resistance has been studied around a point source of SO_2 is the colliery village of Askern in northern England. The source concerned was a smokeless fuel works (now closed) situated in an otherwise relatively clean area. Ayazloo and Bell (1981) made random collections of *L. perenne*, *D. glomerata*, *F. rubra*, and *H. lanatus* from wasteland close to the works and also from the uncultivated verge of a minor road 5 km away. Both populations were screened for acute and chronic SO_2 resistance, as described previously for populations from Helmshore and Philips Park studied by the same workers. Table 2 indicates that three of the species showed a clear differential response to the acute fumigation, with substantially more injury appearing on the clean site than polluted site populations of *D. glomerata*, *F. rubra*, and *H. lanatus*. In the case of the chronic fumigations, there were general indications of resistance in the polluted site populations of all four species, but in only a few cases were there, however, any significant SO_2/population interactions, this being attributed to the collection of an inadequate number of replicate samples.

TABLE 2. Percentage of leaf length injured by a fumigation with 2020 ppb SO_2 for 6 h, on grass collected from polluted and clean sites at Askern.

Species	Clean site	Polluted site	$p<$
Dactylis glomerata	12.52	2.16	0.001
Festuca rubra	26.31	11.16	0.05
Holcus lanatus	3.01	0.84	0.01
Lolium perenne	46.10	46.58	NS

From Ayazloo and Bell 1981.

Resistance in plants from a site near the Askern works was also studied by Awang (1979) for *Trifolium repens*. Clover seeds were collected from Askern and from an unpolluted site in Essex, in the south of England. Seedlings of the two populations were grown for 18 weeks inside a pair of glasshouses located in an inner suburb of the city of Sheffield. One of the glasshouses was ventilated with charcoal-filtered air, while the other was ventilated with ambient air. Growth reductions were recorded in both populations in ambient compared with filtered air, but the magnitude of this effect was consistently greater in the unpolluted site population. This is a unique experiment in that resistance has been demonstrated by exposing the plants to ambient air pollution rather than an artificial SO_2 fumigation. The 3-weekly mean SO_2 concentration during the experiment varied between 14 and 17 ppb, but the filter will also have removed NO_2 and O_3, although these were not measured. Nevertheless, the response can probably be attributed primarily to SO_2 as this was the only gaseous phytotoxic pollutant emitted at high concentrations by the smokeless fuel works.

The existence of the point source of SO_2 at Askern in an otherwise rural area provided an ideal opportunity to study the degree of resistance in plants collected from the field along a gradient of air pollution at different distances away from the works. If a corresponding gradient in resistance could be demonstrated, then this would provide a very high level of proof that evolution had occurred under different degrees of selective pressure imposed by a range of ambient SO_2 concentrations. This was investigated by Wilson and Bell (1986), who collected *Dactylis glomerata* and *Festuca rubra* plants from eight sites on uncultivated land spaced out along a 2.7-km transect downwind of the works. Each population was screened for resistance to acute and chronic injury, as described previously. In the case of the acute fumigations a repeat screening was performed on the same genotypes. *Dactylis glomerata* did not display any clear pattern of relationship between resistance and distance from the works from either of the acute screenings. In contrast, *F. rubra* showed a distinct cline of SO_2 resistance, diminishing along the transect away from the works, which remained consistent over the two screenings (Fig. 2). The chronic fumigations, however, revealed no clear relationship between growth reductions and distance, thereby indicating an apparent absence of differential resistance to this form of injury, at least on the scale of the transect concerned.

The only other study around a point source was carried out by Westman et al. (1985) who fumigated with SO_2 *Bromus rubens* (an annual) plants grown from seed collected from an SO_2-polluted site near an oil-refinery/chemical complex on the coast of California and compared their performance with a population collected from a clean air site in the Santa Monica Mountains. After 5 weeks a significantly greater amount of acute injury was recorded on the clean site than the polluted site plants. This

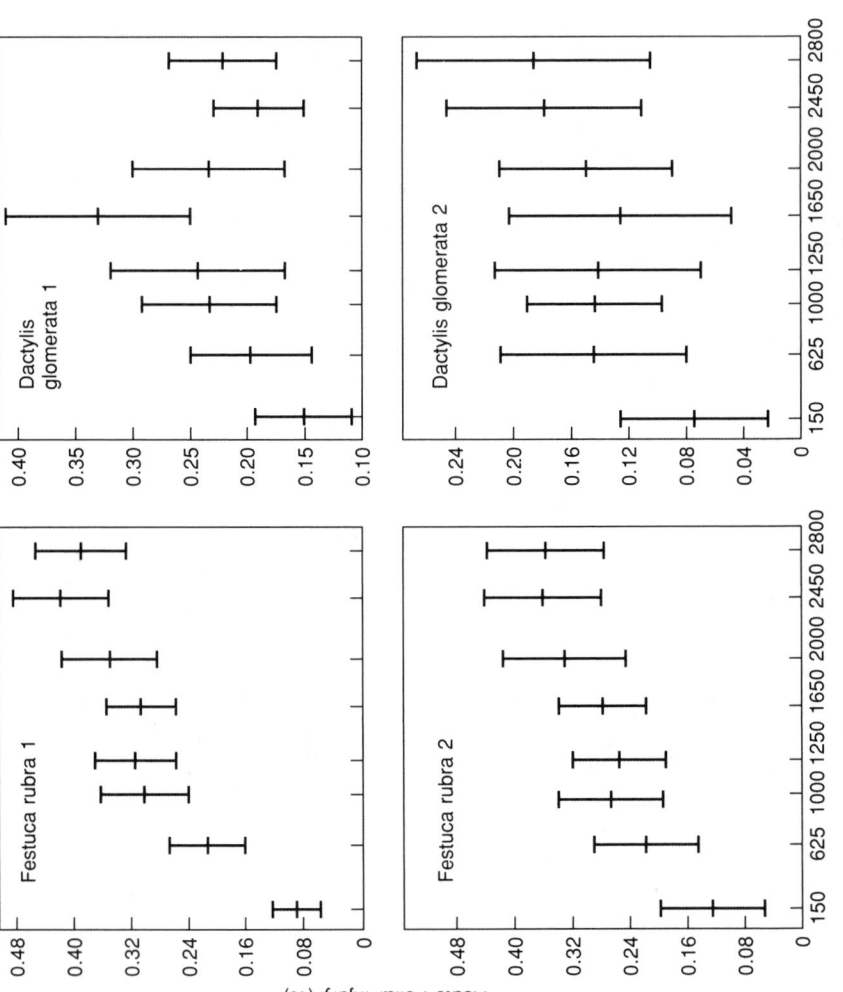

FIGURE 2. Acute injury (percentage after angular transformation) after replicate fumigations with 1700 to 2850 ppb SO_2, in grass populations collected at different distances from a point source of pollution at Askern (95% confidence limits shown). From Wilson and Bell 1986.

TABLE 3. Estimated rates of development of SO_2 resistance.

Species	Location	Ambient SO_2 (ppb) regime	Time (years)	Reference
Lolium perenne	Helmshore, U.K.	30–60 (winter)	<150	Ayazloo and Bell (1981)
Festuca rubra				
Dactylis glomerata				
Holcus lanatus				
Phleum bertolonii			<25	
Lolium perenne	Manchester, U.K.	44–67 (annual)	<17	Ayazloo and Bell (1981)
Festuca rubra	Askern, U.K.	46–53 (annual)	<50	Ayazloo and Bell (1981)
Dactylis glomerata				
Holcus lanatus				
Trifolium repens				Awang (1979)
Festuca rubra	London, U.K.	75–250 (annual)	<34	Wilson and Bell (1985)
Agrostis capillaris			<22	
Lolium perenne	Manchester, U.K.	44–67 (annual)	3.5 (acute)	Wilson and Bell (1985)
Poa pratensis			4.5 (chronic)	
Phleum pratense			4.5 (acute)	
Lolium perenne	Liverpool, U.K.	75–250 (winter)	200?	Horsman et al. (1979)
Lepidium virginicum	Copper Basin, TN	>500 3 h max. for 25% of days	<75	Murdy (1979)
Geranium carolinianum	Newnan, GA	Unspecified	<31	Taylor and Murdy (1975)
Bromus rubens	Nipomo mesa, CA	900 (mean max. daytime concentration)	<25	Westman et al. (1985)

resistance appeared to be associated with an avoidance mechanism, as the polluted site plants showed stomatal closure in the presence of SO_2, whereas no such response occurred in the clean site population.

Studies on the Rate of Development of Resistance

It is of considerable interest to understand the rate at which SO_2 resistance can develop and the pollutant concentrations which can induce this (Bradshaw and McNeilly, Chapter 2, this volume). Table 3 summarizes the available data from the published literature. It appears that SO_2 concentrations as low as 30 to 45 ppb are sufficiently deleterious to act as a selective force for resistance. In most cases it is difficult to determine with any accuracy the time scale over which resistance has evolved: of the work discussed so far, the shortest time involved is 17 years or less in the case of the sown *Lolium perenne* parkland at Philips Park (Ayazloo and Bell, 1981). However, there have been two further studies by Bell et al. (1982) and Wilson and Bell (1985) which have been aimed specifically at identifying the time over which resistance evolves.

One of these studies involved screening for resistance in lawns which had been established at different times over a period of approximately 200 years, adopting the approach of Wu et al. (1975) in their study of the evolution of copper tolerance in *Agrostis stolonifera*. The site concerned was the grounds of The Temple, a complex of buildings and gardens in central London, dating from medieval times, where annual mean SO_2 concentrations were estimated to have been around 75 ppb from the late 17th century, falling to around 27 ppb at the time of the study by Wilson and Bell (1985). Although there was some difficulty in ascertaining the ages of the earlier of The Temple lawns with any degree of accuracy, examination of historical records identified swards which were believed to have been laid down in ca. 1770, 1875, 1946, 1958, and 1977, respectively. In 1979 30 specimens each of *Agrostis capillaris*, *Dactylis glomerata*, and *Festuca rubra* were collected randomly from each lawn. The populations of each species were then subjected to an acute fumigation, prior to assessment of injury. An analysis of variance showed significant differences between populations from lawns of different ages in the case of *A. capillaris* and *F. rubra*, but not *D. glomerata*. *Festuca rubra* displayed a clear relationship with age, with the three youngest populations being the least resistant. However, *A. capillaris* did not show such a clear pattern because, while the 1958 and 1977 lawns contained plants which were considerably more susceptible to SO_2 than the 1875 and 1946 lawns, the oldest population also displayed a low degree of resistance. This study suggests that resistance can evolve within 22 to 34 years, but does not permit a more precise identification of the time scale concerned. Furthermore, the results should be viewed with some caution, in view of uncertainties over the age of the lawns and the origins of their populations.

The second study was carried out by Bell et al. (1982) and Wilson and Bell (1985) on populations in a polluted area over a much shorter time scale, with full knowledge of their origins, thereby overcoming the uncertainties associated with The Temple study. This involved sampling a set of experimental plots containing monocultures of five species in Philips Park, Manchester, situated alongisde the area studied by Ayazloo and Bell (1981). The plots were established in 1975 by the Sports Turf Research Institute as part of a study aimed at assessing the performance of different turf grass species and cultivars in a polluted area. The species studied for evolution of tolerance were *L. perenne* cv. S23, *L. multiflorum* cv. RvP, *Poa pratensis* cv. Baron, *Phleum pratense* cv. S48, and *F. rubra* cv. Engina. Randomly selected tillers of each species were sampled annually from 1976 to 1982, and screened for resistance to acute injury alongside individuals that had been grown from the original source of seed sown at Philips Park in 1975. In addition, a single chronic fumigation was carried out on all five species for the 1980 collection, with an extra screening for the 1979 collection of *L. perenne*, only.

No differences in the degree of acute injury were detected between the two populations for each species until 1979, when *L. perenne* collected that year showed significantly less injury than its corresponding original seed population (Table 4). This differential response was maintained in the 1980 collection, at which point the collections of *Poa pratensis* and *Phleum pratense* also showed the development of resistance. In 1981, resistance was not detected in either *L. perenne* or *Poa pratensis*, but remained in *Phleum pratense*. By 1982 there were no significant differences from the original seed populations in any of the species sampled. This work provides

TABLE 4. Percentage leaf injury on Philips Park (P.P.) and Original Seed (O.S.) populations after acute fumigations.

Species	Population	1976	1978	1979	1980	1981	1982
Lolium perenne	O.S.	20.5	23.5	9.7	42.8	4.8	5.5
	P.P.	18.5	20.1	4.3**	33.0*	4.9	4.8
Lolium multiflorum	O.S.	50.7	48.0	20.0	56.0	7.3	23.6
	P.P.	52.5	44.7	14.9	64.1	12.1	18.8
Poa pratensis	O.S.	6.1	9.4	3.4	44.5	12.7	
	P.P.	5.6	8.8	4.4	35.0*	10.7	
Festuca rubra	O.S.	4.5	4.4	4.1	24.5	9.8	17.6
	P.P.	4.4	3.9	8.1	20.7	9.7	23.2
Phleum pratense	O.S.	59.8	67.0	5.8	67.2	4.6	29.8
	P.P.	61.4	62.5	9.0	54.3	0.8*	35.5

From Wilson and Bell 1985.
* $p < 0.05$; ** $p < 0.01$

strong evidence for the selection of resistance to acute SO_2 injury at a polluted location where mean ambient annual concentrations were generally 40 to 60 ppb, in a period as short as 4 to 5 years. Furthermore, the genetic control of this resistance is suggested by its being maintained in plants grown from tillers of the 1979 *L. perenne* population after 5, 16, and 20 months, respectively, from the date of sampling (data not shown). While the *L. perenne* 1979 population also showed the development of resistance to acute injury, this was not detected for chronic injury until 1980 and was manifested in the form of a smaller reduction in live shoot and total dry weights than in the original seed population.

The Philips Park study throws up a number of interesting questions. First, why had resistance to acute injury apparently disappeared in all three species by 1982? An examination of local air pollution monitoring data reveals that this coincided with a marked fall in mean SO_2 concentration at Philips Park to 23 to 30 ppb. Wilson and Bell (1985) argued that this had reduced the selective pressure of SO_2, such that less resistant individuals were at a competitive advantage in that they have been demonstrated as invariably showing faster growth than resistant individuals when grown in clean air (Bell 1985). In this respect an interesting comparison can be drawn with the loss of industrial melanism with falling smoke levels (Askew et al. 1971). Second, it is surprising that resistance to acute injury appears to have evolved more rapidly and in more species than resistance to chronic injury, in view of the prevailing SO_2 concentrations at Philips Park. This raises the possibility that some form of invisible acute injury (Wilson and Bell 1985) may be taking place where the subcellular disturbance which leads to acute injury occurs but is insufficient to result in visible symptoms. It was also suggested by Wilson and Bell (1985) that this injury might have resulted from a SO_2 frost interaction and, indeed, that the resistance that developed at Philips Park might be a response to a combination of these stresses. It would seem extremely probable that this "invisible" acute injury would result in marked growth reductions, as Roose et al. (1982) considered that a fall in productivity of 5 to 10% per year in sensitive genotypes in the presence of air pollution would be necessary for evolution to occur on a time scale of years as opposed to decades. Indeed, growth reductions larger than this have been demonstrated elsewhere in Manchester over periods of less than 1 year in experiments where the performance of *L. perenne* in chambers ventilated with ambient air was compared with a clean air control (Bleasdale 1973). Finally, it should be stressed that the experimental plots at Philips Park were mown regularly, thereby minimizing the possibility of reproduction by seed or of gene flow into the population. As such, selection for tolerance must have been acting primarily via competition between genotypes which were present at sowing. The considerable variation in response to SO_2 demonstrated in many of the studies described in this section clearly indicates the potential for this to take place.

Ozone

Ozone (O_3) is a major pollutant which causes visible leaf injury, reduces crop yield, and affects forest vitality in North America, western Europe, and other regions of the world. The effects of O_3 on vegetation are experienced over large areas, and thus its overall impact is frequently greater than that of other pollutants which have a more local distribution. The existence of variation in sensitivity to O_3 between bred cultivars has been demonstrated in many studies; this variation has had some commercial value, e.g., in reducing O_3 damage to tobacco crops. Studies have shown a genetic basis for this tolerance in a range of crop species (e.g., Engle and Gableman 1966; Butler et al. 1979) and one tree species (Karnosky 1977).

Despite the clear potential for natural selection in the large areas in which adverse effects of O_3 have been demonstrated, there is very little good evidence of the evolution of tolerance in the field. No studies exist of the temporal changes in O_3 sensitivity in the field, and the only substantive evidence is drawn from comparisons of populations collected from different locations.

The first report of evolution of resistance was that of Dunn (1959) who investigated smog damage to different genetic lines of *Lupinus bicolor* in the Los Angeles basin. This work was carried out before the chemical constitutents of the smog which were responsible for the observed damage had been identified; however, it is likely that ozone was the major compound involved. Dunn (1959) found that a population based on seed collections made in Los Angeles was less susceptible than those derived from other parts of the California coast or inland sites; the population from Los Angeles was able to set seed successfully, whereas the others were not.

Subsequent studies have followed a similar procedure of comparing the O_3 sensitivity of populations collected from different locations. The most convincing evidence of evolution of O_3 resistance in populations from different locations is the work of Berrang et al. (1986a, 1989), with *Populus tremuloides*. They claim to have detected natural selection for O_3 resistance in populations of *Populus tremuloides* from five different national parks in the United States. Continuous records of O_3 concentrations were available within 100 km of each park. The O_3 sensitivity of individuals vegetatively propagated from root collections made in each park was initially assessed by a single fumigation with 180 ppb O_3 for 6 h 1 and 2 years after collection. This material was then used to establish a field planting at a site in New York State subject to relatively high levels of ambient O_3, with the extent of ozone injury being assessed at the end of the next two summers.

In both the acute fumigation and the field trial, little difference between populations was observed after 1 year. In the second year, however, significant differences were found between the populations in the amount of O_3 injury in both the laboratory fumigation and field trial. In particular, more injury was found in both experiments on populations from the two parks with annual mean O_3 concentrations below 30 ppb compared with populations from the three parks with annual mean O_3 concentrations above 40 ppb. A much wider range of visible injury between clones within each population was found in the field trial than in the laboratory fumigation. Within each population, there were clones showing no visible injury in the field trial; the major difference was the greater degree of leaf injury in the most sensitive clones of the populations from parks with low ambient O_3 concentrations, indicating that sensitive clones are less common in populations from areas with higher O_3 concentrations. Berrang et al. (1989) suggest that such genotypes are not killed directly by O_3, but are eliminated though intraspecific competition for light.

A similar study of *Acer rubrum* populations collected from nine national parks was also carried out by Berrang et al. (1986b). Populations from the different parks did develop significantly different amounts of leaf injury after fumigation with 250 ppb O_3 for a total of 24 h, but there was no clear relationship to the current ambient O_3 levels at, or close to, the nine parks. *Acer rubrum* is less sensitive to O_3 than *Populus tremuloides* in terms of visible leaf injury, and this may explain the less convincing evidence of evolution of O_3 resistance for this species. Alternatively, O_3 selection pressure may operate over longer timescales for the more resistant species; in this case O_3 sensitivity of populations may be more closely related to ambient concentrations over several decades than to current levels. However, in the absence of historical records of ambient O_3 concentrations at the locations used by Berrang et al., this possibility cannot be rigorously evaluated.

The absence of data on ambient O_3 concentrations makes other claims of evolution of O_3 resistance based on comparison of populations impossible to evaluate critically. Thus, studies such as those of Johnston et al. (1983), who compared the O_3 sensitivity of *Festuca arundinacea* clones from North America, western Europe and north Africa, and Ernst et al, (1985), who compared the O_3 sensitivity of *Silene cucubalus* populations from East and West Germany, cannot be accepted as evidence of the evolution of O_3 resistance because of the absence of data on O_3 concentrations at the collection sites. A further difficulty arises with a study by Dueck et al. (1987) who compared the O_3 sensitivity of populations from the north and south of the Netherlands. The ranking of the collection sites differed according to whether long-term mean concentration or peak concentration was used as the index of O_3 levels. In the absence of a clear understanding of the relative importance of long-term mean O_3 concentrations or peak O_3 concentrations as selective forces in the field, it is impossible to judge which index of O_3 exposure is the most appropriate.

However, there is another important problem with studies which rely on a comparison of populations from different locations to establish the existence of selection of O_3 tolerance. The existence of such variation in sensitivity to O_3 is not proof that O_3 was the selective force causing this difference between populations from different geographical areas. No study provides evidence of the evolution of a mechanism of resistance specific to O_3. In the only study providing good data on O_3 concentrations at the collection sites, that of Berrang et al., the sensitive and tolerant populations were separated by distances of 1000 km or more, and must have experienced many other different selective forces. Thus, the evidence of natural selection of O_3 resistance in the field can only be described as unconvincing.

Nitrogen Oxides and Ammonia

There is increasing concern about the impact of excess nitrogen input on forests and a range of other ecosystems. Although the additional nitrogen may in itself initially act to stimulate plant growth, it may also tend to increase the susceptibility of the plant to secondary stresses such as insect attack, frost, or pathogen invasion. At high concentrations, nitrogen oxides and ammonia may have direct toxic effects on vegetation. Thus, natural selection pressures may act both in favor of genotypes able to respond positively to additional nitrogen input and those with tolerance of higher toxic concentrations.

There is only one study of the evolution of resistance to NO_2 in populations close to an industrial works emitting high levels of nitrogen oxides. Taylor and Bell (1988) examined the sensitivity of populations of *Dactylis glomerata* and *Lolium perenne* collected from close to a nitrogen fertilizer factory, at Rainham in east London, where mean NO_2 concentrations were 50 to 100 ppb, and peak concentrations may have been considerably higher. Populations of *Dactylis glomerata* from Rainham showed less leaf injury when exposed to acute doses of NO_2 than did plants of the commercial cultivar S26 grown at a rural site: however, *Lolium perenne* showed no consistent difference between the Rainham population and a commercial cultivar.

Taylor and Bell (1988) also compared the responses to lower concentrations of NO_2 of the Rainham population of *Lolium perenne* and a commercial cultivar (S23), when grown in soils containing either 100 mg kg^{-1} N or 10 mg kg^{-1} N. Exposure to 200 ppb NO_2 for 11 weeks caused a significant increase in growth in both populations and on both soils. There was a significant difference between populations only in terms of leaf senescence and root/shoot ratio. The proportion of dead leaf at final harvest was higher in the Rainham population in filtered air, but lower in the Rainham population in the presence of NO_2. In the case of root/shoot ratio, the populations did not differ in NO_2, but the root/shoot ratio was lower in the Rainham population in filtered air. These results indicate the evolution of a greater requirement for nitrogen, which can partly be met by atmospheric NO_2, in the Rainham population.

No other study has reported on the responses of populations from areas with defined differences in NO_2 concentration. This may partly reflect the paucity of industrial sources of nitrogen oxides alone; most combustion processes emit both SO_2 and nitrogen oxides, and in general, elevated concentrations of NO_2 are associated with elevated concentrations of SO_2. Evidence of resistance to NO_2 in situations where evolution of resistance to SO_2 may also be occurring is considered below. There has also been no study reported on the development of tolerance to NO, although the steeper concentration gradients of this primary pollutant, compared with those of NO_2, might offer better opportunities for investigation.

Another nitrogen-containing gas, NH_3, is found in high concentrations close to intensive animal rearing units. The only study of this pollutant in the literature (Dueck et al. 1987) provided no clear evidence of differential responses to NH_3 in populations from different regions of the Netherlands. The ecological impact of NH_3 and other nitrogen gases may be considerable, and it would be valuable to have better understanding of their selective pressure in the field.

Mixtures of Gaseous Pollutants

Air pollutants rarely occur in isolation. In most situations, two or more pollutants are present in combination in the atmosphere (McLaughlin and Norby, chapter 4, this volume). There is good evidence that the impact of such pollutant mixtures may be greater than that of the individual pollutants (Ashenden and Mansfield 1978; Mooi, 1984), and this suggests that pollutant mixtures may exert a greater selective pressure. A number of possibilities exist in terms of the evolution of resistance to air pollution:

1. In situations in which one pollutant is dominant in its effects, the presence of other pollutants may simply alter the selective pressure exerted by the major pollutant.
2. Where several pollutants are present in potentially damaging concentrations, common mechanisms of pollutant avoidance (e.g., changes in growth, form, leaf characters, and stomatal conductance) may evolve.
3. Alternatively, true tolerance of several pollutants might evolve simultaneously. However, this would involve several different physiological changes, and seems intuitively less likely than (2).

To analyze which of these possibilities have actually occurred in the field, one needs at minimum a good description of the pollutant concentrations experienced in the field; experimental exposure to realistic concentrations of those pollutants; and a comparison of the effects of different individual pollutants and pollutant combinations. Here we have concentrated on work which satisfies these conditions in at least one aspect.

Nitrogen dioxide and sulfur dioxide occur in similar concentrations in many parts of western Europe, and there is evidence for synergistic effects of these pollutants. Several studies have compared the sensitivity of populations to these two pollutants and to their mixture. In most cases, these populations are thought to have evolved in response to SO_2. For example, Taylor and Bell (1988) reported on the sensitivity to an acute dose of 4700 to 6200 ppb NO_2 plus 1400 to 1500 ppb SO_2 of populations from Philips Park, which had been shown in some cases to develop tolerance to SO_2 (Wilson and Bell 1985); Taylor (1986) also examined the responses of these populations to a similar concentration of NO_2 alone. The results showed variability between species and collection dates and a lack of consistency in successive fumigations of the same population. Nevertheless, a consistent pattern emerged that the populations from the polluted site at Philips Park, where evolution of resistance to acute SO_2 injury has been shown in *Lolium perenne*, were more sensitive to both NO_2 and the SO_2/NO_2 mixture. This result strongly suggests that selection for resistance to acute SO_2 does not confer resistance to NO_2 or NO_2/SO_2 mixtures.

Differential sensitivity to SO_2, NO_2, and their mixture has also been found in populations of *Plantago major*. Populations from Hyde Park (in central London) and from Ascot (a rural site 40 km from London) were exposed to acute doses of SO_2 (800 ppb for 6 h). NO_2 (3300 ppb for 6 h), and $SO_2 + NO_2$ (700 ppb + 1600 ppb for 6 h) by Taylor (1986). The London population showed less injury in SO_2, but more injury in NO_2, while in the SO_2/NO_2 mixture no significant difference between the populations was found.

A comparison of the sensitivity of grass populations to chronic fumigation with SO_2, NO_2, and $SO_2 + NO_2$ was made by Ashmore et al. (1984). Seed of six populations of *Lolium perenne* was collected from southern Poland. The pollutant concentrations at these sites were not defined, but it is likely that concentrations of SO_2 were much greater than those of NO_2. Exposure to 140 ppb $SO_2 + 200$ ppb NO_2 reduced plant growth, but there was no clear difference between the three populations from highly polluted areas and the three from less polluted areas; in contrast exposure to SO_2 alone reduced the growth of populations from highly polluted areas by 37% and the others by 63%. This conclusion was reinforced by data on dry weight allocation within the plants. Exposure to NO_2 alone had little significance for plant growth overall. These results suggest that in a region where SO_2 is likely to be the most important air pollutant, evolution of resistance to chronic doses of SO_2 alone does not necessarily confer resistance to $SO_2 + NO_2$ mixtures.

This conclusion is consistent with the work of Dueck (1987), who reported the response of a nominally SO_2-tolerant ecotype of *Silene cucubalis* to a mixture of 30 ppb O_3, 14 ppb SO_2, and 30 ppb NO_2 for 12 weeks, with that of nominally metal-sensitive, Cu-tolerant, and Zn-tolerant ecotypes on different soils. The results provide little evidence that the impact of the pollutant mixture was reduced on the SO_2-tolerant ecotype compared with the others.

The conclusion that evolution of resistance to SO_2 alone does not confer resistance to NO_2 or to NO_2/SO_2 mixtures, and may indeed lead to increased sensitivity to these pollutants, is supported by studies of a SO_2-resistant clone of *Lolium perenne*. This clone (S23 Res), which was identified from a fumigation experiment with SO_2 alone, was shown by Ayazloo (1979) to be more resistant to acute doses of SO_2 than the commercial cultivar S23. However, Taylor and Bell (1988) found that fumigation for 6 h with either 4800 ppb NO_2 or 4800 ppb + 1400 ppb SO_2 produced more injury on the SO_2-resistant genotype than on the commercial cultivar.

The sensitivity of populations to SO_2, NO_2, and $SO_2 + NO_2$ has also been compared in one situation in which NO_2, rather than SO_2, is thought to have been the dominant air pollutant in the field. Taylor and Bell (1988) demonstrated that a population of *Dactylis glomerata* collected close to a nitrogen fertilizer factory at Rainham in east London, were more resistant to fumigation with 4700 ppb NO_2 for 6 h than plants grown from the seed of a commercial cultivar. This population was also more resistant to fumigation with 4700 ppb NO_2 + 200 ppb SO_2 than the same commercial cultivar. Longer-term fumigation with 200 ppb NO_2 + 140 ppb SO_2 also provided evidence that the Rainham population was more resistant to adverse effects of $SO_2 + NO_2$, as manifested by increased dead leaf and reduced flowering. However, there was no evidence that this population was more resistant to SO_2 alone (Taylor 1986).

There is little information on the evolution of resistance to other pollutant mixtures. The responses of two populations of *Silene cucubalis* originating from a rural site in West Germany and an industrial site in East Germany to 55 ppb SO_2, 35 ppb O_3, or their mixture, have been compared by Ernst et al. (1985). The industrial population exhibited tolerance to the effects of SO_2 alone, with increased biomass compared with the control. Biomass was significantly reduced in both populations in $SO_2 + O_3$, but the reduction was greater in the rural population (73%) than the industrial population (56%). However, the industrial population had a greater reduction in biomass in O_3 alone than did the rural population. Sulfur content of foliage was increased to a similar extent in both populations in SO_2 alone, but the sulfur content of plants exposed to $SO_2 + O_3$ did not differ significantly from those receiving filtered air. It appears that in this case an increased resistance to SO_2 in the industrial population appears to confer resistance to $SO_2 + O_3$, but not to O_3 alone. A wider range of pollutant mixtures were studied by Dueck et al. (1987), who reported on the sensitivity of populations of *Agrostis capillaris* from four locations in The Netherlands to mixtures of $SO_2 + NH_3$, $SO_2 + O_3$, and $SO_2 + NO_2 + O_3$. The highly variable character of their results make it difficult to draw any firm conclusions, and no significant gas/population interactions were found in any of the experiments which they reported.

In summary, there seems to be no evidence from the small amount of available data of evolution of resistance specifically to pollutant mixtures, as opposed to individual pollutants. Equally, there seems to be no evidence from the very few experiments in which responses to individual pollutants, as well as their mixtures, have been examined, of an evolution of resistance to all components of the pollutant mix. Rather, the picture is of variable responses to individual pollutants (in some cases populations resistant to SO_2 are more sensitive to NO_2 or O_3), with responses to the pollutant mix perhaps depending on the dominant pollutant in the mixture.

However, with the very limited amount of relevant data available these conclusions can only be tentative. Furthermore, none of the studies used more than one set of concentrations. The results observed may depend on the relative concentrations of the pollutants used, and how characteristic these are of those actually experienced in the field. There is a clear need for more rigorous examination of responses of populations growing in different pollutant mixes in the field, when exposed to a range of pollutant concentrations under controlled experimental conditions. However, this would require extensive resources to carry out very large multifactorial experiments with sufficient replication that differences in population responses could be accurately assessed.

Wet Deposition

Wet deposition of acidity, or of sulfate, nitrate or ammonium ions may have wide ecological impact. However, the secondary nature of these forms of pollutant, and their time scale of formation, mean that, as in the case of O_3, there are no consistent sharp gradients in the rate of wet deposition. Thus, good field situations for the study of evolution of resistance are not available.

One attempt has been made to compare the responses of populations from different localities to simulated acid rain. Hodgkin & Briggs (1981) collected seed of *Senecio vulgaris* from Cambridge, in eastern England, and from the west coast of Wales. Application of pH 2.5 rain caused a significantly smaller reduction in leaf growth and flower bud production in the Cambridge population, which was adduced as evidence of tolerance.

However, data from recent national surveys of acid deposition suggest that there is little difference in hydrogen ion deposition between the two locations.

Conclusions

The evolution of changes in sensitivity of vegetation in response to atmospheric pollution is a surprisingly understudied subject, particularly in view of the extensive literature on intraspecific differences in resistance which suggests the potential for selection in the field. In contrast, numerous studies have been performed on the development of resistance to heavy metals, which range from the subcellular to the population levels. Such differences in emphasis may result from the generally less obvious effects of ambient air pollution on plant performance at most locations, compared with the very clear damaging influence of heavy metals on mine and smelter wastes, together with the interest in developing resistant material for the revegetation of such places.

Only in the case of SO_2 is it possible to be confident that such a phenomenon is widespread, at least in grassland species. While many of the studies involving SO_2 can be criticized on a number of grounds, the evidence indicates strongly that resistance has evolved in the field and that marked genetic shifts in populations can occur rapidly in response to changing air quality. In the case of other pollutants, information is very sparse and, in the case of fluorides, nonexistent.

Only a small number of parameters have been used to assess the evolution of resistance to air pollutants. The great bulk of studies have used vegetative characters for growth studies, or the measurement of acute injury in short-term high concentration fumigation experiments. In only a few studies have any reproductive characters been assessed. This is an unfortunate omission as, clearly, resistance to adverse impacts on the reproductive process is a likely selective response to air pollution. Physiological and biochemical responses have been equally neglected, and have scarcely been studied at the population level as opposed to selected resistant genotypes. The demonstration of resistance in key physiological processes would greatly enhance the value of studies where resistance has been demonstrated for growth parameters, particularly if a causal link between these can be established with some degree of confidence.

There is also an extremely unequal distribution of studies between different species, indeed between broad categories of plant type. For SO_2 research has concentrated on perennial pasture grasses; studies on trees are completely absent, lower plants have been ignored, and there are only three studies on annual species. For other pollutants, the paucity of information means that only a very small number of species have been studied. At the moment, we have little understanding of how a species' life history may influence the evolution of resistance to air pollutants; the only clear conclusion at the present time is that resistance can evolve in the absence of reproduction, presumably by competitive exclusion within, for example, grass swards or forest canopies. Knowledge of the responses of a wider range of plant species would assist in furthering this understanding.

The quality of the evidence for evolution of resistance presented in many studies is poor. Few studies have demonstrated changes in resistance to air pollutants through time, and most have relied on comparisons of populations collected from different locations experiencing different pollutant concentrations. This approach is particularly problematical with pollutants, such as O_3, which are regionally distributed, and require populations to be sampled at widely separated locations. With the accumulating evidence of the impact of dry and wet deposited pollutants over wide areas remote from pollutant soures, there is an urgent need for better designed experiments to determine the extent of evolution of resistance to air pollution in such areas.

Some of the possible pitfalls of using comparisons of the responses of populations collected from different locations at one point in time to infer evolution of resistance to an air pollutant are well illustrated by the recent work of Baxter et al. (1989a, 1989b) on the effects of bisulfite on the moss *Sphagnum cuspidatum*. Populations of this ombrotrophic bog species were collected from two areas of the British Isles—the southern Pennines, an area from which *Sphagnum* has almost entirely disappeared since the Industrial Revolution, primarily as a result of atmospheric pollution; and north Wales, a relatively unpolluted area in which *Sphagnum* species still dominate the mire vegetation. Treatment in the laboratory with the SO_2 solution product, bisulfite, resulted in more visible injury and greater reductions in extension growth and photosynthetic rate in a population of *Sphagnum cuspidatum* collected from north Wales than in a population collected from the southern Pennines (Baxter et al. 1989a). However, it proved wrong in this case to conclude that resistance to bisulfite toxicity had developed in the remaining southern Pennine populations; more detailed studies of the mechanisms of bisulfite resistance (Baxter et al. 1989b) indicate that the critical factor is the presence of higher concentrations of extracellular iron in the southern Pennine population. Iron is known to catalyze the oxidation of bisulfite to sulfate, an anion which is much less toxic to *Sphagnum*. Bisulfite is lost from solution much more rapidly in the presence of *Sphagnum* from the southern Pennines than in the presence of *Sphagnum* from north Wales. Resistance to bisulfite can be conferred on the north Wales population by treatment with iron in artificial rainwater, and can be lost from the southern Pennine population by treatment with EDTA (which removes iron bound to the moss surface). It seems likely that the higher concentrations of iron bound to *Sphagnum*, and in bog pool water, in the southern Pennines, as compared with north Wales, are due to higher deposition rates arising from industrial activity.

Thus, the apparent resistance of *Sphagnum* to the SO_2 solution product bisulfite is primarily due to the presence of higher concentrations of iron at the more polluted sites. The "resistance" is conferred by another environmental factor (in this case another pollutant) rather than the pollutant which might have naively been identified as the selective force in the absence of a detailed study of the mechanism of resistance to bisulfite in the two populations.

It is particularly unfortunate that field studies on the evolution of resistance to O_3 are fraught with such considerable difficulties. A useful approach to this pollutant could be a field trial, of the type performed for SO_2 by Wilson and Bell (1985), where plots would be established in a high O_3 location, using material obtained from a clean site, and changes in tolerance studied over a time course. Alternatively, it might be possible to screen for sensitivity wild material which had been collected and then stored as seed or propagated vegetatively from the same area at different times during a period when O_3 concentrations progressively rose. In view of the extreme sensitivity of many plant species to O_3 and its demonstrated importance in the field, including the induction of interspecific changes in community composition, this is an important and potentially fruitful topic for investigation.

It is increasingly recognized that impacts on vegetation may arise from the effects of several pollutants acting together, or from an increased sensitivity to other biotic and abiotic stresses as a result of subtle physiological changes induced by pollutants. The direct effects of high concentrations of individual pollutants are becoming of less importance due to more effective pollution control measures. Furthermore, the measures being taken to control emissions may lead to decreasing concentrations of major air pollutants in many areas over the coming decades.

These major changes in the distribution and concentrations of air pollutants raise important questions concerning the evolution of resistance. There are few studies of resistance to pollution mixtures representative of those currently found in the field; however, there is evidence that the evolution of resistance to SO_2 leads to increased sensitivity to NO_2, and it is clear that co-evolution of resistance to all pollutants present may not necessarily occur. Thus, the relative concentrations of different pollutants may be important for determining how resistance evolves. We have no information on the importance of secondary effects of pollutants as a mechanism in the evolution of resistance in the field, although increasing sensitivity to frost damage has been proposed as a reason for the rapid evolution of resistance to SO_2 in grasses (Wilson and Bell 1985). The importance of episodic exposure to high concentrations of pollutants, as opposed to chronic exposure to lower concentrations, is also unclear. Finally, the rates of change of resistance induced by changing air quality require investigation, in order to establish how rapidly different types of vegetation may evolve in response to the complex changes in air pollution concentration patterns which are currently occurring. In particular, changes in resistance in response to decreasing ambient concentrations are poorly understood at present.

It is thus clear from this review that there is a vast field of investigation that awaits the involvement of air pollution research workers and funding agencies. Evolution of resistance to air pollution is an important process occurring in the field and the study of it is essential if the true impact of air pollution on wild and minimally managed ecosystems is to be understood fully.

References

Ashenden TW, Mansfield TA (1978) Extreme pollution sensitivity of grasses when SO_2 and NO_2 are present in the atmosphere together. Nature 273:142–143

Ashmore MR, Bell JNB, Godzik S (1984) Differential tolerance to SO_2, NO_2 and their mixture in *Lolium perenne* L. populations from southern Poland. Bulletin of the Polish Academy of Science, Biological Sciences 32:339–345

Askew RR, Cook LM, Bishop JA (1971) Atmospheric pollution and melanic moths in Manchester and its environs. Journal of Applied Ecology 8:247–256

Awang MB (1979) The effects of sulphur dioxide pollution on plant growth with special reference to *Trifolium repens*. Ph.D. thesis, University of Sheffield

Ayazloo M (1979) Tolerance to sulphur dioxide in British grass species. Ph.D. thesis, University of London

Ayazloo M, Bell JNB (1981) Studies on the tolerance to sulphur dioxide of grass populations in polluted areas. I. Identification of tolerant populations. New Phytologist 88:203–222

Baxter R, Emes MJ, Lee JA (1989a) Effects of bisulphite ion on growth and photosynthesis in *Sphagnum cuspidatum* Hoffm. New Phytologist 111:457–462

Baxter R, Emes MJ, Lee JA (1989b) The relationship between extracellular metal accumulation and bisulphite tolerance in *Sphagnum cuspidatum* Hoffm. New Phytologist 111:463–472

Bell JNB (1985) SO_2 effects on the productivity of grass species. In: Winner WE, Mooney HA, Goldstein RA (eds) Sulfur dioxide and vegetation. Stanford University Press, Stanford, pp 209–226

Bell JNB, Ayazloo M, Wilson GB (1982) Selection for sulphur dioxide tolerance in grass populations in polluted areas. In: Bornkamm R, Lee JA, Seaward MRD (eds) Urban ecology. Blackwell Scientific Publications, Oxford, pp 171–180

Bell JNB, Clough WS (1973) Depression of yield in ryegrass exposed to SO_2. Nature 241:47–49

Bell JNB, Mudd CH (1976) Sulphur dioxide resistance in plants: a case study of *Lolium perenne*. In: Mansfield TA (ed) Effects of air pollution on plants. Cambridge University Press, Cambridge, pp 88–103

Berrang P, Karnosky DF, Bennett JP (1989) Natural selection for ozone tolerance in *Populus tremuloides*: field verification. Canadian Journal of Forest Research 19:519–522

Berrang P, Karnosky DF, Mickler RA, Bennett JP (1986a) Natural selection for ozone tolerance in *Populus tremuloides*. Canadian Journal of Forest Research 16:1214–1216

Berrang P, Karnosky DF, Mickler RA, Bennett JP (1986b) Population changes in eastern hardwoods caused by air pollution. Proceedings of the North American Forest Biology Workshop, pp 3–10

Bleasdale JKA (1973) Effects of coal-smoke pollution gases on the growth of ryegrass (*Lolium perenne* L.). Environmental Pollution 5:275–285

Bradshaw AD, McNeilly T (1981) Evolution and pollution. Edward Arnold, London

Butler LK, Tibbitts TW, Bliss FA (1979) Inheritance of resistance to ozone in *Phaseolus vulgaris*. Journal of the American Society of Horticultural Science 104:211–213

Dueck TA, Dil EW, Pasman FJM (1987) Adaptation of grasses in the Netherlands to air pollution. New Phytologist 108:167–174

Dunn DB (1959) Some effects of air pollution on *Lupinus* in the Los Angeles area. Ecology 40:621–625

Engle RL, Gabelman WH (1966) Inheritance and mechanism for resistance to ozone damage in onion. Proceedings of the American Society of Horticultural Science 89:423–430

Ernst WHO, Tonneijck AEG, Pasman FJM (1985) Ecotypic response of *Silene cucubalus* to air pollutants (SO_2, O_3). Journal of Plant Physiology 118:439–450

Hodgkin SE, Briggs D (1981) The effects of simulated acid rain on two populations of *Senecio vulgaris* L. New Phytologist 89:687–691

Horsman DC, Roberts TM, Bradshaw AD (1978) Evolution of sulphur dioxide tolerance in perennial ryegrass. Nature 276:493–494

Horsman DC, Roberts TM, Bradshaw AD (1979a) Studies on the effect of sulphur dioxide on perennial ryegrass (*Lolium perenne* L.) II Evolution of SO_2 tolerance. Journal of Experimental Botany 30:495–501

Horsman DC, Roberts TM, Lambert M, Bradshaw AD (1979b) Studies on the effect of sulphur dioxide on perennial ryegrass (*Lolium perenne* L.) I Characteristics of fumigation system and preliminary experiments. Journal of Experimental Botany 30:485–493

Johnson WJ, Haaland RL, Dickens R (1983) Inheritance of ozone resistance in tall fescue. Crop Science 23:235–236

Karnosky DF (1977) Evidence for genetic control of response to sulfur dioxide and ozone in *Populus tremuloides*. Canadian Journal of Forest Research 7:437–440

Kettlewell HBD (1955) Selection experiments on industrial melanism in the Lepidoptera. Heredity 9:323–342

Lees DR, Creed ER, Duckett JG (1973) Atmospheric pollution and industrial melanism. Heredity 30:227–232

Mooi J (1984) Wirkungen von SO_2, NO_2, O_3 und ihrer Mischungen auf Pappeln und einige under Pflanzenarten. Der Forst- und Holzwirt 18:438–444

Murdy WH (1979) Effect of SO_2 on sexual reproduction in *Lepidium virginicum* L. originating from regions with different SO_2 concentrations. Botanical Gazette 140:299–303

Prat S (1934) Die Erblichkeit der Resistenz gegen Kupfer. Berichte Deutsche Botanische Gesellschaft 52:65–67

Roose ML, Bradshaw AD, Roberts TM (1982) Evolution of resistance to gaseous air pollutants. In: Unsworth MH, Ormrod DP (eds) Effects of gaseous air pollution in agriculture and horticulture. Butterworth Scientific, London, pp 379–409

Taylor GE (1978) Genetic analysis of ecotypic differentiation within an annual plant species, *Geranium carolinianum* L. in response to sulfur dioxide. Botanical Gazette 139:362–368

Taylor GE, Murdy WH (1975) Population differentiation of an annual plant species, *Geranium carolinianum*, in response to sulfur dioxide. Botanical Gazette 136:212–215

Taylor HJ (1986) Tolerance of plant species to nitrogen dioxide and sulphur dioxide. M.Phil., University of London

Taylor HJ, Bell JNB (1988) Studies on the tolerance to SO_2 of grass populations in polluted areas. V. Investigations into the development of tolerance to SO_2 and NO_2 in combination and NO_2 alone. New Phytologist 110:327–338

Westman WE, Preston KP, Weeks LP (1985) SO_2 effects on the growth of native plants. In: Winner WE, Mooney HA, Goldstein RA (eds) Sulfur dioxide and vegetation. Stanford University Press, Stanford, pp 264–280

Wilson GB, Bell JNB (1985) Studies on the tolerance to SO_2 of grass populations in polluted areas. III. Investigations on the rate of development of tolerance. New Phytologist 100:63–77

Wilson GB, Bell JNB (1986) Studies on the tolerance to sulphur dioxide of grass populations in polluted areas. IV. The spatial relationship between tolerance and a point source of pollution. New Phytologist 102:563–574

Wu L, Bradshaw AD, Thurman DA (1975) The potential for evolution of heavy metal tolerance in plants. III. The rapid evolution of copper tolerance in *Agrositis stolonifera*. Heredity 34:165–187

4
Atmospheric Pollution and Terrestrial Vegetation: Evidence of Changes, Linkages, and Significance to Selection Processes[1]

SAMUEL B. MCLAUGHLIN and RICHARD J. NORBY

The industrial and urban growth of the past 50 years have produced significant evidence that the earth's atmosphere, soil, and waters are not open systems. This evidence comes in the form of changing concentrations of chemicals that affect physical, chemical, and biological processes (National Academy of Sciences, 1986). In some cases there is evidence that some of these processes themselves are being changed. The linkages between pollution and biological changes may be very apparent where concentration gradients between sources and receptors are strong, and induced changes include easily quantifiable alterations in visible appearance or measurable biological performance. Of potentially greater concern, are the more subtle chronic changes occurring over large regions that may take many years to develop. A major source of that concern has been the effects of atmospheric pollutants on terrestrial biota.

Evaluating the potential effects of atmospheric pollution on terrestrial ecosystems represents a particularly significant research challenge. The spatial scales of exposure range from local and regional to global, and the pathways from sources to receptors often involve formation or induction of secondary pollutants. For this reason terrestrial systems have been exposed to changes in multiple pollutants from diverse industrial sources and widespread geographical origins. These pollutants have a wide range of potential effects, including pollutant interactions. The effects themselves are strongly influenced by natural environmental stresses and may produce symptoms that mimic or amplify those that occur naturally. Responses may also be bimodal since some of the pollutants may be utilized as nutrients under some conditions and cause toxic reactions under others. For these reasons many of the specific pathways linking pollution emissions and

[1] Publication No. 3598, Environmental Sciences Division, ORNL

environmental damage are poorly characterized and poorly understood.

Although there are still many uncertainties regarding specific linkages, there is also a considerable amount of evidence to indicate that chronic pollution stress has had and will continue to have an adverse effect on terrestrial vegetation. Such effects may ultimately have significant implications for genetic selection of species based not only on sensitivity of species to pollution per se, but because pollution stress may alter plant sensitivity to a wide array of secondary stresses. This chapter will examine, in sequence, chemical indicators that atmospheric pollution has increased in recent decades; indications that adverse changes in the condition of vegetation, principally forests, have occurred over widespread areas; and mechanistic evidence to evaluate the linkages between these changes and physiological mechanisms of effect on plants. These discussions will be framed in the context of providing an evaluation of the potential of pollution-derived stresses to exert significant selective pressures on plant growth and development.

The scope of the coverage will be limited to atmospheric changes that have the strongest potential for regional scale impacts on vegetation. These include sulfur and nitrogen oxides and associated acidic deposition, tropospheric ozone, CO_2, and ultraviolet radiation (UVB). Discussion of mechanistic linkages will emphasize responses that indicate that air pollution at *current ambient levels* may be a significant new selective force in our environment.

Criteria for Emphasis on Selected Pollutants as Potential Contributors to Environmental and Genetic Selection Processes

Pitelka (1988) has discussed some of the current uncertainties regarding effects of anthropogenic pollutants on evolutionary processes in plants, focusing primarily on the genetic and physiological characteristics of plants and plant populations. It is the interplay of those properties, the chemical and physiological toxicity of pollutants, and their ambient concentrations, that will determine whether chronic pollution stress represents a potentially significant selective force.

Two processes noted by Pitelka (1988) also play an important role in the extent to which selective pressures are translated into alterations in plant species, populations, and communities. These are plant competition, which determines the net effect of pollution pressure on plant success in mixed species communities, and plant repair mechanisms, which determine costs to the plant either in terms of energy expended to restore cellular or biochemical integrity or in terms of loss in these attributes if repair does not occur.

TABLE 1. Some pollutant-derived changes in plant growth environment.

Influencing factors	Principal effects
SO_x, NO_x, acid deposition	Plant nutrition, metal mobilization, and reproductive physiology
Toxic metals	Plant nutrition, root growth
Ozone	Carbohydrate assimilation and allocation, membrane integrity
CO_2	Carbon assimilation and allocation, water and nutrient use
UVB radiation	Biochemical integrity of cytological and genetic systems

The pollutant stresses discussed in this chapter were included based on current evidence that changes in community structure are already occurring in areas where levels are highest or because they are widely distributed and possess high potential to alter plant communities through either direct or indirect effects on plant growth, physiology, and reproduction. In addition to phytotoxic pollutants, CO_2 and atmospheric nitrogen are included as atmospheric inputs that may either counteract or amplify plant responses to other forms of stress.

The pollutants and other atmospheric additions and principal bases of concern for their effects are shown in Table 1.

The significance of atmospheric pollution as a selective pressure in plant growth and development will be discussed after considering evidence that (1) pollutants have increased in regional environments in recent decades and (2) plants respond to increased pollutant levels in ways that may alter their normal growth and development and hence provide a basis for altered selection pressures.

Evidence of Changes in Deposition of Atmospheric Pollutants to Terrestrial Systems

Atmospheric Emissions

Unfortunately region-wide monitoring of atmospheric pollutants has been restricted to the past 10 years in the United States so there is no direct long-term record of pollution histories for regional pollutants such as acid deposition and ozone. However studies based on either direct measurements of releases from industrial sources or indirect calculations based on historical fuel consumption (Husar 1986) provide a long-term record of the emissions of precursors of these pollutants and hence an indirect record of relative changes in long-term deposition patterns. Although emissions patterns vary somewhat from region to region in the United States, the past 30 years represent the period of most rapid increase over most of the country for combined emissions of SO_x and NO_x, constituents which contribute approximately 90% of the acidity in acid rain (Cogbill and

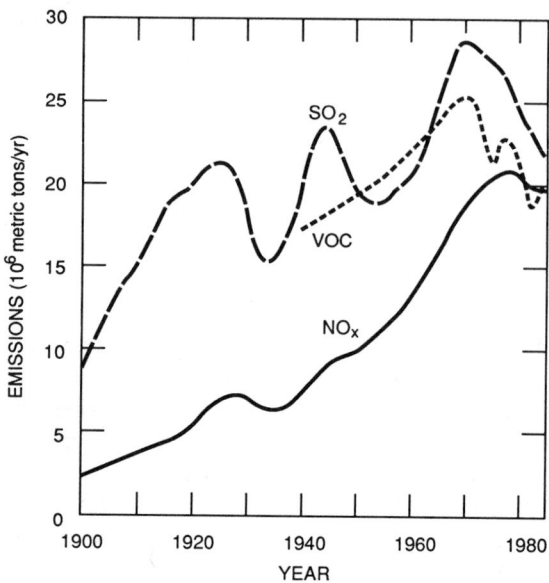

FIGURE 1. Generalized trends in annual anthropogenic emissions of SO_2, NO_x, and volatile organic carbon (VOC) in the United States from 1900 to 1980 (NAPAP 1987).

Likens 1974). In Fig. 1 the combined emission patterns for these pollutants demonstrates an increase in the Northeast for SO_x of approximately 120% from 1900 to 1950 followed by an additional increase of 36% from 1950 to 1970. For the southeast, the increase to 1950 was smaller (40%), but from 1950 to 1970 emissions approximately doubled in the Southeast (McLaughlin 1985; Husar 1986). Emissions of NO_x approximately doubled in the Northeast and tripled in the Southeast during the 1950–1970 interval. European emissions of SO_x increased by approximately 40% from 1900 until 1950 and approximately doubled in the 1950–1970 interval.

While NO_x has generally trended upward since 1970, emissions of SO_2 in the northeast United States during the past 10 to 15 years have decreased from peak levels in the mid-1970s to levels only slightly above those in 1960. In the Southeast, SO_2 emissions have remained approximately the same or increased since 1970 (Husar 1986).

Currently, mean annual rainfall acidity in the United States ranges from a pH of 5.5 and above in rural areas of the west to 4.2 in the northeastern U.S. (Stensland et al. 1986). However, acid deposition on mountaintop sites in the eastern United States may be as much as twice that at nearby lowland sites (Lindberg 1989) largely due to strongly acidic cloud events. Southern coastal California appears to be a special case where fogs with pH values as low as 2.0 have been reported (Walderman et al. 1982).

For ozone, which is formed principally from nitrogen oxides, hydrocarbons from a wide variety of natural and anthropogenic sources, and sunlight, the relationship between anthropogenic emissions and atmospheric concentrations is less clearly defined. Regional emissions of volatile organic carbon (Fig. 1) considered in concert with nitrogen oxides suggest that precursors for ozone formation have increased by 50 to 75% in the United States during the interval between 1940 and 1980. Ozone precursor emissions during the interval from 1970 to 1980, by comparison, declined by approximately 5% (U.S. Environmental Protection Agency, EPA 1984). Recent analyses by Lefohn and Pinkerton (1988) indicate that there was considerable year to year variability and no clearly defined trend in ozone concentrations for the eastern United States over the interval 1978–1985. Longer term records from Europe indicate that there has been an approximate doubling of ozone levels during the past 20 years (Penkett 1984).

Within the United States annual average growing season mean (7 h) ozone concentrations vary by a factor of approximately 1.7, from the lowest (Pacific Northwest = 31 ppb) to the highest (Eastern Piedmont/Mountain/Ridge Valley = 51 ppb) areas (Lefohn and Pinkerton 1988). Evidence from recent studies of atmospheric chemistry on eastern mountain tops indicates that the mean levels of ozone are both higher and less variable over the daily cycle on mountain tops than at low elevations (Mohnen 1988). The absence of local sources of nitrogen oxides, which destroy ozone at night, are thought to reduce the chemical destruction of ozone that normally occurs at night near urban sites and at lower elevations.

Certainly not all of regional differences in ozone formation can be attributed to differences in precursor emissions since solar radiation levels, as well the stability of regional air masses play an important role in ozone formation. Nevertheless, it seems evident that for various reasons plants have to contend with levels of this strong oxidant that may vary by a factor of 50 to 100% on an annually averaged basis across their natural ranges.

Global Carbon Dioxide

The increase in the concentration of CO_2 in the atmosphere over the past century is probably the best documented and most profound change in the global atmospheric environment resulting from human activity. As a primary substrate of photosynthesis and plant growth, as well as a "greenhouse" gas, the potential importance of changes in CO_2 are obvious. The best record of this change comes from the continuous monitoring of the atmosphere at the Mauna Loa observatory in Hawaii since 1958 by Keeling et al. (1982) (Fig. 2). Since 1958 the annual average CO_2 concentration has risen from 315 ppm to 353 ppm in 1990, an increase

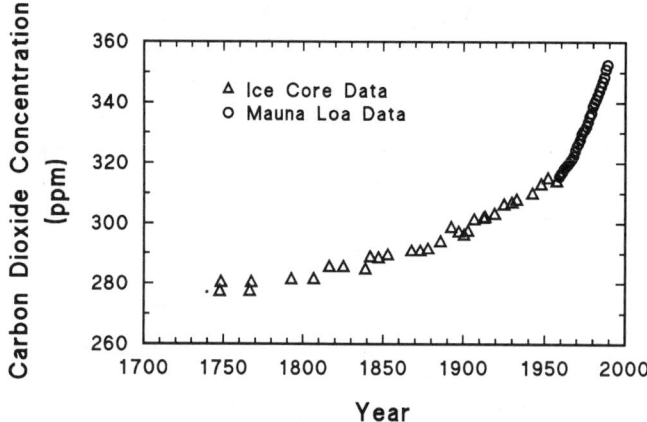

FIGURE 2. Long-term trends in atmospheric CO_2 concentrations derived from analysis of ice cores (Neftel et al. 1985) and atmospheric monitoring at Mauna Loa, Hawaii (Keeling et al, 1982).

of 0.35% per year. The record of atmospheric CO_2 concentrations prior to 1958 is less certain, but a number of independent measurements and historical records have led to a consensus that the preindustrial CO_2 concentration (1800) was in the range of 260 to 285 ppm (Gammon et al. 1985). The best record of CO_2 over the past two centuries comes from analysis of an ice core from Siple Station, Antarctica (Fig. 2), from which the 1800 CO_2 concentration was determined to be 280 ppm (Neftel et al. 1985). It is noteworthy that during the past few million years of plant evolution, the CO_2 concentration has been lower than it is today, fluctuating between 200 and 270 ppm during the glacial-interglacial phases (Gammon et al. 1985). The rate of increase in the past century lies at the extreme of all known natural variability (Gammon et al. 1985).

The increase in atmospheric CO_2 concentration since industrialization can be ascribed with reasonable certainty to human activities—primarily the burning of fossil fuels. The annual release of carbon as CO_2 from fossil fuels has increased from 0.09 Gt in 1860 to 2.3 Gt in 1958 when the Mauna Loa record begins to 5.1 Gt in 1982 (Rotty and Masters 1985) (1 Gt = 10^{15} g). In addition, changes in land use (e.g., deforestation) releases about 0.6 to 2.6 Gt of carbon annually (1980 data) through the burning and decay of vegetation and loss of soil organic carbon (Houghton et al. 1985). The apparent air-borne fraction, that is, the fraction of CO_2 released from fossil fuels and land use activities that remains in the atmosphere, is about 0.58 (Trabalka et al. 1985), the remainder being taken up by the oceans or by plants. Hence, these two anthropogenic sources of CO_2 can account for the increase in atmospheric CO_2 completely, and this represents strong circumstantial evidence for a cause-and-effect relationship.

The concentration of CO_2 in the atmosphere will certainly continue to increase into the indefinite future. The predicted rates of increase are highly uncertain, depending on such diverse factors as world population growth, economic and social policy, energy technology and policy, and current uncertainties in the global carbon cycle. Under various scenarios of fossil fuel emissions and with different assumptions about the contribution of a terrestrial source of CO_2 the model-calculated atmospheric CO_2 concentrations range from 370 to 400 ppm in 2000, 470 to 840 ppm in 2050, and 510 to 1550 in 2075 (Trabalka et al. 1985).

Ultraviolet Radiation

Although UV radiation (wavelengths <400 nm) comprises only 7% of the solar radiation that reaches the earth's surface, UV radiation can be quite damaging to plants, and its biological significance greatly exceeds its relative contribution to the solar spectrum (Caldwell 1981). In fact, UV radiation was probably a major obstacle in the early stages of development of terrestrial life before there was oxygen (or ozone) in the atmosphere (Caldwell 1979).

Ozone forms in the upper atmosphere when oxygen absorbs UV radiation from the sun, and ozone provides a very effective shield against UV radiation. This ozone filter truncates the solar spectrum at 290 nm, absorbing all of the UVC radiation (200–280 nm) and much of the less damaging UVB radiation (280–320 nm). Recent human activity, however, is causing a reduction in the total atmospheric ozone concentration and will increase UVB radiation (National Research Council 1982). Chlorofluorocarbons (CFCs), and to a lesser extent, nitrogen oxides and other trace gases, migrate to the stratosphere where they catalyze the destruction of ozone (Cicerone 1987). The concentration of CFCs in the stratosphere is increasing at a rate of 5 to 8% per year, and even if their industrial production remains limited according to the 1987 Montreal Protocol, the abundance of active chlorine is expected to more than triple by 2050 (Brasseur and Hitchman 1988).

The reduction of the ozone layer caused by CFCs was first predicted in 1974 (Molina and Rowland 1974). Predictions based on a two-dimensional model of the atmosphere (Brasseur and Hitchman 1988) suggest that a 3.3-fold increase in active chlorine causes a net ozone depletion varying from 2.6% at the equator to 5.5% at the poles. A concurrent doubling of CO_2 however, moderates the effects of CFCs on ozone because CO_2 causes cooling of the stratosphere, which slows ozone destruction. Models of stratospheric ozone dynamics did not predict the large (>40%) decrease in springtime ozone abundance over the Antarctic that was first reported by Farman et al. (1985) and soon thereafter confirmed by satellite observations (Stolarski et al. 1986). Explanations of this Antarctic ozone "hole" invoke natural processes as well as involvement of CFCs, and the

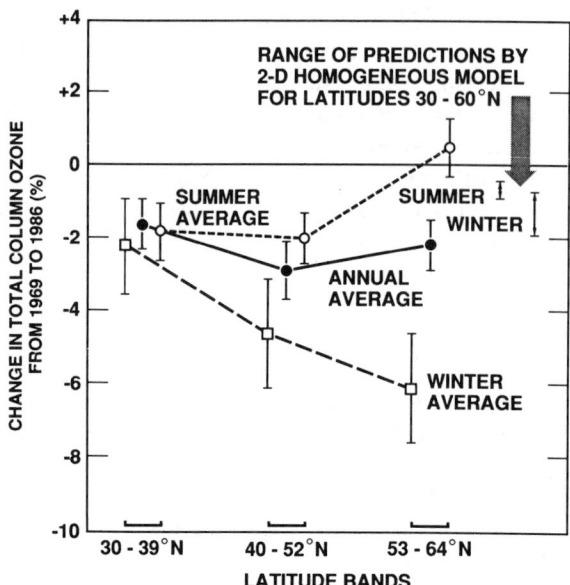

FIGURE 3. Changes in seasonal distribution of total global ozone (Dobson measurements), at northern latitudes over the interval 1969–1986. Data reflect total column ozone (stratospheric and to a lesser extent tropospheric ozone) based on LANDSAT sensing and show the discrepancy between predicted and measured changes by season.

question of how the ozone hole relates to global changes in ozone has not been resolved (Cicerone 1987). Satellite ozone data from the Total Ozone Mapping Spectrometer (Fig. 3) have indicated that decreases in ozone have not been confined to the Antarctic, are global in extent, and occur in all seasons of the year (Bowman 1988). The consensus of the Ozone Trends Panel of NASA was that CFCs are primarily responsible for the Antarctic ozone hole and that stratospheric ozone has decreased globally, but there is not yet complete agreement that CFCs have caused global ozone depletion (Kerr 1988).

The measurement of UVB irradiance is complex, and there is no consensus yet that ozone depletion has led to increased UVB. Ground-level monitoring with Robertson-Berger meters of UVB radiation in the 290 to 330 nm waveband detected no increases in UVB radiation from 1974 to 1985 in the United States (Scotto et al. 1988), but measurements in the Swiss Alps indicated that there has been an increase of 1% per year since 1981 (Blumthaler and Ambach 1990). Nevertheless, the linkage between ozone depletion and increased UVB has a firm theoretical basis, and changes in UVB irradiance can be calculated (Green et al. 1980). Reductions in ozone concentration would most affect irradiance at the shorter wavelengths of the UVB region, and the Robertson-Berger meter underestimates the effect of ozone depletion on UVB radiation. With overhead sun and typical ozone amounts, a 10% decrease in ozone results in a 20% increase in irradiance at 305 nm, a 250% increase at 290 nm, and a 500% increase at 287 nm (Cutchis 1974). Although the increase in total solar flux between 280 and 400 nm is small, the increase in the effective flux of radiation that might inhibit photosynthesis or damage DNA would be much larger (Caldwell 1981).

Atmospheric Pollutants

Ambient Gaseous Pollutants and Thresholds for Plant Response

There is substantial genetically based variation in sensitivity of plants to air pollutants (Roose et al. 1982) as well as potentially wide environmentally based variation in phytotoxicity of a given pollutant dose to plants that are genetically and phenologically very similar. Pollutant toxicity depends on the pollutant uptake potential, biochemical reactivity, and the pollutant exposure dose (concentration × time) to which the plant is exposed. The toxicity of a given exposure dose may also vary widely based on the total amount of pollutant actually entering the plant (internal dose), the rate of entry, the presence of other pollutant species, plant phenology, and environmental conditions. For these reasons, and the fact that relatively few long-term studies relating plant responses to exposure dose have been done, it is impossible to define a single threshold for pollutant phytotoxicity. However, it is useful to evaluate the lower limits of reported damage relative to ambient pollutant levels.

In Table 2 we have contrasted pollutant concentrations reported in the United States and in the Federal Republic of Germany (FRG) relative to "lower limit" concentrations associated with growth, visual injury, and interactive effects with other pollutants. Contrasts between remote (rural regional) and industrial concentrations provide some perspective on concentration gradients to which plants are exposed in these countries. An expected generality that emerges from this table is that gradients in point source pollutants, such as SO_2 and NO_x, are much sharper (sevenfold or more) between remote and industrial areas, than they are for O_3, a more regionally formed pollutant. Of these three gaseous pollutants, only O_3 in the United States, and SO_2 and O_3 in the FRG occur in the regional atmosphere at concentrations that correspond to the levels at which adverse effects have been documented. Effects on growth and interactive effects involving pollutant mixtures may occur at levels that are, in general, less than half of those producing injury. Of the three gaseous pollutants described in Table 2, ozone stands out as exceeding its phytotoxic threshold most significantly over the largest regions. This conclusion is well

TABLE 2. General concentration ranges for SO_2, O_3, and NO_2 in industrial and rural areas of the USA and the Federal Republic of Germany (FRG) shown in relationship to approximate thresholds for plant response.[a]

	USA		FRG				
	Remote[c]	Industrial[c]	Remote[c]	Industrial[c]		Plant response threshold	
SO_2							
Annual	3–6	10–20	5–15	25	G	15–25[c]	Annual
Hour	63	380	10–725	(190)	I	500	1 H
					I+	25	Hours
O_3							
Annual	30–50		30–45		G	30	3 Months[b]
Hour	50–125	100–340	100–200		I	50–80	Hours
					I+	50	Hours
NO_2							
Annual	2–10	25–75	1–2	15–30	G	160	10 Days
Hour	15–40		18 (daily)		I	320	10 Days
					I+	100	35 Days

G = growth; I = injury; I+ = interaction.
[a] For sources of pollutant concentrations and thresholds for plant responses see McLaughlin (9185).
[b] Mean daytime ozone concentration for ambient exposure producing growth responses for loblolly pine seedlings shown in Fig. 5 (after McLaughlin et al. 1988).
[c] Pollutant concentrations (ppb).

supported by numerous studies with seedlings of forest tree species (Pye 1988) and by regional analyses of crop responses to ozone in open top chambers (Heck et al. 1984). A general threshold for ozone effects on native species is 50 ppb and for agricultural crops 36 ppb (Taylor and Norby 1985).

The most systematic survey of crop yield losses attributed to regional ozone levels was the National Crop Loss Survey (NCLAN) (Heck et al. 1984). It provided equations for predicting yield losses for nine crop species. The mean predicted yield loss at 50 ppb O_3 across species was 7.7% but varied widely both across species and for varieties within species, as shown in Table 3.

These responses indicate that ozone is present over widespread areas at levels high enough to adversely affect plant growth and that the differences in response between species could be a significant factor in competitive interactions in mixed species plant communities.

Evidence of Regional Scale Changes in Forest Growth and Vigor

In contrast to studies with seedling trees and crops relatively little is known about the exposure levels at which mature forests respond to regional air

TABLE 3. Predicted yield losses of crop plants to chronic ozone exposure.[a]

Species (cultivars)	Predicted yield loss (%) at two ozone levels	
	50 ppb	60 ppb
Barley (1) (*Hordeum vulgare* L.)	0.2	0.5
Kidney bean (2) (*Phasedus vulgaris* L.)	4–18	9–25
Corn (2) (*Zea mays* L.)	4–3	1–5
Cotton (5) (*Gossypium hirsutum* L.)	3–21	4–31
Peanut (1) (*Arachis hypogaea* L.)	12	19
Sorghum (1) (*Sorghum vulgare* Pers)	2	3
Soybean (11) (*Glycine max* L.)	3–18	5–24
Tomato (2) (*Lycopersicom esculentum* Mill)	2–18	4
Wheat (4) (*Triticum aestivum* L.)	6–38	11–48

Adapted from Heck et al. 1984.
[a] Predicted yield loss at two seasonal 7-h mean ozone concentrations based on results of open top chamber experiments conducted by the National Crop Loss Assessment Network.

pollutants. Yet forests in many industrial countries are showing signs of regional scale stress that may be attributable in part to chronic pollution stress. The implications of region-wide changes in forest vigor for the potential role of pollution-forced genetic selection in natural communities are significant. In this section we examine the symptoms and processes by which these changes in forest condition have been evaluated.

EUROPEAN FORESTS

Much of the early interest in widespread pollution effects on forests came as a consequence of reports of forest damage from western Europe. Beginning in the early 1970s the occurrence of an unexplained foliar disease on Norway spruce (*Picea alba*) was noted in West Germany (FRG 1982). The disease appeared initially on older trees and over widespread areas, some of which were remote from centers of industrial activity. Concern mounted rapidly as the symptoms spread to Norway spruce (*Picea Abies*), a major component of German forests (Schütt and Cowling 1985).

The range of effects noted has been quite diverse, including yellowing and premature loss of needles from conifers that resulted in a thinning of the canopy from the inside out, changes in growth patterns of shoots, reduced numbers of fine roots and associated mycorrhizae, a reduced amount and distribution of annual radial growth, and increased mortality (Schütt and Cowling 1985). In addition a number of the symptoms are of the type that can be produced by hormonal imbalances, including loss of apical dominance, occasional shedding of green leaves and shoots, and abnormal branching patterns.

The two principal theories originally advanced to explain the regional scale decline were (1) soil acidification (Ulrich et al. 1980; Ulrich 1983) associated with atmospheric inputs of sulfuric and nitric acids and (2) individual or interactive effects of gaseous pollutants principally ozone (Krause et al. 1983). The mechanisms of action for soil acidification effects were loss of base nutrients and release of metals such as iron and aluminum to levels toxic to root growth. Gaseous pollutants were considered to act mainly through alteration in carbon assimilation and allocation resulting in a general loss of vigor. More recent analyses of the European forest decline embrace multiple pollutant stresses interacting with natural stresses to cause the variety of symptoms observed (Prinz 1987; Hinrichsen 1986). Drought, frost, harmful biotic organisms, soil fertility, and silvicultural treatments are thought to interact in varying degrees as modifying factors with air pollution to produce the observed symptoms.

UNITED STATES FORESTS

In the U.S., symptoms of forest decline have been neither as diverse nor as well developed as those in the forests of West Germany. Regional scale decline of white pine (*Pinus strobus* L) within the industrialized regions of the northeast (Berry and Hepting 1964) was one of the first symptoms ultimately attributed to chronic pollution stress (Gerhold 1977). The loss of sensitive individual *P. strobus* trees sometimes exhibiting "chlorotic dwarf" symptom from within a several hundred square kilometer area of East Tennessee was documented by Ellertsen et al. (1972). Ultimately controlled studies reviewed by Gerhold (1977) revealed that atmospheric pollutants, principally O_3 and SO_2, can be primary causal agents in the development of foliar symptoms of the type observed on pollutant-sensitive white pine trees throughout much of the northeastern United States. Approximately fivefold differences in response of sensitive and resistant *P. strobus* clones have been found following exposure to low levels to SO_2 and O_3 (Houston 1973).

Although forest community simplification has been documented around smelting operations (Freedman and Hutchinson 1980), there has been little documentation of forest community changes that could be specifically associated with regional atmospheric pollution levels. Noteworthy among results from studies of responses of natural forest community responses to air pollution are forest community simplification along a gradient of pollution levels along an industrial corridor in Ohio (McClenahen 1978); reduced vitality of *P. strobus* and reduced growth of herbs and tree seedlings in ambient air of the Shenandoah Mountains of Virginia, in which ozone was the principal phytotoxicant (Skelly et al. 1983); and the multidisciplinary studies of changes in physiology, growth, and disease resistance of oxidant-stressed coniferous forests in the mountains of southern California (Miller 1983).

With the recognition that substantial areas of the northeastern United States have been exposed to strongly acidic rainfall for at least 2 to 3 decades (Likens and Butler 1981), there has been strong interest in determining whether acidification of forest soils may have occurred and in evaluating the significance of such changes for regional forest communities. The decline of red spruce (*P. rubens*) and its implications for community composition and stability at high elevation sites in the Northeast has been a focal point of concern as a possible consequence of acid deposition. Red spruce has had a long history of disturbance episodes accompanied by mortality (Weiss 1985), but the intensification of mortality first detected in the early 1960s (Siccama et al. 1982) appears to have been unique in its duration, synchrony, and severity (AH Johnson et al. 1988). While the specific causes of decline of *P. rubens* have not yet been determined, a considerable amount has been learned about the patterns and hence the potential causes.

The patterns of mortality (Johnson and Siccama 1983) were found to be more severe in northern sites than southern and to intensify with increasing elevation. Abrupt decreases in radial growth which typically preceded the heaviest mortality were initially thought to be exclusive to northern sites. However, subsequent studies (Adams et al. 1985; McLaughlin et al. 1987) found growth decreases in Tennessee, North Carolina, and Virginia as well. Although the abrupt decreases in radial growth appeared typically in the late 1950s and early 1960s in the North, this growth decline was delayed 5 to 10 years in South, and in contrast to the North, was confined to high elevation sites, typically above 1500 m (McLaughlin et al. 1987; Cook 1988).

A common feature of decline of *P. rubens* in both North and South is the lack of any consistent relationship to tree age or size (Johnson and McLaughlin 1986) or stand stocking levels (McLaughlin et al. 1987). In addition, a shift in the strength of the relationship between growth and climate has been found in the past 25 years by testing the stability of ring width-climate models before and after 1960 (McLaughlin et al. 1987; Cook 1988). Those models suggest that spruce is now growing more slowly than would have been predicted based on pre-1960 growth patterns. The observed shifts in growth across the region indicate that regional scale stress began to intensify at high elevation sites about 25 to 30 years ago (Figure 4). The shifts in climate-growth relationships suggest either the addition of another regional stress, such as air pollution, that has sensitized trees to climate, or alternatively, and not necessarily mutually exclusively, climate itself has changed. These changes occurred at the same time that emissions of sulfur and nitrogen oxides were increased markedly across the region (Fig. 1).

Beginning in the early 1960s, reports of winter damage to foliage of *P. rubens* in the Northeast increased abruptly in an unprecedented manner and this damage may have played a role as an inciting or predisposing factor in the synchronized patterns observed in the Northeast (AH Johnson et al. 1988). This increase in winter damage appears anomalous

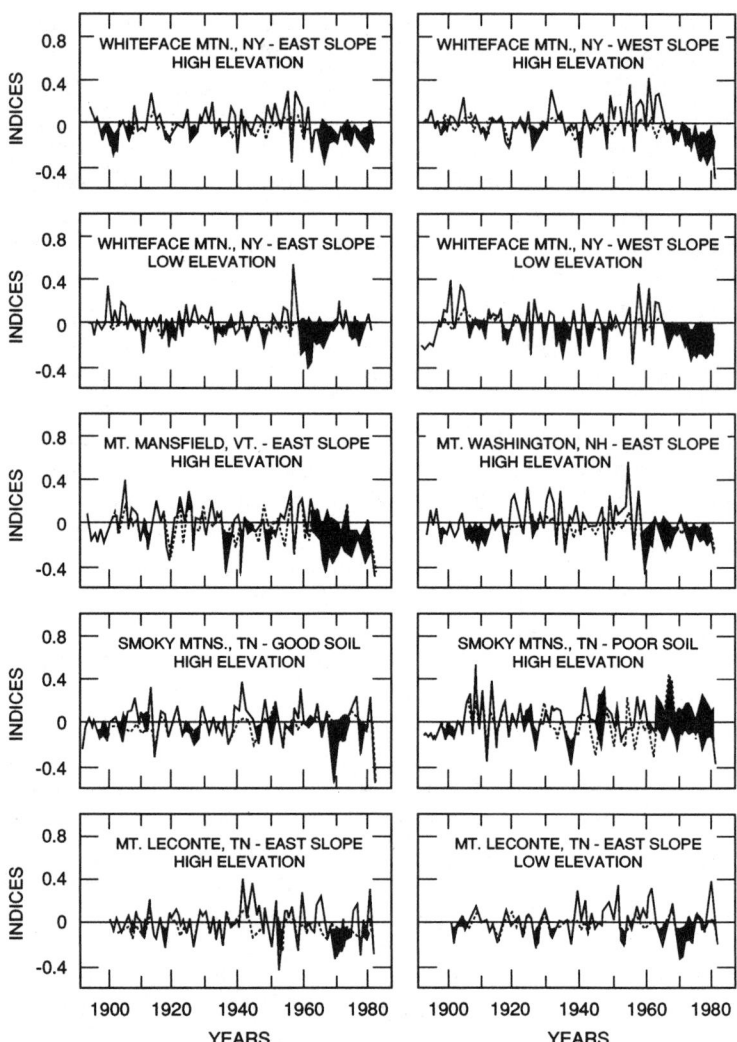

FIGURE 4. Growth trend data from red spruce (*P. rubens*) at many locations in the eastern United States show slower than expected radial growth during the past 20 to 30 years. In this figure discrepancies between actual growth (*solid line*) and growth predicted from the modeling relationship between growth and temperature over the interval 1900–1940 are highlighted to reveal regional patterns for stands in the Northeast and Southeast (after McLaughlin et al. 1987).

since weather records do not show that the winters over this time period were unusually severe (Diaz 1978) and raises the question of possible sensitization of foliage to winter damage. Both exposure to ambient ozone and changes in foliar nutrition are plausible contributors to increased sensitivity of foliage to low temperatures, as will be discussed later.

In addition to the region growth decline of *P. rubens*, during the past 5 years, a slowdown of radial increment has also been reported for southern pines (Sheffield and Cost 1987). Analyses of data from continuous forest inventory plots across a wide diversity of sites within the region showed that diameter growth of southern pines has dropped by 30 to 50% below expectation across the region. Earlier reports indicated that the radial slowdown was more pronounced in the Piedmont region (Sheffield and Knight 1983) than in coastal areas. However, subsequent analyses indicated a similar overall magnitude of reduction during the last three decades (Sheffield and Cost 1987). More recent analyses of the role of climate and competition in the observed growth responses indicated that a substantial portion of the observed decline cannot be explained by these variables, but has been due to intensification of other stresses throughout the region during this time (Zahner et al. 1990). Such evidence provides inferential support but does not prove the involvement of atmospheric pollution in the observed growth declines. For this reason mechanistic studies that explore specific processes by which trees respond to pollutants and other stresses from an important part of efforts to evaluate both the causes and potential consequences of forest decline.

Process Level Effects and Mechanism of Action

Pollutants can alter physiological processes at many levels within the plant and resistance or susceptibility may be controlled at multiple points. For example, Taylor et al. (1986b) found that the evolution of resistance of *Geranium carolinianum* in the field under chronic SO_2 stress was associated with differences in net photosynthesis, internal leaf resistance, photoassimilate partitioning, and growth. In general pollutants affect plant growth and development through altering availability or utilization of carbon, nutrients, or water and therefore may change plant sensitivity to a wide variety of natural stresses.

CARBON ASSIMILATION, ALLOCATION, AND GROWTH

At present there is very good evidence from controlled studies under both laboratory and field conditions that ambient levels of ozone are detrimental to the growth of many species of vegetation (NAS 1977; EPA 1984; Reich and Amundson 1985; Pye 1988). Laboratory studies have shown that photosynthesis of tree species is inhibited at levels which occur during the summer over much of the eastern United States, while field chamber studies have documented growth enhancement produced by filtering ambient air to remove 50% or more of ambient ozone, as well as SO_2 and other acidic pollutants. Loblolly pine has been examined in several studies (Shafer et al. 1987; McLaughlin et al. 1988; Adams et al. 1988) and frequently shows reduced growth under ambient ozone regimes. McLaughlin et al. (1988) found that over 90% of 42 loblolly pine families examined responded negatively in height growth to ambient ozone and the average height reduction across all families was 26%. The range and consistency of those effects is illustrated in Fig. 5. The reductions in the growth of nine species of agricultural crops by ambient ozone levels in the field (Heck et al. 1984) are shown in Table 3.

The high innate phytotoxicity of ozone is related to its capacity to attack chemical double bonds altering molecular structure and hence the chemical and functional integrity of tissues and organelles (Heath 1980). A primary physiological effect of ozone is its disruption of carbon allocation pathways (McLaughlin 1989) including both reduction of photosynthesis (Reich and

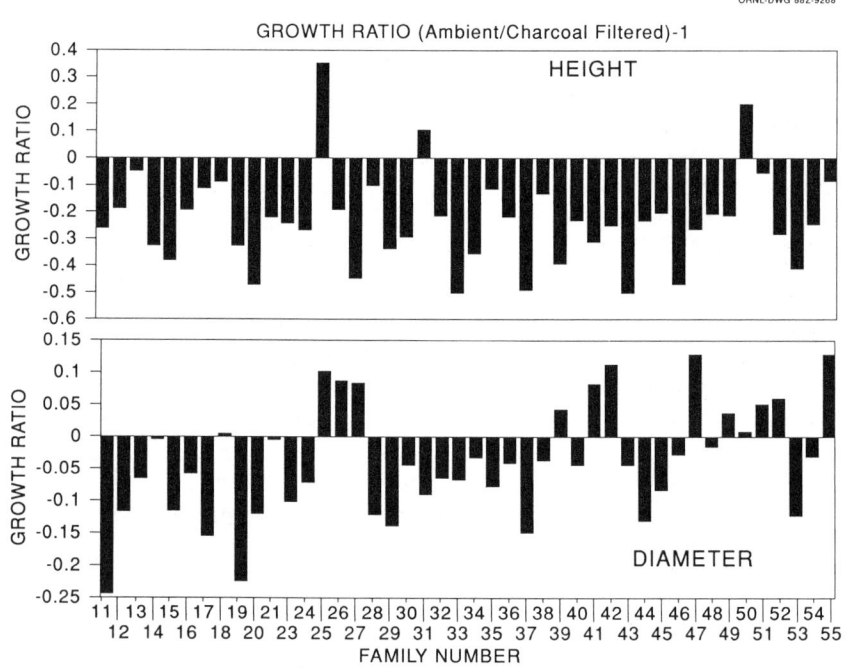

FIGURE 5. Growth responses of seedlings of 44 loblolly pine (*P. taeda*) families to 3 months of exposure to ambient ozone levels in east tennessee. A value of 0.1 or −0.1 represents a change of +10% or −10%. Responses highlight variation in sensitivity across 44 loblolly pine families of commercial importance in the southeastern United States (after McLaughlin et al. 1988).

Amundson 1985), increased dark respiration (Barnes 1972; Skärby et al. 1987; McLaughlin et al. 1982), and reduced transport of photosynthate to root systems (Tingey 1978; Tingey et al. 1976; McLaughlin and McConathy 1983).

The increase in dark respiration of leaves is evidence of higher maintenance costs associated with tissue damage by ozone and that evidence is further supported by 14-C labelling studies that show higher retention of 14-C photosynthate in foliage of beans (*Phasolus vulgaris* L.) (McLaughlin and McConathy 1983) and loblolly pine (*Pinus taeda* L.) (Adams et al. 1988) exposed to ozone. The combination of reduced photosynthesis, increased respiration, and reduced transport of photo-synthate from foliage has been shown to occur in the field in mature *P. strobus* trees showing enhanced foliar symptoms typical of ozone damage, reduced radial growth, and reduced responsiveness to climatic fluctuations (McLaughlin et al. 1982).

Effects of acid deposition on terrestrial systems must be evaluated both in terms of direct effects on growth and indirect effects associated with changes in soil nutrient availability. In contrast to studies with ozone, there is currently little evidence that acid rain at ambient levels has a direct adverse effect on carbon assimilation of tree species (Reich et al. 1986; Taylor et al. 1986a; Hanson et al. 1988). A wide range of field studies involving irrigation of seedling and sapling trees in the field with acid rain in both Scandinavia and the United States have typically produced growth stimulation attributed to fertilizing effects of nitrogen and sulfur (see review by McLaughlin 1985). In general, growth reductions in field grown crops have not been found at pH levels above 3.0 and increased growth of crop species has been reported at pH levels as low as 3.0 and 2.8 in greenhouse and field experiments, respectively (Jacobson 1980).

ALTERATIONS IN AVAILABILITY OF NUTRIENTS AND TOXIC METALS

Concern for chronic adverse effects of atmospheric pollution on nutrient cycles has focused primarily on the input of sulfuric and nitric acids that comprise acid deposition. The problems of aluminum toxicity to crop species under acidic soil conditions have long been recognized and alleviated with liming. Thus, it is likely that acid deposition effects on agricultural soils would be effectively countered by normal liming practices. The major concern regarding acid deposition effects on soils is in natural communities. For forests there is good evidence from nutrient cycling studies that natural cation leaching rates can be substantially (2 to 3 times) increased by acid deposition (Johnson et al. 1985). Historical and resurvey data from Sweden appear to confirm accelerated acidification (0.5 to 0.7 pH units) of soils sampled 50 years earlier in an area of high acid deposition (Tamm and Hallbacken 1988). In addition to the long-term depletion of basic nutrients from soil (see review; Abrahamson 1984) a secondary effect of soil acidification is mobilization of trace metals such as iron and aluminum to levels toxic to root growth (Ulrich et al. 1980; Ulrich 1983).

A major hurdle in conducting and interpreting controlled experiments with acid deposition is the difficulty of representing chronic changes in soil plant nutrient status with typically acute treatment conditions. The problem is that availabilities of nutrients and toxic metals are not linearly related to inputs of soil acidity. Nutrients such as calcium and magnesium will initially be mobilized in the soil, and, with the addition of nitrogen, more favorable nutrient status may be induced initially followed by less favorable conditions over time as predicted by Abrahamson (1980). Thus, a bimodal growth pattern would be predicted, and this is exactly what has been measured in some controlled studies with trees (Chappelka et al. 1988; McLaughlin et al. 1988). As soil acidity approaches a pH of about 4.0 there is also a dramatic reduction in availability of base nutrients such as calcium and increases in availability of iron, aluminum, and other toxic metals (Reuss and Johnson 1986). At this stage in the soil acidification process, the loss in base cations and increases in aluminum and calcium become much more strongly influenced by atmospheric inputs of strong anions such as sulfate and nitrate. With crop species, the bimodal growth response may also by concave with maximum growth inhibition at near ambient pH levels as reported by Johnston and Shriner (1985).

Although loss of base cations is a major concern from a nutritional standpoint, the mobilization of aluminum and other trace metals is stimulated by the same reactions that ultimately deplete these nutrients (Ulrich 1983). Although much of the attention has been focused on aluminum toxicity, a variety of other metals are also mobilized with aluminum that have high potential toxicity to plant roots (Hutchinson 1980). Some of those metals have shown historical increases in both soil (Friedland et al. 1982) and tree rings (Baes and McLaughlin 1984, 1987; Bondietti et al. 1989a,b) in the regional environment. Both the high potential phytotoxicity and the long residence times of many of these metals (Hutchinson 1980) raise the level of concern for cumulative degradation of soils within which native pools are mobilized.

Toxicity from metals may be a consequence of direct enzymatic inhibition or interference with uptake of nutrients (Foy et al. 1978). Recent evidence from several studies suggests that the ratio of Ca to Al is critical to the physiological effects of aluminum. Among those effects are decreased uptake of Ca, Mg, and P (Christianson and Foy 1979), slowed growth of apical meristems, and increased dark respiration. Low levels of Ca uptake can have far-reaching physiological consequences. Increases in dark respiration have been associated with a loss of membrane integrity resulting from reduced availability of Ca (Bangerth 1979). Cossett et al. (1977) reported that both the amount and pattern of translocation of photosynthate of soybean (*Glycine max*) was reduced by low Ca (50 µmol) in nutrient solutions. Rost-Siebert (1983) found that the toxicity to Norway spruce roots of aluminum in soil solutions was dependent on the Al:Ca ratio exceeding a value of 1.0. Shortle and Smith (1988) show data that indicate that this condition is exceeded in root tissues of *P. rubens* at several northeastern sites and discuss the importance of aluminum in interfering with the uptake and utilization of Ca. They report slowed cambial growth under conditions where Ca:Al ratios in roots were below 1.0.

Inferential evidence for an adverse effect of acid deposition on growth of red spruce at high elevations in the Smoky Mountains of Tennessee comes from several sources including (1) the high natural acidity and low buffering capacity of soils at high elevations (DW Johnson et al. 1988); (2) the very high deposition levels of SO_x and NO_x by wet and dry deposition (Lindberg 1988); (3) Al:Ca ratios in soil solution (DW Johnson et al. 1988) that frequently exceed values reported to be toxic to Norway spruce roots in nutrient solution; (4) shifts to higher Al:Ca values in spruce tree rings that occurred during the past 30 years as regional emissions of SO_2 and NO_x increased within the region (Bondietti et al. 1989a); and (5) increasing Al:Ca in foliage, decreasing growth, and altered carbon allocation patterns of red spruce saplings with increasing elevation in the area (McLaughlin et al. 1990). The decreasing levels of foliar calcium and magnesium with increasing elevation reported by McLaughlin et al. (1990), have also been reported for *P. rubens* foliage from several northeastern mountain sites examined by Friedland et al. (1988). Whether lower foliar nutriennt levels are a consequence of higher levels of acid deposition at these sites is not known. The lower levels of bases present at this time, however, should make these sites more susceptible to the effects of additional cation leaching and aluminum mobilization by strong anions deposited by wet and dry deposition. This susceptibility includes both losses from the soil discussed above and enhanced foliar leaching of base cations as reported by Joslin et al. (1988) for a high elevation *P. rubens* site in the southern Appalachians.

Responses to Secondary Stresses

Many of the physiological responses of plants to air pollutants will resemble or amplify those induced by natural stresses that alter the supply of carbon, water, or nutrient resources. The reasons for this are that pollutants may act directly or indirectly to alter the acquisition and/or efficiency of utilization of those resources, or they cause structural changes that the further amplified by natural stresses. Three types of stress for which these interactions are particularly relevant are those induced by water deficits, low temperature, and disease.

MOISTURE

There are several general mechanisms by which air pollutants can alter plant-water relations, including reducing the capacity for water uptake through roots and loss of control of water loss through leaves (see review by McLaughlin 1985). Reduced water supply capacity can be a direct effect resulting from release of aluminum and other metals toxic to fine roots by strong anions in acid deposition (Ulrich et al. 1980). Alternatively it may result from an imbalance of root mass relative to shoot mass. Reduced root shoot ratios may be a secondary result of reduced transport of carbohydrates to roots such as has frequently been noted with ozone exposure, or a consequence of preferential stimulation of shoot growth producing an altered root shoot ratio (McLaughlin 1988).

Preferential stimulation of shoot growth and associated altered shoot physiology associated with increased nitrogen inputs to forest canopies has been a principal tenet of the "nitrogen hypothesis" of forest decline. Evidence that increased drought sensitivity of *P. rubens* seedlings can be produced under greenhouse conditions by simulated acid rain at near ambient conditions (pH 4.1 rain and pH 3.6 mist) has been presented by Norby et al. (1986c). In this case increased sensitivity to drought was attributed to stimulation of transpirational loss from increased needle biomass, not altered physiological resistance to drought. Norby et al. (1989) have also shown that atmospheric nitrogen oxides present at ambient levels can be utilized directly by foliage of high elevation red spruce through induction of nitrate reductase. This provides a mechanism for foliar fertilization and a basis for altering root to shoot ratios, susceptibility to water stress, and, possibly, sensitivity to winter injury, as will be discussed below.

The other principal pathway for altering plant-water relations is through changes in transpirational loss from leaves. The effects of gaseous pollutants on stomatal function have been well documented in laboratory studies under controlled conditions (see Mudd and Kozlowski 1975). Although high levels of either SO_2 or O_3 may cause stomatal closure reducing water loss, low levels of SO_2 stimulated stomatal opening in *Vicia faba* even in previously water-stressed plants (Unsworth et al. 1972). Temple et al. (1988) have also reported increased water stress in alfalfa (*Medicago sativa* L.) grown in the field in the presence of ambient ozone levels. Finally, altered structure of cuticular waxes, particularly those around stomates, in polluted ambient air has been calculated to provide increased moisture stress, particularly during dry periods (Fowler 1980).

LOW TEMPERATURES

The increase in severity of winter temperature and in reports of winter damage to shoots in the northeastern United States in the early 1960s has been implicated as a possible contributing factor in recent declines of high elevation red spruce in that region (AH Johnson et al. 1988). Predisposition of high elevation *P. rubens* to winter injury as a consequence of atmospheric deposition and associated nitrogen-induced delays in cold hardening has been suggested by Friedland et al. (1985) based on cytological examination of damaged tissues collected from the field. However, experimental addition of nitrogen to seedlings that were initially nitrogen-deficient has been shown to reduce rather than enhance sensitivity to freezing injury in laboratory studies (Klein et al. 1989). To date the preponderance of experimental evidence of pollution-induced sensitivity to cold injury comes from experiments with ozone. Experiments at near ambient ozone levels with *P. abies* (Barnes and Davison 1990) and red spruce (Cumming et al. 1988) indicate that sensitivity of foliage to winter cold damage was enhanced by exposure to ozone during the previous growing season. Possible mechanisms for this sensitization included both altered carbohydrate metabolism (Cumming et al. 1988) and increases in membrane damage (Barnes and Davison 1990).

DISEASE

Air pollutants may significantly affect development of plant diseases either through effects on the pathogen itself or through changing resistance of the host plant to initiation and development (see review by Smith 1981). A wide variety of responses have been observed including both stimulation and retardation of disease. The response depends on the species and developmental stage of pathogen and host and the concentration and type of pollutant involved. Perhaps the most useful generalization for today's chronic air pollution regimes is that pollutant combinations that result in reduction in plant vigor will act as predisposing stresses (Manion 1981) to a variety to facultative parasites that affect weakened hosts. A principal mechanism of that predisposition will be the reduced capacity of previously stressed trees to resist or repair damage due to reduced energy reserves (McLaughlin and Shriner 1980). Both the mobilization of carbon-based secondary metabolites such as phenols and the balance of foliar carbon to nitrogen are considered likely factors in reducing resistance of some plants to foliar pathogens (Jones and Coleman 1989).

Examples of responses of this type are the increase in pine bark beetle (*Dendroctonus brevicomis*) attack of ponderosa pine in the oxidant-stressed San Bernardino forest associated with reduced oleoresin production (Cobb et al. 1978). Heavy infection by root disease fungi of *P. strobus* trees that also appeared to be sensitive to repeated ozone damage has been reported in the southern Appalachians (Skelly 1980).

There has been relatively little work on acid deposition effects on disease development to date and results have included both pathogen stimulation and reduction (see review by McLaughlin 1985). Interestingly, the root decay fungus *Armillaria mellea* has apparently not played a significant role in *P. rubens* decline. While generally a strong secondary parasite, it was present in *P. rubens* stands at lowest frequency at high elevations where the damage was greatest (Carey et al. 1984).

Although many types of plant-pathogen interactions, both adverse and beneficial, may be envisioned under current regional air pollution regimes, it seems reasonable to assume that plants will become more susceptible to disease under most conditions where growth is currently being adversely affected. Documented differences in susceptibility to growth reduction within and between species can be expected to increase the selective potential of plant disease under these conditions.

Reproductive Processes

Perhaps the stresses induced by air pollution that can be most directly related to effects on selection processes are the effects on reproductive processes, including seedling establishment and pollen germination.

SEEDLING ESTABLISHMENT

Seedling establishment is generally more sensitive to soil acidity than seedling growth and generally decreases rapidly below a soil pH of 4.4 (Abrahamson 1980). A wide range in sensitivities of seeds of northeastern U.S. forest species to substrate acidity was demonstrated by Raynal et al. (1982). Maximum germination for *P. strobus*, for instance, occurred in the 3.0 to 2.4 pH range. Germination of red maple (*Acer rubrum*) and yellow birch (*Betula alleghaniensis* Britton) was inhibited, while hemlock was unaffected by pH levels in the 3.0 to 4.0 pH range, which is not uncommon for pH of throughfall below the canopy (Miller 1983). Mechanisms discussed by Raynal et al. (1982) as potentially responsible for the observed response were associated with osmotic effects of the substrate solution as well as its ionic characteristics. Both of these solution properties can lead to leaching of metabolically active substances from seeds, with possible adverse physiological effects.

POLLEN GERMINATION

A review of the effects of air pollution on pollen (Wolters and Martens 1987) indicates that pollen germination and tube growth of a variety of plant species can be affected in vivo by concentrations of SO_2, NO_2, and acid rain that occur in the ambient atmosphere. In vitro assays to determine the sensitivity of pollen of 13 Canadian forest species to moisture droplets acidified with dilute H_2SO_4 indicated that the LD_{50} for pollen germination of broad-leaved canopy species during 3-h exposures was in the pH range 3.95 to 3.6 (Cox 1983), well within the range of episodic ambient rainfall events. Pollen tube elongation for an herbaceous forest species, *Oenothera parviflora* L., was found to be more sensitive than germination to inhibition by simulated rainfall, with significant reduction in elongation being produced by pH levels as high as 4.6 (Cox 1984). Subsequent studies of forest vegetation indicated that broad-leaved canopy species were most sensitive, conifers least, and understory species intermediate in sensitvity to in vitro effects of media pH on pollen germination (Cox 1988). As in the previous study, pollen germination of most species examined in these studies was inhibited by pH levels representative of ambient rain of eastern Canada. Similar results (pH 3.4 to 4.2) have been reported for five northeastern U.S. tree species (Van Ryan et al. 1986). These levels of acidity are well below those that occur in the ambient atmosphere, particularly at high elevations where wetting from clouds is frequent and initial pH values may be below pH 3.0 (Saxena and Lin). Differential sensitivity of flowering of alpine plant species to artificial mist at pH 3.5 has been suggested as a basis of concern about effects of acid deposition on the composition of alpine communities in Colorado (Funk and Bonde 1986).

Inhibition of conifer pollen germination at ambient levels of gaseous pollutants has also been reported in controlled studies following fumigations for 16 h with concentrations of SO_2 as low as 74 ppb, a level exceeded in industrial areas of Europe (Keller and Beda 1984). In the field reductions in seed production, seed weight and pollen germination have been reported for *P. strobus* and red pine (*P. resinosa*) growing in areas of high ambient air pollution (Houston and Dochinger 1977).

Collectively, these studies indicate that there is a strong likelihood that current levels of atmospheric pollution are influencing the reproductive cycle at the flowering stage. The consequences of these processes on selection or performance of surviving pollen and the seed they produce are largely unknown. Where selection in the flower for acid tolerance is increased, the influence of such selection on performance of plants growing on basic soil may be undesirable as noted by Cox (1988). For consideration of additional aspects of air pollution effects on flowering see Chapter 6 (Barrett and Bush) in this volume.

Global Increases in CO_2

Evidence of Plant Growth Responses to CO_2

The responses of vegetation to CO_2 must be analyzed differently from the responses to atmospheric pollutants. CO_2 is a fundamental constituent of the atmosphere essential for plant growth. Our interest is in the responses to the *increases* in CO_2 concentration, not to the chemical itself. There is a wealth of experimental evidence showing that plants respond to increases in atmospheric CO_2 concentration, and several comprehensive reviews have summarized much of this evidence (Lemon 1983; Strain and Cure

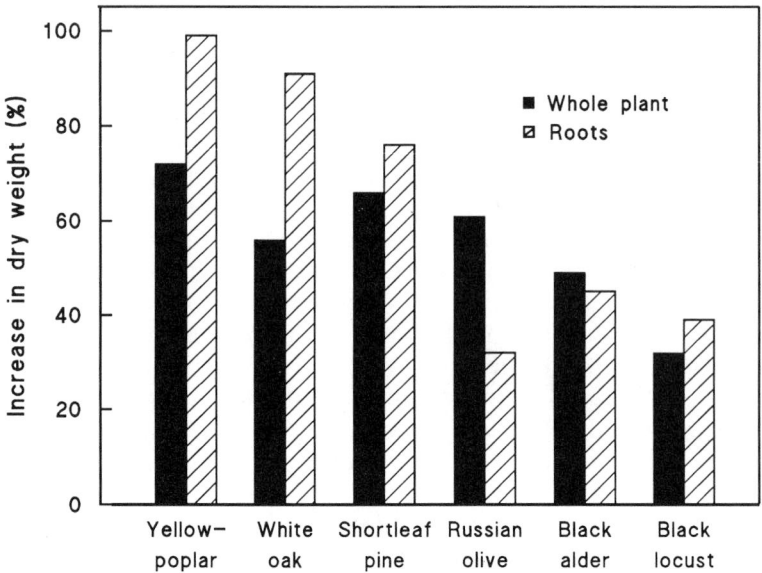

FIGURE 6. The increase in dry weight of whole plants (roots, stems, and leaves) and roots only for tree seedlings grown in 700 ppm CO_2 compared with dry weights of seedlings grown in ambient (350 ppm) CO_2. All of the plants were grown in nutrient-poor soil. Russian olive, black alder, and black locust are nitrogen-fixing species.

1985; Enoch and Kimball 1986). Surveys of the results of many experiments, which are summarized in these reviews, indicate that the growth of C_3 plants increases about 30% or more with a doubling of CO_2 concentration. The increases in C_4 plants usually are less. A typical response to CO_2 enrichment is an increase in mass of all plant organs, with roots gaining proportionately more mass than stems, and stems more than leaves (Fig. 6). The additional dry matter produces longer roots and longer and thicker stems. Increases in leaf mass are associated both with increases in leaf area and thickness.

These data clearly show that plants have the capacity to respond to increases in CO_2, but the questions of whether the CO_2 increases in past decades have impacted vegetation, or whether the increases predicted into the future will have an impact, are much more difficult to answer. There are three types of studies that have purported to document responses of vegetation to recent increases in atmospheric CO_2: dendrochronology, comparison of archived plant material with modern material, and analysis of global atmospheric trends. These studies present indirect evidence at the levels of the individual organism, populations, and large-scale regions, respectively.

The dendrochronology approach to CO_2 response is similar in some respects to the approach used in air pollution studies. That is, tree ring chronologies at a site thought to be sensitive to atmospheric CO_2 are statistically analyzed to separate out the effects of weather, competition, and tree age, and correlate the residual variation in tree ring width to trends in CO_2 concentration. Using these techniques, increases in radial growth in the past few decades not explained by climate or stand conditions have been found for bristlecone pines (*Pinus longaeva*) at two high-elevation sites in California (LaMarche et al. 1984) and for scots pine (*Pinus sylvestris* L.) at high elevations in northern Finland (Hari and Arovaara 1988). Substantial increases (>30% in basal area) were considered to be related to increasing trends in CO_2 in both cases. The importance of site characteristics to the CO_2 sensitivity of tree rings was emphasized in the study of Kienast and Luxmoore (1988). They analyzed tree cores from 34 sites in four different climatic regions and along elevational gradients. Increased ring-width growth trends after 1950 were reported for eight of the sites, all of which were located in the transition from complacent to very sensitive sites to temperature or water stress. On four of these sites the growth increase was unrelated to climatic factors.

The stomatal density of archived leaves that had been collected over the last 200 years was reported to be higher than that of analogous leaves in 1981 (Woodward 1987). The 40% reduction in stomatal density between 1787 and 1981 paralleled the 21% increase in CO_2 concentration during that period. Experiments demonstrated that plants exhibit phenotypic plasticity in stomatal density in response to CO_2 partial pressure (Woodward and Bazzaz 1988). The response is nonlinear with respect to CO_2, so even if this response to increasing CO_2 has occurred in the past, further adjustments in stomatal density as CO_2 continues to rise are less likely.

A global measure of vegetation response to increasing CO_2 may be provided in the amplitude of the seasonal variation in atmospheric CO_2, which is largely determined by photosynthetic uptake of CO_2 in the summer and respiratory release of CO_2 in the winter (Houghton 1987). The seasonal amplitude has been increasing approximately 0.7% per year, with more rapid increases since 1975 (Gammon et al. 1985), and large-scale changes in the growth of boreal forests may account for a large portion of the increase (D'Arrigo et al. 1987).

Mechanisms of Response to Increasing CO_2

The mechanisms of plant response to short-term increases in CO_2 concentration are well understood. The primary effects of increasing atmospheric CO_2 occur as a result of alteration of CO_2 diffusion into leaves and its assimilation into organic compounds (Strain 1985). The increase in net photosynthesis occurs both because of increased carboxylation and decreased photorespiration (Tolbert and Zelitch 1983). Because reduction in photorespiration is an important part of the response to CO_2 enrichment, C_3 plants generally respond more than C_4 plants and are expected to gain a competitive advantage (Bazzaz et al. 1985). High CO_2 concentrations also decrease stomatal apertures, and the ratio of photosynthesis to transpiration (transpiration efficiency) increases.

The proper perspective for discussion of selection pressures on plants in response to rising CO_2 is the long-term responses in unmanaged ecosystems. It is critical, therefore, that feedbacks and acclimation processes be considered. At the biochemical level, increased rate of photosynthesis often leads to an accumulation of starch in chloroplasts, which can reduce the photosynthesis through feedback inhibition (DeLucia et al. 1985). High CO_2 also leads to a decline in carboxylation efficiency (Sage et al. 1989). Effects of elevated CO_2 on photosynthesis per unit leaf must be considered in relation to changes in leaf area (Norby and O'Neill 1989). With perennial plants, ecological feedbacks may become the dominant factor controlling their responses to rising CO_2. Increased carbon supply to root systems not only increases root-to-shoot ratio (see Fig. 6) but also can stimulate plant-microbe interactions, leading to enhancements of mycorrhizal establishment (O'Neill et al. 1987) and symbiotic nitrogen fixation (Norby 1987). Alteration of carbon partitioning patterns with CO_2 enrichment can change leaf chemistry such that herbivore interactions (Lincoln et al. 1986) or decomposition and nutrient cycling (Norby et al. 1986b) are affected. The past or future responses of plants to rising CO_2 may be difficult to detect because of the overriding influence of environmental stresses, but interactions between CO_2 and stress responses could provide an important selection pressure. The increase in transpiration efficiency of plants in high CO_2 can integrate over time into an increase in whole-plant water-use efficiency (Norby and O'Neill 1989), which may increase a plant's resistance to drought. Partial stomatal closure in elevated CO_2 reduces the uptake of gaseous pollutants (Carlson and Bazzaz 1982). Woody plants in CO_2-enriched, but nutrient-poor, environments exhibit a decreased physiological demand for some nutrients (e.g., nitrogen) and an increased uptake of other nutrients (e.g., phosphorus) (Norby et al. 1986a). Experimental data at the whole-plant level suggest that the stimulation of plant growth by CO_2 enrichment is roughly the same over wide ranges of light and temperature (Kimball 1986).

The effects of elevated CO_2 on reproduction have not been well studied, but in general, high CO_2 appears to accelerate all events from the dates of anthesis to the date of seed maturation and may effect larger and more numerous flowers and seeds (Strain 1985). Although most plants are likely to increase reproductive output as CO_2 increases, there is a great deal of variability among species in the nature of the response, and this variability provides the potential for changing reproductive success in natural ecosystems (Oechel and Strain 1985).

Ultraviolet Radiation

Growth and Physiological Responses to Ultraviolet Radiation

The reduction in stratospheric ozone concentration is a recent phenomenon, and there is no documentation that plants have responded to increases in UVB radiation caused by anthropogenic emissions of CFCs and other trace gases. Although responses to temporal changes in UVB have not been documented, plant responses to geographic variation in UVB irradiance are well known (Klein 1978). The flux of UVB radiation effective in DNA damage increases with decreasing latitude and increasing elevation, and the gradient from arctic to tropical alpine environments represents more than a seven fold difference in daily DNA-effective UVB irradiance (much more than any change that might result from ozone destruction) (Caldwell 1981). Plants native to high UVB environments exhibit properties that are effective in reducing the penetration of UVB into mesophyll tissue, notably the occurrence of UVB-absorbing phenolic compounds (flavonoids and flavones) in the epidermis (Caldwell 1981). Plants from temperate latitudes that have been introduced into high UVB environments also have low epidermal UVB transmittance, suggesting that some acclimatization occurs (Robberecht et al. 1980). Short-term changes in epidermal transmittance, attributable to pigment synthesis, occur in some plants exposed to enhanced UVB irradiance (Robberecht and Caldwell 1978). Other morphological and pigmentation characteristics of alpine plants have been ascribed to the effects of UVB radiation, but the observations are somewhat equivocal (Klein 1978).

Mechanisms of Response to Ultraviolet Radiation

There have been many experiments in which plants were exposed to enhanced or reduced UVB radiation and various biochemical, physiological, or morphological responses recorded (see reviews by Klein 1978 and Teramura 1983). Unfortunately, in the majority of these studies the UVB exposure was poorly controlled or the amount of photosynthetically-active radiation (PAR) was low. Many of the responses of plants to UVB that are observed under low PAR are much less or non-existent under higher levels of PAR as occur in the field (Teramura 1983) and, therefore, these studies do not address the issue of plant responses to stratospheric ozone depletion. In addition, some of the described responses to UVB radiation may be of no ecological consequence because the primary photoreaction occurs in a waveband that is unimportant in solar radiation, has a low quantum efficiency, or is overwhelmed by its response in the visible spectrum (Caldwell 1981).

With these precautions in mind, and recognizing that the effects of UVB on plant growth are rather subtle in comparison with other environmental parameters, it is still recognized that the consequences of increased UVB radiation on plants are potentially insidious (Caldwell 1981). The two mechanisms by which higher plants are most likely to be affected are reductions in photosynthetic capacity and leaf expansion (Caldwell 1981). The effects on photosynthesis occur primarily through an effect on photosystem II, although the site of action is not clear (Caldwell 1981; Teramura 1983). The effect on leaf expansion is associated with inhibited or delayed cell division in addition to the secondary effects of reduced photosynthate supply (Dickson and Caldwell 1978).

The effect on cell division may be associated with the absorption of UVB by nucleic acids (Caldwell 1981). Nucleic acids are efficient absorbers of UVB radiation, and damage to DNA by UVC and, to a lesser extent, UVB has been well studied (National Research Council 1982). The DNA of higher plants appears to be much better shielded from UVB than is bacterial DNA, and it is unlikely that mortality would result from DNA damage in higher plants (Caldwell 1981). The DNA in pollen, however, may be more vulnerable to UVB damage, and pollen is the most plausible stage where mutations or chromosomal aberrations might occur in germinal cells (Caldwell 1981).

There is a considerable degree of variability between different plant species in their sensitivity to UVB radiation (Teramura 1983). In mixed assemblages of species it can be expected that the competitive balance of plants will shift as the amount of UVB radiation reaching the earth's surface increases (Fox and Caldwell 1978). Even without a reduction in stratospheric ozone, current levels of UVB may be a subtle factor in plant competition (Caldwell 1979).

A critical point for consideration is the degree to which plants can acclimate to increasing UVB. The UVB induction of flavonoid synthesis in epidermal cells is apparently an important mechanism of this type. Such a reaction may occur when they are moved from a greenhouse to outdoors or from a temperate to high-elevation tropical habitat (Caldwell 1981). Whether plants in their natural habitat can acclimate to increased UVB resulting from ozone depletion is a critical uncertainty that will determine whether UVB will exert significant selection pressure.

Synthesis and Conclusions

Responses of Terrestrial Systems to Environmental Pollution

In the preceding sections evidence is reviewed that man has influenced the chemistry of the atmosphere and in turn affected terrestrial plant life. With respect to fossil fuel combustion, the chemical signal produced on the terrestrial environment is unequivocal. That signal includes increasing acidity of precipitation, soils, and water bodies within and downwind of industrialized regions (NAS 1986). The patterns of increase in fossil fuel combustion during the past few decades are reasonably well represented by the changing chemistry of tree rings, lake sediments, and on a global scale by ice cores and increasing atmospheric CO_2 levels. There is at present no proof that acid deposition has caused widespread changes in forest vigor reported in industrialized countries (McLaughlin 1985), however a strong inferential case can be made that acid deposition has been a contributing factor. In support of an acid deposition effect are the strong temporal and spatial correlations between increases in fossil fuel emissions and changes in mortality and/or growth of *P. rubens* in the eastern United States. The changes began at about the same time as emissions began to increase rapidly across the region, appeared first and most strongly at northern sites where emissions were highest, and have been most apparent in both northern and southern sites at highest elevation where atmospheric deposition is highest.

There is strong evidence of the adverse effects of acid inputs on soils including both leaching of base cations and mobilization of toxic metals that can alter root function. These types of effects would be expected to be most pronounced on the more acidic nutrient-poor soils that occur at many high elevation forested sites where decline is occurring. In the southern Appalachians, analyses of soil solution chemistry indicate that Al:Ca ratios may substantially exceed the values for toxicity to root growth determined from laboratory studies. Trends in Al:Ca in tree rings from these areas also show shifts to higher values during the past 30 years that parallel both increases in fossil fuel emissions within the region and the decline in radial growth of the red spruce trees.

Some growth stimulation may be induced through atmospheric nitrogen "fertilization" or mobilization of cations, however the potential risks of depletion of soil cation pools, and the dangers of altering root-shoot balance and increasing foliar succulence with foliar nitrogen additions must be considered as longer term risks to be weighed against those advantages.

A wide variety of laboratory and field studies indicate that ozone at ambient levels adversely effects growth of crops, native herbs, and seedling trees. The primary mechanism for those effects is the alteration of normal carbon allocation patterns including reduction of carbon assimilation,

increased allocation to maintenance and repair, and decreased translocation to and growth of roots. The shift to lower root-shoot ratios induced by
ozone can be expected to amplify natural stresses from a limited supply of
soil moisture or nutrients and susceptibility to some foliar and root
pathogens.

The effects of increasing atmospheric CO_2 on plant growth are
predominantly positive and attributable to increased carbon gain. Several
lines of evidence suggest that plants (particularly trees) have responded to
past increases in CO_2. As the atmospheric CO_2 concentration continues to
increase, selection pressures will probably arise from secondary or tertiary
effects of elevated CO_2, particularly those related to acquisition or
utilization of other environmental resources by plants in competitive
relationships. Laboratory studies have shown that elevated CO_2 can
increase the efficiency at which water and nutrients are used.

Continued increases in destruction of stratospheric ozone and increases
in UVB radiation are predicted based on anthropogenic inputs of
chlorofluorocarbons to the upper atmosphere. Concern regarding effects
of these changes on plants is based on the potential of UVB radiation to
induce cellular and molecular damage leading to reduced photosynthesis
and leaf expansion (cell division). Although depletion of stratospheric
ozone has been documented, there is not yet evidence that UVB radiation
has increased on a global scale or that plants have responded.

Selection Pressures

There is now widespread evidence that environmental pollution can exert
selective pressures that influence plant evolutionary processes and that
detectable evolutionary changes can occur in one or two generations
(Roose et al. 1982; Pitelka 1988; Bradshaw 1975). The general basis of
evolutionary changes is that plants vary in their responses to environmental
stresses thereby making genetic selection possible.

This chapter has examined the evidence that anthropogenic pollution
per se represents a significant selective force in the regional atmosphere.
The nature of today's complex stress regimes makes both the verification of
that relationship and an adequate description of the nature of the net
changes difficult. For example, some anthropogenic changes such as
increasing CO_2 may reduce effects of natural or anthropogenic stresses
while others can amplify those stresses. There are, however, some
generalizations that can be made regarding the types of anthropogenic
stress that are most likely to provide the framework for selection:

1. The effects of gaseous pollutants in reducing overall growth rate of
   sensitive species will likely be most pronounced in mixed species
   communities where differential sensitivity of component species to

growth inhibition will provide a basis for changing competitive potential.

2. The root-rhizosphere system can be expected to be adversely impacted by several aspects of current pollution regimes including reduced transport of carbohydrates below ground (ozone), increased utilization of nutrients above ground (nitrogen fertilization of shoots), and metal toxicity and reduced nutrient supply (acid deposition) to roots. For this reason selection pressures in response to pollution will likely be enhanced on nutrient poor sites and during drought episodes. By contrast, elevated CO_2 also alters nutrient acquisition and utilization, typically stimulating root growth. Those plants with the greatest response will acquire limited nutrients to the detriment of less responsive plants, thus intensifying some aspects of plant competition.

3. Increased importance of differences in species sensitivity to disease can be anticipated with reduced vigor of some species under chronic pollution stress regimes. This assumption is based on predisposition of pollution-stressed plants to attack by some types of plant pathogens.

4. Reproductive physiology, particularly pollen germination and tube growth, is another converging point for several of these stresses at ambient pollutant levels. Ozone, acid deposition (particularly cloud deposition at high elevation sites), the UVB radiation would be expected to act synergistically to adversely influence flowering physiology.

5. Elevated CO_2 concentrations stimulate growth, particularly root growth, and would be expected to counteract some of the negative effects of ozone and acid deposition in more polluted areas. In areas more remote from pollutants, improved growth and competitive potential might be expected with species for which the root-shoot ratio and nutrient or water uptake are most limiting growth.

6. There is sufficient variation in the response of different plant species to UVB radiation that shifts in competitive potential can be expected with increasing UVB radiation. Biochemical adaptations through mechanisms such as flavonoid synthesis may increase resistance.

The net effect of selection for resistance to chronic pollution levels cannot be predicted based on current information on mechanisms and specificity of past responses. Certainly plant breeding programs, particularly with reference to crops, innately incorporate some components of resistance to air pollutants when they select for superior growth under regional pollutant levels (Reinert et al. 1982). In the case of forest trees, while variations in the extent of growth reduction by pollutants provide an indication of the basis of selection, there is some evidence that the individual trees affected most are those that were originally the fastest growing individuals in the population (Oleksyn 1988; Pye 1988). If this is a general property of selection for pollution resistance, the new definition of "the fittest" may have potentially significant implications for productivity and species composition in natural plant communities.

References

Abrahamson G (1980) Acid precipitation, plant nutrients, and forest growth. In: Tablos, D, Tollen, A (eds) Ecological impact of acid precipitation. Proceedings of An International Conference, SNSF Project, Oslo, Norway, pp 56–63

Abrahamson G (1984) Effects of acidic deposition of forest soil and vegetation. Philosophical Transactions of the Royal Society London 305:369

Adams HS, Stevenson SL, Blasing TJ, Duvick DN (1985) Growth-trend declines of spruce and fir in mid-Appalachian subalpine forest. Environmental and Experimental Botany 25:315–325

Adams MB, Kelly JM, Edwards NT (1988) Growth of *Pinus taeda* L. seedlings varies with family and ozone exposure level. Water, Air, and Soil Pollution 38:137–150

Baes CF, McLaughlin SB (1984) Trace elements in tree rings: evidence of recent and historical air pollution. Science 224:494–497

Baes CF, McLaughlin SB (1987) Trace metal uptake and accumulation in trees as affected by environmental pollution. In: Hutchinson TC, Meema KM (eds) Effects of atmospheric pollutants on forests, wetlands and agricultural ecosystems. Springer-Verlag, New York

Bangerth F (1979) Calcium related disorders in plants. Annual Review of Phtopathology 17:97–122

Barnes JD, Davison AW (1990) The influence of ozone on winter hardiness of Norway spruce *Picea abies* (L. Karst). New Phytologist 108:159–166

Barnes RL (1972) Effects of chronic exposure to ozone on photosynthesis and respiration of pines. Environmental Pollution 3:133–138

Bazzaz FA, Garbutt K, Williams WE (1985) Effect of increased atmospheric carbon dioxide concentration on plant communities. In: Strain BR, Cure JD (eds) Direct effects of increasing carbon dioxide on vegetation. DOE/ER-0238, U.S. Department of Energy, Washington, D.C., pp 155–170

Berry CR, Hepting GH (1964) Injury to eastern white pine by unidentified atmospheric constituents. Forest Science 10:2–13

Blumthaler M, Ambach W (1990) Indication of increasing solar ultraviolet-B radiation flux in Alpine regions. Science 248:206–208

Bondietti EA, Baes CF III, McLaughlin SB (1989a) Radial trends in cation ratios in tree rings as indicators of the impact of atmospheric deposition on forests. Canadian Journal of Forest Research 19:586–594

Bondietti EA, Baes CF III, McLaughlin SB (1989b) The potential of trees to record aluminum mobilization and changes in alkaline earth availability. Proceedings of the National Research Council Workshop on Markers of Air Pollution Effects on Forest Trees. Little Switzerland, NC, April 1988, pp 281–292

Bowman KP (1988) Global trends in total ozone. Science 239:48–50

Bradshaw AD (1975) Pollution and evolution. In: Mansfield TA (ed) Effects of air pollution on plants. Cambridge University Press, London

Brasseur G, Hitchman MH (1988) Stratospheric response to trace gas perturbations: changes in ozone and temperature distributions. Science 240:634–637

Caldwell MM (1979) Plant life and ultraviolet radiation: some perspective in the history of the earth's UV climate. BioScience 29:520–525

Caldwell MM (1981) Plant response to solar ultraviolet radiation. In: Lange OL, Nobel PS, Osmond CB, Ziegler H (eds) Physiological plant ecology. I. Responses to the physical environment. Springer-Verlag, Berlin, pp 169–197

Carey AC, Miller EA, Geballe GT, Wargo PM, Smith WH, Siccama TG (1984) Armillaria mellea and decline of red spruce. Plant Disease 68:794–795

Carlson RW, Bazzaz FA (1982) Photosynthetic and growth response to fumigation with SO_2 at elevated CO_2 for C_3 and C_4 plants. Oecologia 54:50–54

Chappelka AH, Chevone BI, Burk TE (1988) Growth responses of green ash to ozone, sulfur dioxide, and simulated acid rain. Forest Science 34:1016–1029

Christiansen MN, Foy CD (1979) Fate and function of calcium in tissue. Communications in Soil Science and Plant Analysis 10:427–443

Cicerone RJ (1987) Changes in stratospheric ozone. Science 237:35–42

Cobb FW Jr, Wood DL, Stark RW, Parmeter JR Jr (1978) Theory on the relationships between oxidant injury and bark beetle investation. Hilgardia 39:141–151

Cogbill CV, Likens GE (1974) Acid precipitation in the northeastern United States. Water Resources Research 10:1133

Cook E (1988) A tree ring analysis of red spruce in the Southern Appalachian Mountains. In: Van Deusen PC (ed) Analysis of Great Smoky Mountain red spruce tree ring data. USDA–Forest Service General Technical Report SO-69. Southern Forest Experiment Station, New Orleans, LA, pp 6–20

Cox RM (1983) The sensitivity of forest plant reproduction to long range transported air pollutants: In vitro sensitivity of pollen to acidity. New Phytologist 95:269

Cox RM (1984) Sensitivity of forest plant reproduction to long range transported air pollutanats: in vitro and in vivo sensitivity of Oenothera parviflora L. pollen to simulated acid rain. New Phytologist 97:63–70

Cox RM (1988) The sensitivity of pollen from various coniferous and broad-leaved trees to combinations of acidity and trace metals. New Phytologist 109:193–301

Cumming JR, Alscher RG, Chabot J (1988) Effects of ozone on the physiology of red spruce seedlings. Proceedings of U.S./German Symposium on the Effects of Atmospheric Pollutants in The Federal Republic of Germany and the Eastern United States, Burlington, VT, October 1987, U.S. Forest Service, pp 355–364

Cutchis P (1974) Stratospheric ozone depletion and solar ultraviolet radiation on earth. Science 184:13–19

D'Arrigo R, Jacoby GC, Fung IY (1987) Boreal forests and atmospheric-biospheric exchange of carbon dioxide. Nature 329:321–323

DeLucia EH, Sasek TW, Strain BR (1985) Photosynthetic inhibition after long-term exposure to elevated levels of atmospheric carbon dioxide. Photosynthesis Research 7:175–184

Diaz HF (1978) A long record of weather observations at Cooperstown, New York 1854–1977. National Climatic Center, Asheville, NC

Dickson JG, Caldwell MM (1978) Leaf development of Rumex patientia L. (Polygonaceae) exposed to UV irradiation (280–320). American Journal of Botany 65:857–863

Ellertsen BW, Powell CJ, Massey DL (1972) Report on a study of diseased white pine in East Tennessee. Mitt Forstl Bundes-Versuchsanst (Wien) 97:195–208

Enoch HZ, Kimball BA (eds) (1986) Carbon dioxide enrichment of greenhouse crops, vol 2 Physiology, yield, and economics. CRC Press, Boca Raton, FL

Farman JC, Gardiner BG, Shanklin, JD (1985) Large losses of total ozone in Antarctica reveal seasonal ClO_x/NO_x interaction. Nature 315:207–210

Federal Republic of Germany, Federal Minister of Food, Agriculture, and Forestry, 1982. Forest damage due to air pollution. The Situation in the Federal Republic of Germany. Bonn, p 63

Fowler D, Cape JN, Nicholson IA, Kinnaird JW, Patterson IS (1980) The influence of a polluted atmosphere on cuticle degradation in Scots pine (*Pinus sylvetris*). In: Drablos D, Tolln A (eds) Ecological impact of acid precipitation. Proceedings of an International Conference, SNSF Project, Oslo, Norway, 1980, p 146

Fox FM, Caldwell MM (1978) Competitive interaction in plant populations exposed to supplementary ultraviolet-B radiation. Oecologia 36:173–190

Foy CD, Chaney RL, White MC (1978) The physiology of metal toxicity in plantas. Annual Review of Plant Physiology 29:511–566

Freedman B, Hutchinson TC (1980) Long-term effects of smelter pollution at Sudbury, Ontario, on forest community composition. Canadian Journal of Botany 58:2123

Friedland AJ, Gregory RA, Karenlampi L, Johnson AH (1985) Winter damage to foliage as a factor in red spruce decline. Canadian Journal of Forest Research 14:963–965

Friedland AJ, Hawley GJ, Gregory RA (1988) Red spruce (*Picea rubens Sarg*) foliar chemistry in northern Vermont and New York, USA. Plant and Soil 105:189–195

Friedland AJ, Johnson AH, Siccama TG (1982) Accumulation of trace meetals in the forest floor in Vermont: spatial and temporal patterns. Water, Air, and Soil Pollution 21:161–170

Funk DW, Bonde EK (1986) Effects of artificial acid mist on growth and reproduction of two alpine plant species in the field. American Journal of Botany 4:524–528

Gammon RH, Sundquist ET, Fraser PJ (1985) History of carbon dioxide in the atmosphere. In: Trabalka JR (ed) Atmospheric carbon dioxide and the global carbon cycle. DOE/ER-0239, U.S. Department of Energy, Washington, D.C., pp 25–62

Gerhold HD (1977) Effects of air pollution on *Pinus strobus* L. and genetic resistance. Corvallis Environmental Research Laboratory, USEPA 600/3-77-002

Gossett DR, Egli DB, Leggett JE (1977) The influence of calcium on translocation of photosynthetically fixed C14 in soybeans. Plant and Soil 48:243–251

Green AES, Cross KR, Smith LA (1980) Improved analytic characterization of ultraviolet skylight. Photochemistry and Photobiology 31:59–65

Hanson PJ, McLaughlin SB, Edwards NT (1988) Net CO_2 exchange of *Pinus taeda* shoots to variable ozone levels and rain chemistries in field and laboratory settings. Physiologia Plantarum 74:635–642

Hari P, Arovaara H (1988) Detecting CO_2 induced enhancement in the radial increment of trees. Evidence from northern timber line. Scandinavian Journal of Forest Research 3:67–74

Heath RL (1980) Initial events in injury to plants by air pollutants. Annual Review of Plant Physiology 31:395

Heck WW, Cure WW, Rawlings JO, Zaragoza LJ, Heagle AS, Heggestad HE, Kohut RJ, Kress LW, Temple PJ (1984) Assessing impacts of ozone on agricultural crops. II. Crop yield functions and alternative exposure statistics Journal of the Air Pollution Control Association 34:810–817

Hinrichsen D (1986) Multiple pollutants and forest decline. Ambio 15:258–265

Houghton RA (1987) Terrestrial metabolism and atmospheric CO_2 concentrations. BioScience 37:672–678

Houghton RA, Schlesinger WH, Brown S, Richards JF (1985) Carbon dioxide exchange between the atmosphere and terrestrial ecosystems. In: Trabalka JR (ed) Atmospheric carbon dioxide and the global carbon cycle. DOE/ER-0239, U.S. Department of Energy, Washington, D.C., pp 113–140

Houston DB (1973) Response of selected *Pinus strobus* L. clones to fumigations with sulfur dioxide and ozone. Canadian Journal of Forest Research 4:65–68

Houston DB, Dochinger LS (1977) Effects of ambient air pollution on cone, seed, and pollen characteristics in eastern white and red pines. Environmental Pollution 12:1–5

Husar RB (1986) Emissions of sulfur dioxide and nitrogen oxides and trends for eastern North America. Acid deposition: long term trends. National Research Council. National Academy Press, Washington, D.C.

Hutchinson TC (1980) Acid precipitation impact on terrestrial and aquatic ecosystems. Proceedings of the Symposium on Effects of Air Pollution on Mediterranean and Temperate Forest Ecosystems. USDA Technical Report PSW-43, pp 158–164

Jacobson JS (1980) The influence of rainfall composition on the yield and quality of agricultural crops. In: Tablos, D, Tollen A (eds) Ecological impact of acid precipitation. Proceedings of An International Conference, SNSF Project, Oslo, Norway, pp 41–46

Johnson AH, Cook ER, Siccama TG (1988) Climate and red spruce growth and decline in the northern Appalachians. Proceedings of the National Academy of Science USA 85:5369–5373

Johnson AH, McLaughlin SB (1986) The nature and timing of the deterioration of red spruce populations in Appalachian forests. Monitoring and assessing trends in acidic deposition. National Academy of Sciences, Washington, D.C., pp 200–230

Johnson AH, Siccama TJ (1983) Acid deposition and forest decline. Environmental Science Technology 17:294

Johnson DW, Friedland AJ, Miegroet HV, Harrison RB, Miller E, Lindberg SE, Cole DW, Schaefer DA, Todd DE (1988) Nutrient status of some contrasting high-elevation forests in the eastern and western United States. Proceedings of the U.S./German Symposium on the Effects of Atmospheric Pollutants in the Federal Republic of Germany and the Eastern United States. Burlington, VT, October 1987. U.S. Forest Service, pp 453–460

Johnson DW, Richter DD (1983) Effects of atmospheric deposition on forest nutrient cycles. TAPPI 676:311

Johnson DW, Richter DD, Lovett GM, Lindberg SE (1985) The effects of atmospheric deposition on potassium, calcium, and magnesium cycling in two deciduous forests. Canadian Journal of Forest Research 15:773–782

Johnston JW, Shriner DS (1985) Responses of three wheat cultivars to simulated acid rain. Environmental and Experimental Botany 25:349–353

Jones CG, Coleman JS (1989) Biochemical indicators of air pollution effects in trees: unambiguous signals based on secondary metabolites and nitrogen in fast growing trees. Biologic markers of air pollution stress and damage in forests. Proceedings of a National Research Council Symposium, April 1988. National Academy Press, pp 261–274

Joslin JD, McDuffie CM, Brewer PF (1988) Acidic cloud water and cation loss from red spruce foliage. Water, Air and Soil Pollution 39:355–363

Keeling CD, Bacastow RB, Whorf TP (1982) Measurements of the concentration of carbon dioxide at Mauna Loa Observatory, Hawaii. In: Clark WC (ed) Carbon dioxide review: 1982. Oxford University Press, New York, pp 377–385

Keller T, Beda H (1984) Effects of SO_2 on the germination of conifer pollen. Environmental Pollution 33:237

Kerr RA (1988) Stratospheric ozone is decreasing. Science 239:1489–1491

Kienast F, Luxmoore RJ (1988) Tree-ring analysis and conifer growth responses to increased atmospheric CO_2 levels. Oecologia 76:487–495

Kimball BA (1986) CO_2 stimulation of growth and yield under environmental restraints. In: Enoch HZ, Kimball BA (eds) Carbon dioxide enrichment of greenhouse crops, vol 2. Physiology, yield, and economics. CRC Press, Boca Raton, FL, pp 53–67

Klein RM (1978) Plants and near-ultraviolet radiation. Botanical Review 44:1–127

Klein RM, Perkins TD, and Myers HL (1989) Nutrient status and winter hardiness of red spruce foliage. Canadian Journal of Forest Research 19:754–758

Krause GHM, Prinz B, Jung KD (1983) Forest effects in West Germany. In: Davis DD (ed) Air pollution and the productivity of the forest. Proceedings of Symposium, Izaak Walton League, Washington, D.C., pp 297–332

LaMarche VC, Graybill DA, Fritts HC, Rose MR (1984) Increasing atmospheric carbon dioxide: tree ring evidence for growth enhancement in natural vegetation. Science 225:1019–1021

Lefohn AS, Pinkerton JE (1988) High resolution characterization of ozone data for sites located in forested areas of the United States. Journal of the Air Pollution Control Association 38:1504–1511

Lemon ER (ed) (1983) CO_2 and plants: the response of plants to rising levels of atmospheric carbon dioxide. AAAS Selected Symposium 84, Westview Press, Boulder, CO

Likens GE, Butler TJ (1981) Recent acidification of precipitation in North America. Atmospheric Environment 15(7):1103–1109

Lincoln DE, Couvet D, Sionit N (1986) Response of an insect herbivore to host plants grown in carbon dioxide enriched atmospheres. Oecologia 69:556–560

Lindberg SE (1988) Atmospheric deposition of sulfur to forests in the Integrated Forest Study. p 3–12 In: Lindberg SE, Johnson DW (eds) 1987 Group leaders reports of the Integrated Forest Study. Oak Ridge National Laboratory Report ORNL/TM-11052

Manion PD (1981) Tree disease concepts. Prentice-Hall, Englewood Cliffs, NJ

McClenahen JR (1978) Community changes in a deciduous forest exposed to air pollution. Canadian Journal of Forest Research 8:432

McLaughlin SB (1985) Effects of air pollution on forests: a critical review. Journal of the Air Pollution Control Association 35:516–534

McLaughlin SB (1988) Whole tree physiology and forest responses to air pollutants. Proceedings of the Commission of European Communities Workshop Interrelationships Between Above and Below Ground Influences of Air Pollutants on Forest Trees. Gennep, The Netherlands, December 1988 pp 8–26

McLaughlin SB (1989) Carbon allocation processes as indicators of pollutant impacts on forest trees. Biologic markers of air pollution stress and damage in forests. Proceedings of a National Research Council Symposium, April 1988. National Academy Press, pp 293–302

McLaughlin SB, Adams MB, Edwards NT, Hanson PJ, Layton PA, O'Neill EG, Roy WK (1988) Comparative sensitivity, mechanisms, and whole plant physiological implications of responses of loblolly pine genotypes to ozone and acid deposition. Oak Ridge National Laboratory Technical Report ORNL/TM-10777

McLaughlin SB, Andersen CP, Edwards NT, Roy WK, Layton PA (1990) Seasonal patterns of photosynthesis and respiration of red spruce saplings from two elevations in declining southern Appalachian stands. Canadian Journal of Forest Research 20:485–495

McLaughlin SB, Downing DJ, Blasing TJ, Cook ER, Adams HS (1987) An analysis of climate and competition as contributors to decline of red spruce in high elevation Appalachian forests of the eastern United States. Oecologia 72:487–501

McLaughlin SB, McConathy RK (1983) Effects of SO_2 and O_3 on allocation of C^{14}—photosynthate in *Phaseolus vulgaris*. Plant Physiology 73:630–634

McLaughlin SB, McConathy RK, Duvick D, Mann LK (1982) Effects of chronic air pollution stress on photosynthesis, carbon allocation, and growth of white pine trees. Forest Science 28:60–70

McLaughlin SB, Shriner DS (1980) Allocation of resources to defense and repair. In: Horsfall JB, Cowling EB (eds) Plant diseases, vol 5. Academic Press, New York, pp 407–431

Miller PR (1983) Ozone effects in the San Bernardino national forest. In: Davis DD, Miller AA, Dochinger L (eds) Air pollution and the productivity of the forest. Isaak Walton League of America, Arlington, VA, 1983, pp 161–193

Mohnen VA (1988) Exposure of forests to air pollutants, clouds, precipitation, and climatic variables. A preliminary assessment. 1987. USEPA, Research Triangle Park, NC

Molina MJ, Rowland FS (1974) Stratospheric sink for chlorofluoromethanes: chlorine atom-catalysed destruction of ozone. Nature 249:810–812

Mudd JB, Kozlowski TT (eds) (1975) Responses of plants to air pollution. Academic Press, New York

National Academy of Sciences (1977) Ozone and other photochemical oxidants. National Academy of Sciences Committee on Medical and Biological Effects of Environmental Pollutants, Washington, D.C.

National Academy of Sciences (1986) Acid deposition: long term trends. National Academy Press, Washington, D.C.

National Acid Precipitation Assessment Program (1987) Interim Assessment. Executive Summary, vol 1. Washington D.C.

National Research Council (1982) Causes and effects of stratospheric ozone reduction: an update. National Academy Press, Washington, D.C.

Neftel A, Moor E, Oeschger H, Stauffer B (1985) Evidence from polar ice cores for the increase in atmospheric CO_2 in the last two centuries. Nature 315:45–47

Norby RJ (1987) Nodulation and nitrogenase activity in nitrogen-fixing woody plants stimulated by CO_2 enrichment of the atmosphere. Physiologia Plantarum 71:77–82

Norby RJ, O'Neill EG (1989) Growth dynamics and water use of seedlings of *Quercus alba* L. in CO_2-enriched atmospheres. New Phytologist 111:491–500

Norby RJ, O'Neill EG, Luxmoore RJ (1986a) Effects of atmospheric CO_2 enrichment on the growth and mineral nutrition of *Quercus alba* seedlings in nutrient-poor soil. Plant Physiology 82:83–89

Norby RJ, Pastor J, Melillo JM (1986b) Carbon-nitrogen interactions in CO_2-enriched white oak: physiological and long-term perspectives. Tree Physiology 22:233–241

Norby RJ, Taylor GE, McLaughlin SB, Gunderson CA (1986c) Drought severity of red spruce seedlings affected by precipitation chemistry. Proceedings of the Ninth North American Forest Biology Workshop, Stillwater, Oklahoma, June 15–18, 1986

Norby RJ, Weerasuriya Y, Hanson PJ (1989) Induction of nitrate reductase activity in red spruce needles by NO_3 and HNO_3 vapor. Canadian Journal of Forest Research 19:889–896

Oechel WC, Strain BR (1985) Native species responses to increased atmospheric carbon dioxide concentration. In: Strain BR, Cure JD (eds) Direct effects of increasing carbon dioxide on vegetation. DOE/ER-0238. U.S. Department of Energy, Washington, D.C., pp 117–154

Oleksyn J (1988) Height growth of different European scots pine *Pinus sylvestris* L. provenances in a heavily polluted and a control environment. Environmental Pollution 55:289–299

O'Neill EG, Luxmoore RJ, Norby RJ (1987) Increases in mycorrhizal colonization and seedling growth in *Pinus echinata* and *Quercus alba* in an enriched CO_2 atmosphere. Canadian Journal of Forest Research 17:878–883

Penkett SA (1984) Ozone increases in European air. Nature 311:14

Pitelka LF (1988) Evolutionary responses of plants to anthropogenic pollutants. Trends in Ecological Evolution 3(9):233–236

Prinz B (1987) Causes of forest damage in Europe. Major hypotheses and factors. Environment 29(9):10–37

Pye JM (1988) Impact of ozone on growth and yield of trees: a review. Environmental Quality 17:347–360

Raynal DJ, Roman JR, Eichenlaub WM (1982) Response of tree seedlings to acid precipitation. I. Effect of substrate acidity on seed germination. Environmental and Experimental Botany 22:377–384

Reich PB, Amundson RG (1985) Ambient levels of ozone reduce net photosynthesis in tree species. Science 230:566–570

Reich PB, Schoettle AW, Amundson RG (1986) Effects of O_3 and acid rain in sugar maple and northern red oak seedlings. Environmental Pollution 40:1–15

Reinert RA, Heggestad HE, Heck WW (1982) Response and genetic modification of plants for tolerance to air pollutants. In: Christiansen MN (ed) Breeding plants for less favorable environments. Wiley, New York, pp 259–292

Reuss JO, Johnson DW (1986) Acid deposition and the acidification of streams and waters. Springer-Verlag, New York

Robberecht R, Caldwell MM (1978) Leaf epidermal transmittance of ultraviolet radiation and its implications for plant sensitivity to ultraviolet-radiation induced injury. Oecologia 32:277–287

Robberecht R, Caldwell MM, Billings WD (1980) Leaf ultraviolet optical properties along a latitudinal gradient in the arctic-alpine life zone. Ecology 61:612–619

Roose ML, Bradshaw AD, Roberts TM (1982) Evolution of resistance to gaseous pollutants. In: Unsworth MH, Ormrod DP (eds) Effects of gaseous pollutants in Agriculture and Horticulture Butterworth Scientific, London, 379–406

Rost-Siebert K (1983) Aluminum toxicity and tolerance of *Picea abies* and *Fagus Silvatica* seedlings. Allg Forst No 26/27:686–680

Rotty RM, Masters CD (1985) Carbon dioxide from fossil fuel combustion: trends, resources, and technological implications. In: Trabalka JR (ed) Atmospheric carbon dioxide and the global carbon cycle. DOE/ER-0239. U.S. Department of Energy, Washington, D.C., pp 63–80

Sage RF, Sharkey TD, Seemann JR (1989) Acclimation of photosynthesis to elevated CO_2 in five C_3 species. Plant Physiology 89:590–596

Saxena VK, Lin NH (1990) Cloud chemistry measurements and estimates of acid deposition on air above cloudbase coniferous forest. Atmospheric Environment 24A:329

Schütt P, Cowling EB (1985) Waldsterben—a general decline of forests in Central Europe: symptoms, development and possible causes. Plant Disease 69:448–558

Scotto J, Cotton G, Urbach F, Berger D, Fears T (1988) Biologically effective ultraviolet radiation: surface measurements in the United States, 1974 to 1985. Science 239:762–764

Shafer SS, Heagle AS, Camberato DM (1987) Effects of chronic doses of ozone on field grown loblolly pine: seedling responses in the first year. Journal of the Air Pollution Control Association 37:1179–84

Sheffield RM, Cost ND (1987) Behind the decline. Journal of Forestry 85:29–33

Sheffield RM, Knight HA (1983) Georgia's forest. USDA Forest Service Research Bulletin SE-73

Shortle WC, Smith KT (1988) Aluminum-induced calcium deficiency syndrome in declining red spruce. Science. 240:239–240

Siccama TG, Bliss M, Vogelmann HW (1982) Decline of red spruce in the Green Mountains of Vermont. Bulletin of the Torrey Botanical Club 109:163

Skärby L, Troeng E, Bostrom C (1987) Ozone uptake and effects on transpiration, net photosynthesis, and dark respiration in scots pine. Forest Science 33:801–808

Skelly JM (1980) Photochemical oxidant impact on Mediterranean and temperate forest ecosystems: real and potential effects. In: Miller PR (ed) Effects of air pollutants on Mediterranean and temperate forest ecosystems, USDA Forest Serice Report PSW-43, pp 38–50

Skelly JM, Yang Y-S, Chevone BI, Long SJ, Nellessen JE, Winner WE (1983) Ozone concentrations and their influence on forest species in the Blue Ridge Mountains of Virginia. In: David DD, Miller AA, Dochinger L (eds) Air pollution and the productivity of the forest. Isaac Walton League of America, Arlington, VA, pp 143–160

Smith WH (1981) Air pollution and forests. Springer-Verlag, New York

Stensland GJ, Whelpdale DM, Oehlert G (1986) Precipitation chemistry. Acid deposition: long term trends. National Research Council, National Academy Press, Washington, D.C., pp 128–199

Stolarski RS, Krueger AJ, Schoeberl MR, McPeters RD, Newman PA, Alpert JC (1986) Nimbus 7 satellite measurements of the springtime Antarctic ozone decrease. Nature 322:808–811

Strain BR (1985) Physiological and ecological controls on carbon sequestering in terrestrial ecosystems. Biogeochemistry 1:219–232

Strain BR, Cure JD (eds) (1985) Direct effects of increasing carbon dioxide on vegetation. DOE/ER-0238. U.S. Department of Energy, Washington, D.C.

Tamm CO, Hallbacken L (1988) Changes in soil acidity in two forest areas with different acid deposition: 1920s to 1980s. Ambio 17:56–61

Taylor GE, Norby RJ (1985) The significance of elevated levels of ozone on natural ecosystems of North America. In: Lee SD (ed) International specialty conference on evaluation of the scientific basis for ozone/oxidant standards. Air Pollution Control Association, Pittsburgh, PA, pp 152–175

Taylor GE, Norby RJ, McLaughlin SB, Johnson AH, Turner RS (1986a) Carbon dioxide assimilation and growth of red spruce seedlings in response to ozone, precipitation chemistry, and soil type. Oecologia (Berlin) 70:163–171

Taylor GE, Tingey DT, Gunderson CA (1986b) Photosynthesis, carbon allocation, and growth of sulfur dioxide ecotypes of Geranium carolinianum L. Oecologia 68:350–357

Temple PJ, Bennoit LF, Lennox RW, Reagan CA, Taylor OC (1988) Combined effects of ozone and water stress on alfalfa growth and yield. Journal of Environmental Quality 17:108–113

Teramura AH (1983) Effects of ultraviolet-B radiation on the growth and yield of crop plants. Physiologia Plantarum 58:415–427

Thorton FC, Shaedle M, Raynal D (1987) Effects of aluminum on red spruce seedlings in solution culture. Environmental and Experimental Botany 27:489–498

Tingey DT (1978) Effects of ozone on root processes. California Air Environment 7:5

Tingey DT, Wilhour RG, Standley C (1976) The effect of chronic ozone exposure on the metabolite content of ponderosa pine seedlings. Forest Science 22:234–241

Tolbert NE, Zelitch I (1983) Carbon metabolism. In: Lemon ER (ed) CO_2 and plants: the response of plants to rising levels of atmospheric carbon dioxide. AAAS Selected Symposium 84, Westview Press, Boulder, CO, pp 21–64

Trabalka JR, Edmonds JA, Reilly J, Gardner RH, Voorhees LD (1985) Human alterations of the global carbon cycle and the projected future. In: Trabalka JR (ed) Atmospheric carbon dioxide and the global carbon cycle. DOE/ER-0239. U.S. Department of Energy, Washington, D.C., pp 247–287

Ulrich B (1983) Soil acidity and its relations to acid deposition. In: Ulrich B, Pankrath J (eds) Effects of accumulation of air pollutants in forest ecosystems. Reidel, Nijmegen, pp 127–146

Ulrich B, Mayer RT, Khana K (1980) Chemical changes due to acid precipitation in a loess-derived soil in central Europe. Soil Science 130:193–199

United States Environmental Protection Agency (1984) Air quality standard for ozone and other photochemical oxidants. EPA-600/8-84-020A Review Draft, vol 3, Chapter 8. Effects of ozone and other photochemical oxidants on natural and agroecosystems. Environmental Criteria and Assessment Office, Research Triangle Park, NC

Unsworth MH, Biscoe PV, Pinckney HR (1972) Stomatal responses to sulfur dioxide. Nature 239:458

Van Ryn DM, Jacobsen JS, Lassoi JP (1986) Effects of acidity on in vitro pollen germination and tube elongation in four hardwood species.Canadian Journal of Forest Research 16:397–400

Walderman JM, Munger JW, Jacob DJ, Flagan RC, Hoffman MR (1982) Chemical composition of acid fog. Science 218:677–680

Weiss MJ, McGeary LR, Millers I, O'Brien JT, Miller-Wooks M (1985) Cooperative survey of red spruce and balsam fir decline and mortality in New Hampshire, New York and Vermont. Interim Rept. USDA Forest Services, P.O. Box 640, Durham, NY

Wolters JHB, Martens MJM (1987) Effects of air pollutants on pollen. Botanical Review 53(3):372–414

Woodward FI (1987) Stomatal numbers are sensitive to increase in CO_2 from pre-industrial levels. Nature 327:617–618

Woodward FI, Bazzaz FA (1988) The response of stomatal density to CO_2 partial pressure. Journal of Experimental Botany 39:1771–1781

Zahner R, Saucier JR, Myers RK (1990) Tree-ring model interprets growth decline in natural stands of loblolly pine in the southeastern United States. Canadian Journal of Forest Research 19:612–621

Evidence of Changes, Linkages, and Significance of Atmospheric Pollution to the Selection Process

T. MIKE ROBERTS

Introduction

McLaughlin and Norby (Chapter 4) have developed criteria for selecting the pollutants which may cause evolutionary changes in plant populations on a regional scale. The criteria are based on (a) evidence for effects on yield or reproductive success, (b) increased concentrations in recent decades, (c) continued exposure to present concentrations or projected increases in concentration. The chapter considers the emissions and effects in North America of sulfur dioxide, nitrogen oxides, hydrocarbons, ozone, acid deposition, carbon dioxide, and ultraviolet radiation. The authors conclude that there is evidence of regional effects of both acid deposition and ozone on seminatural and forested ecosystems. The authors speculate that there may be significant consequences of increased CO_2 emissions, whereas significant effects of ultraviolet radiation are less certain. This Commentary explores the magnitude of air pollution damage to terrestrial ecosystems in Europe and questions whether evolutionary responses have been demonstrated on a regional scale.

Air Pollutants as Stress Factors in Europe

Applying the three criteria used by McLaughlin and Norby in this book to air pollution in Europe gives a slightly different set of priorities. First, emissions of SO_2 are considerably more important. Evidence presented by Bell et al. (Chapter 3) indicate that SO_2 levels in urban and industrial areas in the United Kingdom have induced significant evolutionary effects in grasslands there. However, in many areas of western Europe, SO_2 emissions have declined since the mid-1970s and recent decisions on emission controls will ensure that ground level concentrations will be drastically reduced by the year 2010. In eastern Europe, there are still severe effects of acute SO_2 exposures on forested ecosystems, and it is unlikely that the emissions will be reduced as rapidly as in western Europe.

In eastern Europe, therefore, SO_2 is having a significant effect on terrestrial ecosystems on a regional scale and the evolutionary responses should be considered.

Attention is now focusing on NO_x emissions as these continue to increase for most of western Europe, and the current control strategies will not produce as significant a reduction as for SO_2. Emissions of NO_x may affect terrestrial ecosystems directly in combination with other gases; through formation of phytotoxic concentrations of O_3 in the atmosphere and through indirect effects induced by excess nitrogen deposition.

Considerable progress has been made in understanding the effects of air pollutants on forests in Europe (Blank et al. 1988). The causes of the five main decline types of Norway spruce are reasonably well understood, i.e., (1) acute SO_2 injury in parts of Eastern Europe; (2) magnesium deficiency in upland areas of West Germany induced by acid deposition, tree harvesting, and drought; (3) needle reddening of older stands in southern Germany (induced by needle-cast pathogens); (4) needle yellowing on calcareous soils in the Alps (induced by nutritional factors); and (5) die-back of pine in Holland, Denmark, and northern Germany (induced by excess nitrogen deposition).

Ozone meets some of the criteria used by McLaughlin and Norby to identify pollutants which may cause evolutionary changes on a regional scale in Europe. Concentrations have nearly doubled since the 1950's, although there is some indication of a decrease in recent years at remote sites. Despite emission controls on NO_x and hydrocarbons, it is unlikely that O_3 levels will decrease significantly in the next decade. However, O_3 concentrations are generally lower than those reported in some areas of North America, and the evidence for effects on crops and forests is less equivocal. For example, a European Communities research project has used open-top chambers to determine the effect of filtering pollutants on agricultural crops in ten countries (Bonte and Mathy 1989). For 1987 and 1988, which were low O_3 years in western Europe (7-h/day mean summer values ranged from 15 to 50 ppb), significant effects of air filtration were recorded at only two sites. Nevertheless, in hot summers producing higher O_3 levels, it seems likely that the effects of O_3 on crops and other herbaceous vegetation will occur.

Recent studies also indicate that O_3 is not a significant factor in the nutrient deficiencies associated with some of the recent forest declines in Europe (Brown and Roberts 1988). It had been proposed that O_3 would accelerate the leaching of nutrients from conifers by acid deposition, leading to accelerated nutrient deficiencies. However, recent studies have shown that this interaction was due to anhydrous nitric acid produced as a byproduct of O_3 generation. O_3 may play a role in reduced nutrient uptake through effects on root growth, but this has not been demonstrated experimentally.

As in North America, direct effects of acid deposition on crop yields are unlikely, but acid deposition may affect forested and seminatural ecosystems on a regional scale through accelerated soil leaching. It has been shown that acid deposition plays a significant role, at least in the Type 1 (magnesium deficiency) forest decline in Central Europe (Roberts et al. 1989). Tree harvesting and acid deposition have both contributed to soil acidification and depletion of magnesium in certain areas (about 5% of the forest area in West Germany is affected by Mg deficiency).

Nitrogen deposition is very high (>50 kg/hectare/each year) close to intensive agricultural parts of Holland, Denmark, Belgium, and northern Germany, and this has led to damage in some forest areas (Van Breemen and Van Dijk 1988). Damage to pine stands has been reported through increased susceptibility to pathogens, nutrient deficiencies, and soil acidification. Deposition on a regional basis in Europe is lower (20 to 30 kg/hectare/each year), and the contribution to forest decline in these regions is less certain. Nitrogen deposition may accelerate growth and thereby exacerbate deficiencies of other nutrients (Kenk and Fischer 1988). Indeed, there have been significant increases in the yield of Norway spruce in the Black Forest in the last 20 years which may have added to the magnesium deficiencies that are now seen on a regional scale.

Little attention has yet been paid to the effects of increased CO_2 or ultraviolet (UVB) radiation. If increased CO_2 results in significant climate change, then there will be significant changes of species and communities on a regional scale. In contrast, the effects of increased CO_2 alone may act in part to counteract some of the effects of air pollutants (e.g., by increased water use efficiency or higher root-shoot ratios), although nutrient deficiencies may be exacerbated by accelerated growth. Increases in UVB radiation may act at higher altitudes through changes in cuticle structure or induction of genetic changes. The latter may not be a significant factor as meristems are generally protected from direct radiation. The attenuation of UVB light by plant cuticles is a critical factor and needs investigation. It is possible that the elevated levels of ozone at higher altitudes could interact with elevated UVB radiation to produce significant effects on cuticle structure.

Investigating Evolutionary Responses to Air Pollution Stress

It is apparent that air pollutants constitute a significant environmental stress in many terrestrial ecosystems. However, while research has established that individuals, populations, and varieties of a particular species can differ in their response to pollutants, the importance of evolutionary responses to pollutants on a regional scale remains to be established. There are a number of technical complications which make it difficult to produce definitive evidence.

Roose et al. (1982) drew attention to the possibility of comparing evolution of tolerance to herbicides and metal wastes as comparisons with responses to acute and chronic gaseous pollutants. Table 1 is based on the original tabulation of Roose et al. (1982) which has been extended to include indirect effects of gaseous emissions through changes in soil chemistry by either acid deposition or excess nitrogen deposition. Studies of the evolutionary responses to air pollutants will be more difficult than studies of metal or herbicide tolerance for a number of reasons.

First, a major factor determining how fast resistance can evolve in a population is the strength of selection. There are a few examples where it has been demonstrated that air pollutants could be a strong selective force (e.g., acute SO_2 effects on forests in parts of eastern Europe and around point sources in western Europe and in North America). However, away from point sources, selection by gaseous air pollutants or acid deposition will be weaker. Occasional acute exposures will be less effective as resistance will depend on the comparative performance of resistant genotypes in unpolluted air. In contrast, chronic exposures will act continuously on all phases of the life-cycle, although in general the effects on important competitive parameters, such as growth, are relatively small.

Second, there are no sharp gradients in air pollutant concentrations— with the exception of SO_2, NO_x, and HF emissions from point sources. The pollutants of most concern on a regional scale (e.g., O_3 and acid deposition) cannot readily be investigated by working along gradients. However, improved recognition of the effects of air pollutants on managed ecosystems should allow breeding programs to include a range of field sites which take into account air pollution concentrations as well as aspects of site quality.

Third, there are technical difficulties in developing experimental procedures for screening for resistance under controlled conditions. Studies of metal tolerance were made possible by the development of a rapid routine bioassay test. Screening for resistance to the direct effects of gaseous pollutants has until recently been limited by the availability of large exposure chambers. It is also difficult to determine the concentration to be used in the screening trial which will give significant differentiation among genotypes. This is particularly so when the plant species is long-lived and the mechanism of action of the pollutant is cumulative.

Fourth, as shown in Table 1, the effect of gaseous pollutants may be indirect through changes in sensitivity to other environmental stresses or through changes in soil chemistry. In the former case the appropriate screening test may have to involve a combination of factors (e.g., pollutant-cold stress or pollutant-drought stress). In other cases, the deposition of the pollutant in dry or wet form may result in soil acidification or nutrient deficiency. In this case, selection results from

TABLE 1. Comparison of factors affecting evolution of resistance to metal wastes, herbicides, gaseous air pollutants, and soil acidification.

	Metal wastes	Herbicides	Gaseous pollutants	Acid (or excess N) deposition
			Severity	
	Severe	Severe	Moderate	Moderate/Severe
Selection factor				
Temporal nature	Continuous	Intermittent	Intermittent	Continuous
Nature of gradients	Sharp	Sharp	Gradual	Gradual
Biological response				
Variation in populations	Rare	Rare	Common	Common
Costs of resistance	Large	Large	?	Large (?)
Rate of evolutionary response	Fast	Fast	Probably slow	Slow

Developed from Roose et al. 1982.

either aluminum toxicity or deficiency of essential nutrients such as magnesium.

With the exception of effects around point sources and some areas of eastern Europe, chronic effects of air pollutants on a regional scale do not appear to be causing significant changes in species diversity. The potential effects of CO_2 emissions are much greater, depending upon the magnitude of the resultant climatic change. Significant changes in the distribution of species and communities could result, with consequent evolutionary changes within species. The effects of increased UVB radiation need to be determined at the physiological and biochemical levels before implications for genetic diversity can be considered.

Conclusions

McLaughlin and Norby (Chapter 4, this volume) outline the criteria for identifying pollutants which may act as significant selective factors on a regional scale. They provide evidence of regional-scale changes in plant growth and vigor in response to O_3 and acid deposition in North America. In a European context, the effects of O_3 are less severe but acute effects of SO_2, and chronic effects of excess nitrogen and sulfur deposition—soil acidification and nutrient depletion—are significant. Attention is also drawn to the fact that the mechanism of action of air pollutants may be both direct and indirect. Indirect effects act through increased sensitivity to other stresses (e.g., drought, cold stress, disease, and pests) or through changes in soil chemistry.

The severity and temporal fluctuations in the selection factor is of critical importance in determining the rate of evolution. Continuous exposure to stress results in fitness differences accumulating over time and eventually becoming quite large. In contrast, intermittent stress means that growth and survival of resistant genotypes depend upon their response to both the polluted and normal environment, thus slowing the evolution of resistance.

The degree of genetic variability for a particular stress may limit evolution and this seems to be the case for resistance to heavy metal and herbicides because resistance is absent in many species. In contrast, variation in resistance to air pollutants appears to be common in most species. In part this difference probably reflects the severity of the stress imposed: resistance to heavy metals and herbicides is identified by survival at concentrations which kill a large proportion of the population, whereas resistance to air pollution is studied at dosages causing differential injury or growth reduction.

References

Blank LW, Roberts TM, Skeffington RA (1988) New perspectives on forest decline. Nature 336:27–30

Bonte J, Mathy P (eds) (1989) European Commission research project on open top chambers. Results on Agricultural Crops 1987–88. European Commission, Brussels

Brown KA, Roberts TM (1988) Effects of ozone on foliar leaching in Norway spruce. Environmental Pollution 55:55–73

Kenk G, Fischer H (1988) Evidence from nitrogen fertilisation in the forests of Germany. Environmental Pollution 54:199–218

Roberts TM, Skeffington RA, Blank LW (1989) Causes of type 1 Spruce decline in Europe. Forestry 62:179–222

Roose ML, Bradshaw AD, Roberts TM (1982) Evolution of resistance to gaseous air pollutants. In: Unsworth MH, Ormrod DP (eds) Effects of gaseous air pollution in agriculture and horticulture. Butterworth Scientific, London, pp 379–409

Van Breemen N, Van Dijk HFG (1988) Ecosystem effects of atmospheric deposition of nitrogen in the Netherlands. Environmental Pollution 54:249–274

5
Genetics of Response to Atmospheric Pollutants

MIKEAL L. ROOSE

Introduction

Human activities have altered the composition of the atmosphere on both a local and global scale (Chapter 4, this volume). For many plant species, these environmental changes represent a novel stress to which they may or may not adapt. The alternatives to adaptation include range reduction and local or global extinction. The capacity of plant populations to adapt to an environmental stress is considerably influenced by several aspects of their genetic system. Adaptation can only occur if the population possesses or generates heritable genetic variation influencing fitness in the new environment. During the period of adaptation, phenotypic characteristics of the population (such as plant health and density) will also be influenced by the frequency and pleiotropic effects of the genes being selected. The purpose of this paper is to consider these genetic issues. It does not include a comprehensive review of the literature on the genetics of pollution resistance, but rather compares the various approaches which can be taken to characterize the genetics of resistance in natural populations.

It is commonly stated that natural populations have little genetic variation for fitness (Falconer 1981) and therefore, it is inferred that they are at an evolutionary equilibrium. This is sometimes seen as a consequence of Fisher's fundamental theorem of natural selection which states that the rate of increase of mean fitness is equal to the additive genetic variance in fitness (Fisher's proof applies only to single locus viability models). Thus, if there is genetic variance in fitness, the population will evolve and (ignoring mutation) the genetic variance will eventually disappear. Imposition of anthropogenic stresses clearly provides the population with a new "target" to evolve toward, and therefore we have the opportunity to observe adaptive evolution.

The Role of Intrinsic Variation

The first prerequisite for plant populations to evolve increased resistance to atmospheric pollutants is appropriate genetic variation: genes which, compared with their alternate alleles, increase net fitness in the new (polluted) environment. The ultimate source of all such variation is mutation (at least in a broad sense). Mutational events range from simple base substitutions to deletions, duplications, and gene fusions. Additional variation can also be introduced by introgression with other species. The frequency and type of mutational events which occur in a population are now being accorded a more central position in evolutionary theory (Endler and McLellan 1988).

One major issue concerns the relative importance of new mutations vs. existing variation in allowing adaptation of plants to novel environments. A few years ago, most geneticists would have dismissed new mutations as unimportant, but our view of mutation has been changing rapidly. We now recognize the potential significance of transposable elements in generating mutation, and it is clear that they can substantially increase response to selection in some situations (MacKay 1988). Transposable elements may be particularly important for evolutionary responses to atmospheric stresses because at least some are activated by stress events (McClintock 1984; Nevers et al. 1986). Activation results in a burst of new mutation. As with other types of mutation, most new mutations with strong phenotypic effects are deleterious, but some may not be. Essentially nothing is known about the selective value of transposable element induced mutations with minor effects on phenotype. Transposable element induced mutations may play an important role in altering protein sequences and their expression (Saedler and Schwarz-Sommer 1987). Such mutations, as well as the creation of new genes by exon shuffling, may be important in adaptation to novel stresses (McDonald 1983).

The number of plant species in which transposable elements have been identified is small but growing, and many more species have mutable loci which suggest the presence of transposable elements (Nevers et al. 1986). In species in which they occur, the proportion of plants in wild populations having active transposable elements—those currently generating new mutations—is not known. Despite these caveats, it is unwise to assume that new mutations do not contribute to adaptation. This is an area clearly ripe for innovative new research at both the molecular and population levels.

More generally, many quantitative geneticists are reconsidering the view that spontaneous mutations make little contribution to long-term selection response (Hill 1982; Hill and Keightley 1988). To a considerable extent, this reassessment is prompted by difficulty in explaining the continued response to artificial selection after 50 or more generations in artificial selection experiments. Although too little is known about mutation rates for genes influencing quantitative traits to conclude that mutation plays a major role in such long-term selection experiments, it is also clear that our knowledge of mutation rates for quantitative traits is rather meager.

In classical neo-Darwinian models, the major source of genetic variation for evolution already exists in populations. Such variants are generally seen as more important for rapid evolution than new mutations because they may already exist at substantial frequencies in populations. Why such variation is present in populations is, of course, a matter of considerable debate among population geneticists. The principal competing explanations are that most such variants are selectively neutral and are maintained by a balance between mutation and genetic drift (Turelli et al. 1988), or mutation and stabilizing selection (Lande 1975), or that the variants are subject to some type of balancing selection through overdominance, frequency dependence, environmental heterogeneity, or pleiotropic overdominance (Turelli 1988). The resolution of this debate is not in sight. This issue has important implications for evaluating adaptation to pollutants because the two models make quite different predictions about secondary effects of adaptation. If variants responsible for pollution resistance are selectively neutral in unpolluted environments, then adaptation to pollution involves no "costs," and the resulting plants should remain equally well adapted to unpolluted environments. On the other hand, if selection is important in maintaining these variants, then increasing the frequency of alleles for resistance will have "costs" and will reduce adaptation to unpolluted environments.

Genetic variation in natural populations has been measured for both single genes and quantitative traits. The frequency of allelic polymorphisms at essentially random loci has been determined using molecular markers, primarily isozymes, and more recently RFLP analysis and DNA sequence data. Isozyme studies reveal extensive polymorphism in nearly all outcrossing species and many predominantly selfing species (Brown 1979). Although only a few such studies have yet been reported, it is clear that analysis at the DNA level reveals much more variation. How the variants detected with these techniques are maintained in populations, and whether or not they could contribute to adaptive evolution are still matters of much debate.

A smaller number of studies have used the methods of quantitative genetics to measure genetic variation in natural populations (Schoen 1982; Mitchell-Olds 1986; Roach 1986; Shaw 1986). In many cases, the traits measured in these studies are related to fitness, and therefore it is more difficult to dismiss variation in such traits as unimportant to current or potential adaptation. These studies have also found genetic variation affecting the traits examined, but there may often be negative genetic correlations among traits, presumably reflecting pleiotropic effects of the genes involved.

Overall, both molecular and quantitative genetics studies reveal that natural populations of most organisms include a vast array of genotypes. Whether this variation is maintained by selection or mutation and genetic

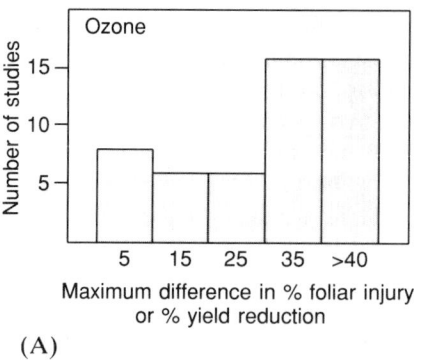

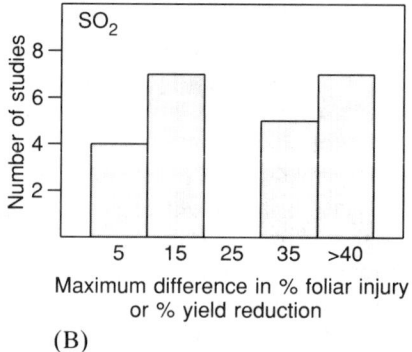

FIGURE 1. (A, B) Distributions of the maximum difference among genotypes (cultivars, clones, or families) within plant species in response to ozone (A) or SO_2 (B). Five ozone and three SO_2 studies reported percent yield reduction, all other studies reported only percent foliar injury (from Roose et al. 1982, Tables 18.3 and 18.4).

drift, it may be the raw material for evolutionary response to atmospheric pollution. Since pollutants represent a novel stress, it is quite possible that even genes which are selectively neutral in an unpolluted environment will confer a selective advantage in a polluted environment. If genetic variation influencing adaptation to polluted environments is maintained by selection in unpolluted environments, then when the frequency of these genes changes in response to pollution stress, the adaptation of the population to those aspects of the environment which previously determined the frequency of these genes may also change. Ecologically, this would be seen as a change in the niche or range of the population. This argument is a genetic perspective on the "costs" of resistance to pollution.

For response to air pollutants, there is no evidence concerning how variation in resistance is maintained in populations from unpolluted environments. Neither are there studies of the effects of pollution on fitness components of natural populations from truly unpolluted environments. For CO_2, the opportunity for such studies is already lost since all populations may well have been influenced by increasing CO_2 levels.

Pollutant resistance has been studied in many crop species (reviewed Roose et al. 1982; Karnosky 1985), and variation is usually observed (Fig. 1), but a single natural population may have somewhat less genetic diversity than a collection of cultivars as is typically included in such surveys. In any case, few of these studies evaluate fitness components— they emphasize variation in such measures of resistance as leaf chlorosis, growth, and yield. Furthermore, many of these studies involve exposure to higher pollutant levels than typically occur in natural populations. The function relating fitness variance to pollutant levels is not known, but it is unlikely to be monotonic. Fitness variance in natural populations is typically low in the environment to which they are adapted (Falconer 1981). As pollution stress increases, variance should increase, but must eventually decrease because few genotypes survive. Despite these limitations in our understanding of the genetic details, it is clear that populations often contain variation influencing response to atmospheric pollutants.

Mode of Inheritance

When the objective is to understand and predict the evolutionary response of natural populations to pollutant stress, a genetic analysis should attempt to answer the following questions:

1. To what extent is variation in fitness components in polluted and unpolluted environments determined by environmental vs. genetic factors?
2. To what extent can this variation be explained by segregation at a single gene locus?
3. Do genes influencing response to pollutant stress also influence other traits?

Genotype vs. Environment

The first problem is distinguishing the relative contributions of genetic and environmental causes to variation in response to pollution. It is fairly simple to design and conduct appropriate experiments for annual plants in controlled environments, but much more difficult for perennials or plants growing in natural environments. Controlled experiments comparing injury, growth, and yield of different crop cultivars or clones from natural populations generally reveal substantial genetic variation for pollution resistance (Roose et al. 1982; Karnosky 1985).

Gregorius (1990) describes a general approach for distinguishing genetic and environmental effects on phenotypic varation in which pairs of neighboring individuals which apparently differ in resistance are evaluated for allozyme gene frequencies. If gene frequencies differ significantly between resistant and sensitive classes, then it can be concluded that the phenotypic differences have some genetic basis. Unfortunately, predictions of evolutionary response require a more detailed description of genetic control of the trait than can be provided by partitioning phenotypic variation into genetic and environmental components.

Detection of Single Genes for Resistance

Most studies of inheritance of resistance to atmospheric pollutants have found that the trait is quantitatively inherited (reviewed by Roose et al.

1982). In fact, most such studies have *assumed* that resistance is quantitatively inherited and made no attempt to identify major genes. In part this reflects the nature of the traits studied. For leaf chlorosis or necrosis, it may be possible to distinguish distinct phenotypic classes at a given level of pollutant and thereby to fit data to Mendelian models. However, there is abundant evidence that even these measures of resistance are influenced by environmental conditions which are difficult to apply uniformly to all plants in a segregating population. For example, degree of injury or growth reduction from SO_2 can be influenced by wind speed (Ashenden and Mansfield 1977), nutrients (Cowling and Koziol 1982), irradiance (Davies 1980), and temperature (Jones and Mansfield 1982). When single plants are scored, misclassification of genotypes could result from variation in any of these environmental factors. In many cases, such environmental variation could transform a single locus segregation into apparently continuous variation in resistance. It is not argued that most cases of variation in resistance are attributable to single genes but rather that adequate testing of this simple hypothesis has not been conducted before moving to quantitative analysis. A variety of methods are available for detecting single genes with major effects on a trait which is also influenced by the environment (Thoday 1961; Wehrhahn and Allard 1965; Elston and Stewart 1973; Fain 1978; Elston 1984). Admittedly, most of these methods require inbred lines and therefore are difficult to apply to natural populations of outcrossing species, but none have been applied to pollution resistance data.

A different approach to detecting single genes with major effects on a trait also influenced by environmental variation involves testing for associations between marker loci (typically isozymes or RFLPs) and the trait of interest in segregating populations. This approach has been used to detect genes influencing insect resistance (Nienhuis et al 1987), water use efficiency (Martin et al. 1989), and soluble solids content (Osborn et al. 1987) of tomatoes, and yield and other traits of maize (Edwards et al. 1987). However, identifying a sufficient number of informative markers to have a reasonable probability of detecting genes with strong effects can be difficult unless the parents are quite divergent (Ellis 1986). Clearly this approach could be applied to pollution resistance using the genetic markers already developed in tomato, maize, brassicas, Arabidopsis, and other crop plants.

The approach described by Gregorius (1989) and mentioned above is conceptually similar but more easily applied to natural populations. Muller-Starck (1989) compared genotypes of adjacent healthy and declining trees at 9-17 isozyme loci in six stands of beech exposed to a mixture of air pollutants in Europe. Allele frequencies at several loci differed between the apparently resistant and sensitive genotypes, and resistant genotypes were generally more heterozygous. Bergmann and Scholz (1989) also found some alleles associated with resistance in Norway spruce. Either the loci studied are directly involved in response to pollutants, or they are in linkage disequilibrium with genes responsible for differences in resistance. Linkage disequilibrium is generally rare in populations of outcrossing plants (Brown 1979, but see Epperson and Allard 1987), suggesting that the loci studied may contribute to resistance.

In a few cases pollution resistance does seem to be simply inherited. Ozone resistance in onion (Engle and Gabelman 1966), copper resistance in *Mimulus* (MacNair 1983) are examples. Koziol et al. (1986) observed that the RUBISCO from a ryegrass genotype resistant to SO_2 had greater resistance to inhibition by sulfite than that from a sensitive genotype. RUBISCO subunits are specified by a single gene in the chloroplast genome and a small family of nuclear genes, but the enzyme may be modified by other gene products after synthesis. Inheritance of resistance from this genotype has not been reported, but one gene could account for much of its resistance.

Several studies report that activities of enzymes or metabolites induced by pollution stress differ between sensitive and resistant genotypes (Lee and Bennett 1982; Tanaka et al. 1985). While these studies suggest that enzymes causing pollution resistance have been identified, genetic tests showing that resistance cosegregates with the enzyme activity have not yet been reported. Even if such tests show association, unless resistance cosegregates with the gene specifying the enzyme, it cannot be assumed that this enzyme is responsible for resistance. Resistance may be conferred by a gene which modifies the structure or expression of many different proteins, and activity of all such proteins will cosegregate with resistance.

Resistance as a Quantitative Trait

Although detection of single genes involved in pollution resistance has rarely been attempted, the general belief that variation in resistance is a quantitative trait is probably sound. This, however, tells us little about the genetic architecture of resistance, rate of evolution of resistance, or its effects on other characteristics of the population. The most common statistic used to quantify genetic variation in resistance is its heritability: the ratio of genetic variance to total phenotypic variance. Two types of heritability estimates are relevant to evolutionary biologists: the narrow-sense heritability is appropriate for sexually reproducing populations, while the broad-sense heritability is appropriate for apomictically or clonally reproducing populations. The difference between the two is that the numerator of the narrow-sense heritability contains only the additive genetic variance component, that portion of the genetic variance which determines response to selection in a sexually reproducing population. Since the progeny of clonally reproducing genotypes are genetically identical to the parent, all genetic variation can contribute to response to selection, and therefore all types of genetic variance are included in the numerator. These heritabilities relate to Bradshaw's concept of various stages of evolution (Chapter 2). Broad-sense heritability is related to variation in survival and reproduction of the plants in the initial population (stages 1 and 2), while narrow-sense heritability determines the long-term response of the population over several generations (stage 3).

When evaluating heritability estimates for crop plants, evolutionary biologists should note the unit for which the heritability was calculated. Plant breeders frequently calculate heritability estimates based on evaluation of plots of related individuals, and select among genotypes using a similar basis. Such heritabilities are appropriate for selection on a plot basis, but will often be substantially larger than the single-plant heritabilities appropriate for predicting response to natural selection. For crop plants, estimates of environmental variances are likely to be lower (and heritabilities correspondingly higher) than those in natural populations because breeders typically evaluate the plants in environments which are as uniform as possible. Although the magnitude of this effect is difficult to evaluate, it seems likely to be substantial. Thus, considerable caution should be used in attempting to extrapolate from crop plant heritabilities to those in natural populations.

Heritability is considered an important measure of variation in quantitative traits because, for a single trait, it is directly related to the response to selection (rate of evolution). However, in natural populations, selection acts simultaneously on many traits so that the overall rate of evolution is influenced not just by heritabilities of each trait, but also by the genetic covariance or correlation between traits. Genetic correlations arise because two traits are influenced by the same genes (pleiotropy) or by linked genes which are in linkage disequilibrium. The heritabilities and genetic correlations of fitness components must be known to predict response to natural selection.

Experimental Approaches for Quantitative Resistance

In contrast to leaf injury, the stress resistance of single plants cannot be measured for traits such as yield and fitness components because resistance is the deviation of the stressed phenotype from the "normal" phenotype. Both phenotypes cannot be measured on the same plant. There are several approaches to measuring inheritance of such traits. One is to clone each genotype, grow the clonal propagules in each environment, and calculate the heritability of the difference in the value of the trait between environments. Siblings can be used in a similar way to compare resistance of parental genotypes. Stress resistance can be expressed as the ratio (or percentage) of the value of the stress phenotype to the normal phenotype. Use of ratios or percentages further complicates testing the significance of the variation observed.

Another approach is to treat the phenotype in the two different environments as two different characters, calculate the heritability of each character, and the genetic correlation between them (Falconer 1952). This genetic correlation measures the extent to which the traits are influenced by the same (or linked) genes. This general approach can be extended to consider the genetic variance-covariance matrix for various fitness components in both stress and normal environments. This information is necessary to predict the rate of evolution of resistance, because it allows us to evaluate the extent to which the various fitness components are correlated with one another. For example, if the genetic correlation between survival and female fecundity in the polluted environment is negative, say because under stress some genotypes allocate resources to growth rather than flowering, the rate of evolution will be much slower than if this correlation is positive or zero. Air pollution clearly influences multiple fitness components such as survival (Davidson and Bailey 1982), fecundity (Bonte 1982), and pollen viability (Wolters and Martens 1987), but there is little information on genetic correlations among these effects. Similarly, if pollutant stress is episodic, so that plants must be adapted to both polluted and unpolluted environments, the appropriate genetic variance-covariance matrix is necessary to predict the rate of evolution of the population.

What would be necessary to estimate the genetic and environmental variance-covariance matrix in a natural population? The general approach is to measure all traits of interest in individuals varying in degree of relatedness (half-sibs, full-sibs, unrelated). Various statistical methods can then be used to relate phenotypic similarity to expected genetic similarity. Mitchell-Olds and Rutledge (1986) consider the assumptions involved for application of quantitative genetic analysis to plant populations. The problems in obtaining sufficiently precise, unbiased estimates of quantitative genetic parameters are formidable. To obtain estimates useful for predicting response to selection these authors point out that:

1. For designs requiring half-sibs controlled pollinations are necessary because natural families are unlikely to represent half-sibships.
2. If mortality and environmental variation are high in the population, measurement of hundreds or thousands of plants from scores of families is likely to be necessary to obtain reliable estimates.
3. Parents must be equally inbred and progeny not inbred.
4. No genes with large effects on the trait should be present.
5. Genotypes must not interact with new or unmeasured environments.

The last point is particularly important with regard to pollution resistance because it emphasizes the difficulty in interpreting controlled environment experiments on the inheritance of pollution resistance. If plants are exposed to pollutants in an environment far different from that they experience in natural populations, it is likely that genotypes will interact differently with this environment in unpredictable ways. The consequence is that heritabilities and genetic correlations measured in such an environment may bear little relationship to those in the natural population.

Because these conditions are difficult to meet, in some cases it may be simpler to try to measure the results of selection than to predict it. Clearly, if a population evolves to become more resistant to pollutants, the necessary genetic variation must have been present—regardless of how thoroughly we can characterize it. However, if no genetic analysis is undertaken, we gain little predictive ability and little insight into the architecture of resistance.

Quantitative genetic analysis of pollution resistance could be performed directly in natural populations if information on parentage of individuals were available. In principle, sufficiently polymorphic molecular markers could be used to identify genetic relationships among plants in natural populations, but current methods are inadequate for reliable identification of even paternity when the maternal genotype is known (Chakraborty et al. 1988). More powerful statistical methods for identification of parentage, and polymorphic loci having many very rare alleles are necessary for this approach to succeed. If these problems were solved, then information on responses of individuals to pollutants would allow quantitative genetic analyses of pollutant resistance. The data would also be suitable to test for associations between markers and resistance. The technical requirements for such a project are immense, but it is not necessarily impossible.

The Value of Genetic Studies

The rate of evolution of resistance will depend primarily on selection pressure and the magnitudes of genetic and environmental variances and covariances. However, the duration of evolutionary response, and the level of resistance achieved may be quite different for resistance determined by a single gene vs. many genes. With a single gene model, the final level of resistance of the population is unlikely to be better than that of the best plant in the existing population (assuming that homozygotes for resistance exist). In contrast, with resistance determined by several or many genes of approximately equal effect, the most resistant possible genotype probably does not exist in the initial population.

Using a model in which fitness increases exponentially with the value of the character, Lande (1983) compared the rate of evolution due to a major mutation with that from polygenic variation. He concluded "a large evolutionary change in the mean phenotype in a population will occur more rapidly by a major mutation than by polygenic changes only if the initial frequency of the major mutant exceeds a critical value depending on its degree of dominance." In general, conditions favoring fixation of a major mutation are that negative pleiotropic effects of the mutation are small, selection for the mutation is reasonably strong, the initial frequency of the mutation is fairly high (or mutation rate is high), and polygenic variation for the trait is small. It is not clear to what extent this result is dependent on the particular exponential fitness model employed.

Genetic Aspects of "Costs" of Resistance

Jinks and Pooni (1988) point out that selection for high performance (fitness) in a below average environment (stress) will reduce the environmental sensitivity of the population. This occurs because, if mean performance and environmental sensitivity are determined by independently segregating genes, then considering only genotypes with the same genes for performance, those less sensitive to the poor environment will have a higher mean and therefore a greater probability of being selected than genotypes which are more sensitive to the poor environment. These effects can be illustrated by a simple two-locus model with one locus influencing mean performance and the second influencing environmental sensitivity (Fig. 2). Genotypes selected in a poor environment would therefore be expected to have poorer performance in a good environment than those selected in a good environment because of their lower environmental sensitivity, in addition to the fact that they were not selected in this environment. The major assumption in this analysis is that mean performance and environmental sensitivity are at least partially genetically independent traits. Similar logic should apply to natural populations subject to selection under pollution stress, but the direction and magnitude of the genetic correlation between sensitivity and mean fitness is not known.

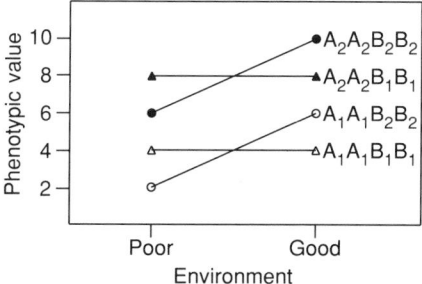

FIGURE 2. Two-locus model illustrating effect of independent genes determining performance and environmental sensitivity. Each *A2* allele increases performance by 2 units, while *B2* causes the genotype to be environmentally sensitive. Among plants with the same genes for performance (genotype at locus *A*), those with the *B2* allele would be selected for in a good environment, but against in a poor environment.

In a theoretical study, Rosielle and Hamlin (1981) show that artificial selection for stress resistance (defined as low reduction in yield in stress environments) will usually reduce performance in nonstress environments unless genetic variation in the stress environment is greater than that in the nonstress environment. While this result is important for breeders working with artificial selection, in natural populations selection acts on fitness variation in the environment experienced by the population. Resistance is an artificial measure relevant to our understanding of the genetic changes which occur in the population, but variation in resistance is not the property subject to selection. This emphasizes again the importance of evaluating fitness variation in the polluted and normal environments and the extent to which these properties are genetically correlated. The heritability of "resistance" cannot be related to response to selection in natural populations.

The importance of genetic correlations is illustrated by the model of Via and Lande (1985) which investigates evolution of mean phenotype in a population exposed to two different environments. If genetic correlations between fitness in the two environments are strongly negative, the population may be poorly adapted to both environments while evolving toward the optimum phenotype. Although this model is for spatial environments, similar models could be developed for temporally fluctuating environments such as episodic air pollution stress.

CO_2 and Other Changes Increasing Resource Availability

Although the literature on variation in response to these factors is rather meager, it does suggest that variation exists (Wulff and Alexander 1985). Genotypes which are better able to make use of the increased resources should increase in frequency. However, increased resource availability is likely to result in secondary selection pressures such as changes in interspecific competition intensity (Zangerl and Bazzaz 1984) perhaps involving new competitors. Increased competition will be accompanied by other potential selection pressures including increased temperatures and altered rainfall patterns. In contrast to the novel stresses imposed by anthropogenic pollutants, these changes alter intensity of stresses to which populations are normally exposed. How rapidly populations will evolve in response to this complex of stresses depends on whether the persistence of populations depends on adaptive genetic variation, or phenotypic plasticity and homoeostasis. While some evolution is no doubt possible, these secondary effects of increased CO_2 do not create novel environments, but rather environments which already occur elsewhere. If populations have the capacity to adapt to these environments, then the species should already occur in them. This analysis suggests that the initial effects of these environmental changes will be ecological (community change), and that much of the adaptive evolution may follow later among those species which survive.

Conclusions

Plant populations frequently contain genetic variation influencing resistance to atmospheric pollutants, but in most cases it is difficult to evaluate the potential for evolution of resistance because the genetics of this variation has not been well characterized. The major factors which have not been adequately investigated are (1) the role of major genes in determining the variation in resistance, (2) the role of new mutations in providing additional variation important for longer-term evolution, and (3) the genetic architecture of fitness in both clean and polluted environments. Methods from biochemical and molecular genetics complement the statistical tools of quantitative genetics and should be helpful in addressing these problems.

References

Ashenden TW, Mansfield TA (1977) Influence of wind speed on the sensitivity of ryegrass to SO_2. Journal of Experimental Botany 28:729–735

Bergmann F, Scholz F (1989) Selection effects of air pollution in Norway spruce (*Picea abies*) populations. In: Scholz F, Gregorius H-R, Rudin D (eds) Genetic effects of air pollutants in forest tree populations. Springer-Verlag, Berlin, pp 143–160

Bonte J (1982) Effects of air pollutants on flowering and fruiting. In: Unsworth MH, Ormrod DP (eds) Effects of air pollution in agriculture and horticulture. Butterworth, London, pp 207–223

Brown AHD (1979) Enzyme polymorphism in plant populations. Theoretical Population Biology 15:1–42

Chakraborty R, Meagher TR, Smouse PE (1988) Parentage analysis with genetic markers in natural populations. I. The expected proportion of offspring with unambiguous paternity. Genetics 118:527–536

Cowling DW, Koziol MJ (1982) Mineral nutrition and plant response to air pollutants. In: Unsworth MH, Ormrod DP (eds) Effects of air pollution in agriculture and horticulture. Butterworth, London, pp 349–375

Davidson AW, Bailey IF (1982) SO_2 pollution reduces the freezing resistance of ryegrass. Nature 297:400–402

Davies T (1980) Grasses more sensitive to SO_2 pollution in conditions of low irradiance and short days. Nature 284:483–485

Edwards MD, Stuber CW, Wendel JF (1987) Molecular-marker facilitated investigations of quantitative-trait loci in maize. I. Numbers, genomic distribution and types of gene action. Genetics 116:113–125

Ellis THN (1986) Restriction fragment length polymorphism markers in relation to quantitative characters. Theoretical and Applied Genetics 72:1–2

Elston RC (1984) The genetic analysis of quantitative trait differences between two homozygous lines. Genetics 108:733–744

Elston RC, Stewart J (1973) The analysis of quantitative traits for simple genetic models from parental, Fl, and backcross data. Genetics 73:695–711

Endler JA, McLellan T (1988) The process of evolution: toward a newer synthesis. Annual Review of Ecology and Systematics 19:395–421

Engle RL, Gabelman WA (1966) Inheritance and mechanism for resistance to ozone damage in onion, *Allium cepa* L. Proceedings of the American Society for Horticultural Science 89:423–430

Epperson BK, Allard RW (1987) Linkage disequilibrium between allozymes in natural populations of lodgepole pine. Genetics 115:341–352

Fain PR (1978) Characteristics of simple sibship variance tests for the detection of major loci and application to height, weight, and spatial performance. Annals of Human Genetics 42:109–120

Falconer, DS (1952) The problem of environment and selection. American Naturalist 86:293–298.

Falconer, DS (1981) Introduction to quantitative genetics. Longman, London pp 340

Gregorius HR (1989) The attribution of phenotypic variation to genetic or environmental variation in ecological studies. In: Scholz F, Gregorius HR, Rudin D (eds) Genetic effects of air pollutants in forest tree populations. Springer-Verlag, Berlin, pp 3–15

Hill WG (1982) Rates of change in quantitative traits from fixation of new mutations. Proceedings of the National Academy of Science USA 79:142–145

Hill WG, Keightley PD (1988) Interrelations of mutation, population size, artificial and natural selection. In: Weir BS, Eisen EJ, Goodman MM, Namkoong G (eds) Proceedings of the Second International Conference on Quantitative Genetics. Sinauer Associates, Sunderland, pp 57–70

Jinks JL, Pooni HS (1988) The genetic basis of environmental sensitivity. In: Weir BS, Eisen EJ, Goodman MM, Namkoong G (eds) Proceedings of the Second International Conference on Quantitative Genetics. Sinauer Associates, Sunderland, pp 505–522

Jones T, Mansfield TA (1982) The effect of SO_2 on growth and development of seedlings of *Phleum pratense* under different light and temperature environments. Environmental Pollution (series A) 27:57–71

Karnosky DF (1985) Genetic variability in growth responses to SO_2. In: Winner WE, Mooney HA, Goldstein RA (eds) Sulfur dioxide and vegetation. Stanford University Press, Stanford, pp 346–356

Koziol MJ, Shelvey JD, Lockyer DR, Whatley FR (1986) Response of SO_2-sensitive and resistant genotypes of ryegrass (*Lolium perenne* L.) to prolonged exposure to SO_2. New Phytologist 102:345–357

Lande R (1975) The maintenance of genetic variation by mutation in a polygenic character with linked loci. Genetic Research (Cambridge) 26:221–235

Lande R (1983) The response to selection on major and minor mutations affecting a metrical trait. Heredity 50:47–65

Lee EH, Bennett JH (1982) Superoxide dismutase—a possible protective enzyme against ozone injury in snapbeans (*Phaseolus vulgaris* L.). Plant Physiology 69:1144–1149

MacKay TFC (1988) Transposable element-induced quantitative genetic variation in *Drosophila*. In: Weir BS, Eisen EJ, Goodman MM, Namkoong G (eds) Proceedings of the Second International Conference on Quantitative Genetics. Sinauer Association, Sunderland, pp 219–235

MacNair MR (1983) The genetic control of copper tolerance in the yellow monkey flower, *Mimulus guttatus*. Heredity 50:283–293

Martin B, Nienhuis J, King G, Schaefer A (1989) Restriction fragment length polymorphisms associated with water use efficiency in tomato. Science 243:1725–1728

McClintock B (1984) The significance of responses of the genome to challenge. Science 226:792–801

McDonald JF (1983) The molecular basis of adaptation: a critical review of relevant ideas and observations. Annual Review of Ecology and Systematics 14:77–102

Mitchell-Olds T (1986) Quantitative genetics of survival and growth in *Impatiens capensis*. Evolution 40:107–116

Mitchell-Olds T, Rutledge JJ (1986) Quantitative genetics in natural plant populations: a review of the theory. American Naturalist 127:379–402

Muller-Starck G (1989) Genetic implications of environmental stress in adult forest stands of *Fagus sylvatica* L. In: Scholz F, Gregorius HR, Rudin D (eds) Genetic effects of air pollutants in forest tree populations. Springer-Verlag, Berlin, pp 127–142

Nevers P, Shepherd NS, Saedler H (1986) Plant transposable elements. Advances in Botanical Research 12:104–203

Nienhuis J, Helentjaris T, Slocum M, Ruggero B, Schaefer A (1987) Restriction fragment length polymorphism analysis of loci associated with insect resistance in tomato. Crop Science 27:797–803

Osborne TC, Alexander DC, Fobes JF (1987) Identification of restriction fragment length polymorphisms linked to genes controlling soluble solids content in tomato fruit. Theoretical and Applied Genetics 73:350–356

Roach DA (1986) Life history variation in *Geranium carolinianum* I. Covariation between characters at differents stages of the life cycle. American Naturalist 128:47–57

Roose ML, Bradshaw AD, Roberts TM (1982) Evolution of resistance to gaseous air pollutants. In: Unsworth MH, Ormrod DP (eds) Effects of air pollution in agriculture and horticulture. Butterworth, London, pp 379–409

Rosielle AA, Hamlin, J (1981) Theoretical aspects of selection for yield in stress and non-stress environments. Crop Science 21:943–946

Schoen, DJ (1982) The breeding system of *Gilia achillefolia*: variation in floral characteristics and outcrossing rate. Evolution 36:352–360

Saedler H, Schwarz-Sommer Z (1987) Transposable elements and their role in plant evolution. In: Von Wettstein D, Chua N-H (eds) Plant molecular biology. Plenum Press, New York, pp 163–166

Shaw RG (1986) Response to density in a natural population of the perennial herb *Salvia lyrata*: variation among families. Evolution 40:492–505

Tanaka K, Suda Y, Kondo N, Sugahara K (1985) O_3 tolerance and the ascorbate-dependent H_2O_2 decomposing system in chloroplasts. Plant Cell Physiology 26:1425–1431

Thoday JM (1961) Location of polygenes. Nature 191:368–370

Turelli M (1988) Population genetic models for polygenic variation and evolution. In: Weir BS, Eisen EJ, Goodman MM, Namkoong G (eds) Proceedings of the Second International Conference on Quantitative Genetics. Sinauer Associates, Sunderland, pp 601–618

Turelli M, Gillespie JH, Lande R (1988) Rate tests for selection on quantitative characters during macroevolution and microevolution. Evolution 42:1085–1089

Via S, Lande R (1985) Genotype-environment interaction and the evolution of phenotypic plasticity. Evolution 39:505–522

Wehrhahn C, Allard RW (1965) The detection and measurement of the effects of individual genes involved in the inheritance of a quantitative character in wheat. Genetics 51:109–119

Wolters JHB, Martens MJM (1987) Effects of air pollutants on pollen. Botanical Review 53:372–414

Wulff RD, Alexander HM (1985) Intraspecific variation in the response to CO_2 enrichment in seeds and seedlings of *Plantago lanceolata* L. Oecologia (Berlin) 66:458–460

Zangerl AR, Bazzaz FA (1984) The response of plants to elevated CO_2. II. Competitive interactions among annual plants under varying light and nutrients. Oecologia (Berlin) 62:412–417

Commentary to Chapter 5

Genetics of the Resistance of Plants to Pollutants

Mark R. Macnair

Introduction

The study of the genetic architecture of resistance to pollutants can be useful in a number of fields. If the nature and magnitude of the genetic variance for resistance is known, it is possible to predict whether, and how fast, truly resistant races of a species can evolve. Only a genetic study can distinguish between physiological adaptation or acclimation and innate evolved resistance. Finally, a knowledge of the number of genes involved in producing a resistant ecotype can assist greatly in thinking about the possible physiological mechanisms of resistance.

In Chapter 5, Roose discussed in detail the problems of extrapolating from relatively simple genetic studies to predicting the rate and course of evolution, and emphasized the necessity to establish the genetic correlations between resistance and fitness components for this exercise to be successful. The objectives of this Commentary are to consider (a) the conditions under which major gene or polygene inheritance is more likely to be involved; (b) the problems inherent in distinguishing between modes of inheritance; and (c) some of the implications that follow from a knowledge of the mode of inheritance of the character.

Polygene or Major Gene Inheritance?

There has been some controversy over the role of major gene mutations in the evolution of higher plants. Gottlieb (1984) and Hilu (1983) have argued that the evolution of certain characters may be more likely to involve major gene mutations. Coyne and Lande (1985) have criticized Gottlieb's analysis and suggested that polygenic inheritance is more usually found. Lande (1983) analyzed the conditions under which a population was more likely to evolve by polygenic variance or major genes and concluded that polygenic adaptation was more likely except where the major gene was at least partially dominant and was already at a moderate gene frequency.

For certain well-studied adaptations, particularly mimicry in animals, insecticide resistance, warfarin resistance in mammals, and industrial melanism, major gene adaptation is far more common than polygenic inheritance. There is a reasonably simple explanation for this, and why we should expect certain sorts of adaptation to be governed by major genes.

For a species to evolve an adaptation, there must be genetic variance for the character under selection. If the variance is not present or is too small, the population will not evolve the adaptation, and, if the adaptation is necessary for existence in a particular environment, the species will go locally extinct. Consider a simple genetic model in which a character is determined by n additive genes of equal effect in which each gene has the following effect:

Genotype	bb	Bb	BB
Mean phenotype	0	a	$2a$
Genotype frequency	q^2	$2pq$	p^2

where a is the additive effect of the B allele, p is the frequency of the B allele, and q the frequency of the b allele ($p + q = 1$).

The mean of the character under such circumstances is $2nap$ and the genetic variance $2npqa^2$. A high genetic variance (a prerequisite for selection to be effective) is achieved when either p and q are roughly equal, or when a is large. So, for a given mean (i.e., for constant $2nap$), a high variance can either be achieved by having many genes of small effect at intermediate frequency, or a few genes of large effect at low frequency (i.e., $p < 0.01$). These two scenarios represent the two contrasting models of polygenic control or major gene control.

Figure 1 illustrates the adaptive problem faced by a population confronted by a novel environment (e.g., a pollution-induced stress). In the normal environment, the population has a mean of $2nap$ and is well adapted to the prevailing environment. This mean value is not well adapted, however, to the novel environment. In order to achieve even minimal adaptation to the new environment, the population has to move x units under selection. The maximum value that the character can achieve under directional selection will be when all the B alleles have spread to fixation, when the mean will be $2na$. Thus, the maximum *response* to selection (i.e., the difference between the mean before selection and the mean after selection has occurred) is

$$2na - 2nap = 2naq,$$

and the maximum response as a function of the mean is

$$2naq/2nap = q/p.$$

This value will be small if p and q are roughly equal in magnitude, but large when p is small. We can consider two contrasting scenarios:

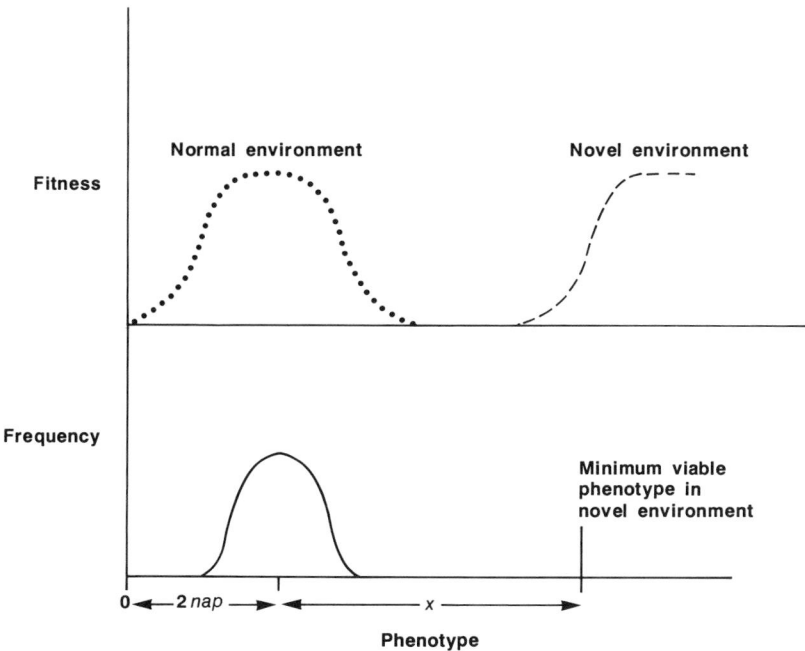

FIGURE 1. Distribution of phenotypes in a population (*bottom*) and their fitness (*top*) in a normal environment (*dotted curve*) or novel environment (*dashed curve*). The population has a mean of $2nap$ in the normal environment; individuals have to have a value of at least x units more to be viable in the novel environment. If x is substantially more than $2nap$ then only a major gene mutation can achieve the minimum viable phenotype; if x is of the same order of magnitude as $2nap$ then polygenic adaptation is probable (see text).

1. x of same order of magnitude as $2nap$. In this case, genes at intermediate frequencies will be able to achieve the response required, and polygenic adaptation is both possible and likely. Most outbreeding species that have been investigated have shown additive genetic variance (which is most probably caused by genes at intermediate gene frequencies) for most characters, including resistance to common air pollutants (Roose et al. 1982; Karnosky, 1985; Roose, Chapter 5, this volume). Thus, in this scenario, the majority of species would be able to achieve the adaptation in time, and it is unlikely that the selective force would lead to widespread local extinction.

2. x much larger than $2nap$. In this case, genes at intermediate frequencies *cannot* achieve the response required, and adaptation can *only* be achieved by the spread of genes initially at low frequency. Only if these genes are of large effect will the genetic variance be large enough for adaptation to occur within a reasonable time period. In this case, it is

unlikely that all populations of all species will have the requisite genetic variance, and only some species will evolve the adaptation, so that it is much more probable that the selective force would lead to many species becoming locally extinct.

The first scenario is the one most frequently studied when considering small selective coefficients, or minor changes in the environment. For instance, when Lande (1983) considered the relative probability of adaptation occurring by polygenes or major genes, his model involved weak selection and did not envisage a situation in which the evolution of a character could be constrained through lack of appropriate genetic variability.

The second scenario is most probable in cases of extreme selection, such as insecticide resistance and heavy metal resistance in plants. This model provides a conceptual framework to understand why only some species have evolved resistance (Bradshaw 1984), and why, when the genetics has been done in detail (Wood 1981; Macnair 1990), major genes are most frequently found.

The problem with the model is that measuring the scale on which x and $2nap$ are located is not possible, and thus there is no way of empirically determining their relative magnitude. In the case of typical metrical characters such as plant height or growth rate, it is likely that $2nap$ will be quite large relative to zero, and so x would have to be enormous for the second scenario to be relevant. But in the case of physiological or biochemical characters, particularly where one is considering an adaptation to a novel anthropogenic pollutant, it is more likely that under normal circumstances $2nap$ is close to zero. It is necessary to use one's judgment and experience to assess which scenario is more probable in any particular situation. In many cases of air pollution, the selection pressure is not very extreme, and perhaps the first scenario is more likely. There are, however, other cases, primarily those where there is a major local source of pollution, for which the second might be more appropriate. It is emphasized that the resistance to chronic and acute pollution stresses is often independent (Bell et al. Chapter 3, this volume), and it is possible that the two forms of stress challenge the plants with different degrees of selection pressure.

How Easy Is It to Distinguish Major Gene from Polygenic Inheritance?

In practice it can be difficult and time-consuming to differentiate between the two genetic models empirically. There are a number of reasons for this, but the primary cause is the difficulty in defining and measuring the character under consideration. This automatically inflates the error and environmental variances in any genetic analysis, making identification of the genetic segregations produced by major genes difficult or impossible. It is a common misconception that, if a continuous or quasicontinuous distribution of phenotypes is found in a population or a series of crosses, this is in itself evidence for polygenic inheritance. Only where the magnitude of the error and environmental terms are substantially less than the genetic effects (as in the sorts of characters studied in classical genetic exercises) does the overall distribution of phenotypes carry any information about the nature of the genetic control.

Relative pollution resistance is frequently measured as a genotype × environment interaction. Plants are exposed to two environments (a polluted and a control environment), and the relative growth of genotypes in the two environments are assessed (Fig. 2). Genotypes which are least affected by the change in the environment are said to be more resistant. However, growth is a character that is inevitably affected by many (polygenic) genes that have nothing to do with resistance to the pollutant under study, and which may show genotype × environment interaction to other environmental factors. In addition, most growth characters show high environmental and error variances, and these are going to be magnified in any combined measure of relative growth. Thus, where relative growth rate is used as the measure of resistance, it is almost inevitable that it will appear to be inherited polygenically, even when there is in fact a single major gene (see Macnair 1990 for an example).

A more subtle problem with the use of relative growth rate to measure tolerance is illustrated in Fig. 2. Two genotypes (G1 and G2) are grown in each of two environments, E1 (control) and E2 (polluted). Where the growth of the genotypes is as shown in Fig. 2a or 2b, the interpretation of G2 as being more resistant is unambiguous. However, in the situation

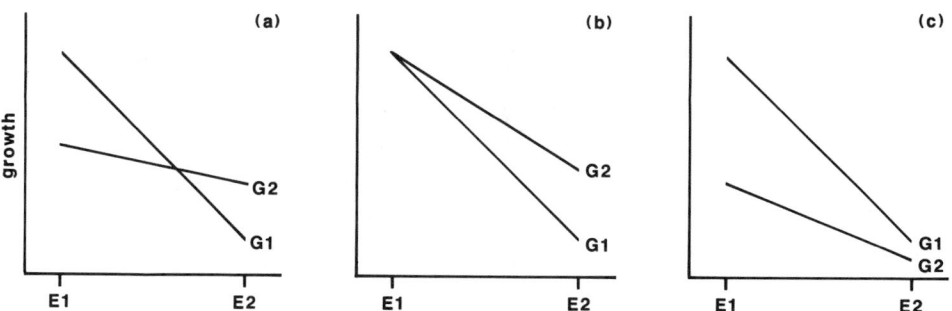

FIGURE 2(a–c). Resistance is frequently measured by the relative growth of genotypes in two or more environments. Two genotypes (G1 and G2) are grown in a control environment (E1) and an environment containing the pollutant (E2). Resistance is then defined as *Growth in E2/Growth in E1*. In all three cases, by this measure G2 has a higher resistance than G1.

illustrated in Fig. 2c, it is much less clear that G2 (though by any measure of relative growth it is less affected by the change of environment) is more resistant, since it is still performing worse in E2 than is G1. In this case, it is possible for G2 to be the more resistant genotype if, because it is under less stress due to the pollution, it is better able to resist other environmental stresses, such as drought or competition. This would mean that the two environments chosen for the experiment did not reflect properly the true conditions in the field, and that, had the two environments been chosen properly, the plants would have grown as in Fig. 2a. This may well be so, but it would be very unwise to assume that it *must* be so, and that where an experiment gives a result as in Fig. 2c, that G2 is necessarily the more resistant. It is for these reasons that Roose (Chapter 5) and Macnair (1990) suggested that relative growth rate is a poor measure of resistance in genetical studies. Roose has suggested using genetic correlations to distinguish between Fig. 2a and 2c (2a will give a negative genetic correlation between growth in the two environments, 2c will give a positive correlation); Macnair (1990) suggests that where the situation is as shown in Fig. 2a or 2b the growth of the genotype in E1 is irrelevant, and growth in E2 should be a valid measure of tolerance on its own. Obviously, this latter approach should only be used after preliminary experiments have shown it to be valid (i.e., the situation is as 2a or 2b).

When tolerance has already evolved, major genes may be difficult to detect if analysis of genetic variance *within* the resistant or susceptible classes is not separated from analysis of the genetic variance *between* classes. There is normally a within-population genetic variance for most characters, including pollution resistance. If resistance evolves polygenically, then the differences between classes will involve the same genes as the differences within them. But if it evolves by major genes, then the genes giving the differences between the resistant and susceptible classes will not be the same as those giving the polygenic variance within classes. In *Mimulus guttatus*, for instance, it is possible to find phenotypic variability for level of copper resistance both within mine and nonmine populations, and at least within the mine population this variability is highly heritable. The basic difference between the populations, however, is determined by a single gene.

There would appear to be two relatively simple approaches to attempting to determine whether a single major gene confers resistance. The first is to attempt to refine a measure of resistance in controlled environments that is minimally affected by error and environmental variances. Classic studies into the genetics of fruit shape in *Capsicum* (see Gottlieb 1984) and dwarfing in wheat (Law and Gale 1979) illustrate how major genes can be detected in the midst of an apparently continuous phenotypic distribution once a suitable assay has been discovered. If the difference between resistant and susceptible plants is large, or if it is possible to define a physiological or biochemical assay, it may not prove too difficult to obtain such a measure of resistance. For the reasons outlined above, the measure should not use relative growth rates. If such a technique can be refined, large numbers of crosses and progenies can be screened, and segregations determined if a major gene is present. It is this sort of technique that has enabled major genes to be detected for copper resistance in *Mimulus guttatus* (Macnair 1983), arsenic resistance in *Holcus lanatus* (unpublished results), and ozone resistance in onions (Engle and Gabelman 1966). However, if the difference between resistant and susceptible individuals is not very large relative to the residual environmental and error variances, it will still be impossible to detect segregation by this method even if there is a single major gene present. Thus, the failure to perfect such a technique should not be taken as evidence of polygenic inheritance.

The second technique relies on the fact that if a single gene is present, families with heterozygous parents will have a large variance than families with homozygous parents (see Mitchell-Olds and Bergelson, 1990). This will result in families with heterogeneous within-family variances. Under polygenic inheritance there should be little heterogeneity of within-family variances. The screening of F_3 families, or even, in outbreeding species, wild pollinated families, should be able to detect such heterogeneity if a single gene is important. As an example, Watkins (1985) collected 24 wild pollinated families of *Agrostis capillaris* from a copper mine, and tested them for copper resistance. No segregation could be detected in this material, but the families appeared to fall into two groups, one with a high family mean and low within-family variance, and one with a low mean and high variance. This result is consistent with a single gene for copper resistance, with the high mean families having a homozygous parent and the low mean families a heterozygous parent. This result does not prove major gene inheritance but is a relatively efficient experiment to perform; a positive result would at least indicate that a more detailed investigation (for instance, by some of the methods outlined by Roose in Chapter 5) would be worthwhile.

What Use Is a Knowledge of the Genetic Architecture of Resistance?

A knowledge of the heritability of resistance is essential if any prediction is to be made about the probability or rate of evolution of resistant populations or species. However, the sorts of experiments outlined in the second section of this Commentary will only be of limited use for this purpose. As emphasized by Roose (Chapter 5) and Tonsor and Kalisz (Chapter 10) predictions about evolution can only be made from heritability studies undertaken under field conditions, where fitness is the character studied (which takes into account genetic correlations between resistance and other attributes) rather than just resistance measured under more or less unrealistic conditions.

This is not to say that the detailed genetic studies outlined above are of no use. First, a knowledge of the number of genes and the nature of the genetics of resistance can help in planning breeding programs for particularly sensitive or important species. If resistance is multifactorial, then only a conventional breeding program, identifying more resistant varieties and selecting from them or their hybrid progeny, will be likely to be effective. If, on the other hand, single genes are of importance in some species, it might be possible to use genetic engineering techniques to isolate and transfer the genes to critical varieties or species.

A knowledge of the genetics can also be of considerable help to physiologists seeking to understand the mechanisms of resistance. There are many different levels at which a pollutant can exercise its phytotoxic effects, and thus potentially many different levels at which resistance could operate (Tinqey and Andersen, Chapter 8, this volume). The conclusion drawn from Chapter 8 is that resistance must be multifactorial and polygenic. This may be so, but it does not follow necessarily, because only some of the interacting physiological processes may have genetic variation in the population studied, and some may have more influence on fitness than others.

For example, many different physiological processes have been implicated in heavy metal resistance. Heavy metals have different phytotoxic effects, disrupting membranes, inhibiting and denaturing enzymes, and disrupting cytokinesis. There have been many studies into the physiology of metal resistance, concentrating on different aspects of the complex interaction between cells and heavy metals. Many physiological differences have been found between genotypes, and this has led some (e.g., Woolhouse 1980; Ernst 1976) to propose that a variety of different processes confer resistance. However, the demonstration that in at least one species there is a single major gene for tolerance has led to a realization that there may well be a single underlying basis for tolerance, and that all the other processes investigated may be secondary effects caused by the fact that healthy resistant cells behave differently from unhealthy (in some cases dead) ones. It is still not clear in any species what the primary mechanism is, but it is highly probable that it is different for different metals.

Conclusion

Genetic analyses can assist greatly in the investigation of pollution resistance in any species, and these ought to be included in a research program at an early stage. However, it is essential to be clear what questions a genetic analysis is addressing. A fairly simple analysis, such as a parent offspring regression or partitioning the variance into within- and between-family components in wild collected seed, can demonstrate that the phenomenon is heritable and that evolution of resistance may take place. It may also be possible to gain some indication of whether a major gene is involved. The information obtained is extremely valuable, and though limited in scope, may be all that is required in a particular investigation. However, further research will need to take one or both of two routes. For detailed information on the genetic architecture of resistance (number of genes, dominance, epistatic relations of the genes) studies will have to be conducted under carefully controlled conditions where the environmental variance can be reduced as much as possible. The information gleaned from such studies is of interest to the plant breeder, the geneticist, and to the plant physiologist. It may not, however, be very useful in predicting the interaction of plants with the pollutant in the wild. For this, experiments in more complex and natural environments will be required, in which the genetic correlation between the physiological characters of resistance with various fitness attributes needs to be determined.

References

Bradshaw AD (1984) The importance of evolutionary ideas in ecology—and vice versa. In: Shorrocks B (ed) Evolutionary ecology. Blackwell, Oxford, pp 1–25

Coyne JA, Lande R (1985) The genetic basis of species differences in plants. American Naturalist 126:141–145

Engle RL, Gabelman WA (1966) Inheritance and mechanism for resistance to ozone damage in onion, *Allium cepa* L. American Society for Horticultural Science 89:423–430

Ernst WH (1976) Physiological and biochemical aspects of metal tolerance. In: Mansfield TA (ed) The effects of air pollutants in plants. Cambridge University Press, Cambridge, pp 115–133

Gottlieb LD (1984) Genetics and morphological evolution in plants. American Naturalist 123:681–709

Hilu KW (1983) The role of single gene mutations in the evolution of flowering plants. Evolutionary Biology 16:97–128

Karnosky DF (1985) Genetic variability in growth responses to SO_2. In: Winner WE, Mooney HA, Goldstein RA (eds) Sulfur dioxide and vegetation. Stanford University Press, Stanford, pp 346–356

Lande R (1983) The response to selection on major and minor mutations affecting a metrical trait. Heredity 50:47–65

Law CN, Gale MD (1979) Cytological markers and quantitative variation in wheat. In: Thompson JN, Thoday JM (eds) Quantitative genetic variation. Academic Press, New York, pp 273–293

Macnair MR (1983) The genetic control of copper tolerance in the yellow monkey flower, *Mimulus guttatus*. Heredity 50:283–293

Macnair MR (1990) The genetics of metal tolerance in natural populations. In: Shaw J (ed) Heavy metal tolerance in plants: evolutionary aspects. CRC Press, Boca Raton, pp 235–253.

Mitchell-Olds T, Bergelson, J (1990) Statistical genetics of an annual plant, *Impatiens capensis*. 1. Genetic basis of quantitative variation. Genetics 124:407–415

Roose ML, Bradshaw AD, Roberts TM (1982) Evolution of resistance to gaseous air pollutants. In: Unsworth MH, Ormrod DP (eds) Effects of air pollution in agriculture and horticulture. Butterworth, London, pp 379–409

Watkins AJD (1985) Within-mine differentiation in degree of copper and arsenic tolerance in *Agrostis capillaris* L (*Agrostis tenuis* Sibth.) Ph.D. Thesis, University of Exeter

Wood RJ (1981) Insecticide resistance: genes and mechanisms. In: Bishop JA, Cook LM (eds) Genetic consequences of man made change. Academic Press, London, pp 53–96

Woolhouse HW (1980) Heavy metals in plants. Chemistry in Britain 16:72–76

6
Population Processes in Plants and the Evolution of Resistance to Gaseous Air Pollutants

SPENCER C.H. BARRETT and ELIZABETH J. BUSH

Introduction

Human activities since the industrial revolution have resulted in considerable pollution of the earth's air, water, and soil, particularly in technologically advanced countries of the world. Pollution is manifested in many ways including alterations in the chemical composition of the atmosphere, and contamination of air, water, and soils by heavy metals, pesticide residues, and toxic chemicals associated with industrial activities and agriculture. The most obvious ecological effect of pollution stress is alteration of the species composition of biological communities through elimination of sensitive species or individuals (e.g., Hutchinson and Meema 1987). There is also evidence that populations of some species have responded to the ecological challenges presented by pollution by evolving the ability to grow and reproduce in contaminated environments (Bradshaw and McNeilly 1981). The best evidence for the evolution by natural selection of plant populations resistant to anthropogenic pollutants involves plants growing in heavy metal contaminated soils (Antonovics et al. 1971) and those that are tolerant to herbicides (LeBaron and Gressel 1982). There are fewer examples of evolutionary responses of plant populations to the effects of gaseous air pollutants, however, the widespread occurrence of genetic variation in sensitivity to air pollutants suggests that a potential for adaptive responses exists in many species (Roose et al. 1982; Hutchinson 1984; Pitelka 1988; Scholz et al. 1989). The purpose of this chapter is to identify features of the population biology of plants that may be important in determining whether evolutionary responses to gaseous air pollutants are likely to occur.

It is often tacitly assumed that observable differences in phenotype that distinguish populations and species are the products of natural selection. Demonstrating this, however, is not a trivial issue because natural selection is a complex process that is notoriously difficult to quantify directly, particularly in long-lived organisms (Endler 1986). Microevolutionary forces largely occur within populations and gradually convert individual genetic differences in fitness to variation among populations, races, and ultimately species. As a result, studies of natural selection require a population-level approach, involving ecological and genetic investigations. Although demonstrating that species level genetic variation exists for a particular trait (e.g., resistance to a pollutant) is of value if the goal is to breed for increased resistance, from an evolutionary perspective, it may tell us little about the potential for evolutionary response in contemporary populations. To do this it is necessary to demonstrate that heritable variation occurs *within* natural populations and that the variation can be transformed into fitness differences among genotypes over generations. If this occurs, natural selection of the trait exhibiting genetic variation will inevitably occur resulting in increased adaptedness of the population.

The nature of evolutionary change depends on the amounts and kinds of genetic variation within natural populations as well as the intensity and type of natural selection that occurs. In addition, however, patterns of evolutionary diversification are also influenced by the intrinsic features of the population biology of organisms. Plant populations exhibit an enormous diversity of life history and of demographic and reproductive characteristics (reviewed by Harper 1977; Solbrig 1980). This variation plays an important role in controlling the organization of genetic variation within and among populations and hence the type of selective responses that occur. Simple selection models developed for many animal populations that assume diploid, unisexual, outbreeding populations of large effective size with nonoverlapping generations are largely invalid for most plant species because of their distinctive features. Restricted recombination systems involving high degrees of inbreeding, limited gene flow, small effective population size, large dormant seed banks, clonal propagation, and high phenotypic plasticity make it difficult to generalize about the outcome of selection in plant populations (Levin 1978; Crawford 1984). Features of some plant populations such as a small effective size and the large environmental component to fitness variation can reduce the intensity of selection and increase the likelihood that genetic drift will be a dominant force in influencing population genetic structure. Although random processes are unlikely to retard the progress of selection where intensities are very high (e.g., in heavy metal contaminated sites), they may be of more significance where selection pressures are weaker and intermittent in nature such as in the case of some gaseous air pollutants.

The main objective of this chapter is to address two general questions that pertain to population-level processes governing the evolutionary responses of plants to gaseous air pollutants. First, how might variation in the life history attributes and reproductive systems of populations facilitate or constrain selective responses to pollution stress? Second, what features of population genetic structure are likely to be important in determining whether resistance to air pollution will evolve and what genetic consequences might result from such a change? While empirical data demonstrating the selective effects of air pollutants are limited, wherever possible the largely intuitive arguments presented here will be buttressed with information gleaned from case studies from the air pollution literature. Since the harmful ecological effects of air pollution on forest communities have been of major concern in recent years, particular attention is given in the discussion to the ecological genetics of tree species and how the population biology of long-lived organisms is likely to effect genetic responses to air pollution.

Life History Characteristics

The interaction of life history variation with anthropogenic pollutants has perhaps no greater relevance than with respect to air pollution stress. Atmospheric pollution is so widespread that no habitat may be entirely protected and consequently all plant life forms are likely to risk exposure to air pollution. This contrasts with the exposure of plants to heavy metal contamination, in which case the contaminated habitat is often discrete, being confined to the boundaries of mining areas or other point sources of metal contaminants. Only a limited number of species can survive on soils contaminated with heavy metals, and the vast majority of cases discovered to date have involved herbaceous perennial plants. An even more restricted subset of plants has evolved herbicide resistance (e.g., weeds of annuals crops) and their life histories represent only a small sample of the existing diversity.

The range of variation in plant life histories adds considerable complexity to attempts to assess the potential for evolution of resistance to air pollution. Longevity in plants covers the spectrum from short-lived annuals to perennials that may live many hundreds of years. This variation creates practical problems in assessing the effects of air pollution stress on life time fitness and makes comparisons among groups with contrasting longevities particularly difficult. A further level of complexity is introduced by variation in the frequency of sexual versus asexual reproduction, within and among plant populations. This character is itself often very plastic but is critical for assessing rates of evolution in populations. Furthermore, there is an array of mechanisms for achieving asexual propagation, by corms, bulbs, rhizomes, and stolons and by fragmentation of stems, runners, and tillers. It seems likely that these contrasting clonal regenerative strategies will vary in the extent to which they are susceptible to air pollution stress, just as we might expect differences among other ecological and life history groupings. However, predicting which sets of life history traits are more likely to confer resistance is a formidable task without comparative experimental work or surveys of the response of plant communities to pollution-induced stresses.

Given the diversity of life history traits in plants it is relevant to ask whether Grime's classification of adaptive plant strategies (Grime 1979) can be employed when considering plant responses to air pollution stress. Grime's initial divisions placed plants into one of three primary categories: ruderals, competitors, and stress tolerators. Further subdivisions into secondary strategies include plants adapted to environments at the interface of each of these three extremes. The importance of Grime's work in the context of responses of plants to air pollution stress is that each strategic category contains plants united by common features of morphology, life history, and physiology. Although this simplifies considerable complexity it does allow us to evaluate whether traits that are believed to be important as adaptations to natural stress conditions may also confer some degree of resistance to anthropogenic stresses, and conversely whether traits associated with disturbed or competitive environments may be disadvantageous in polluted environments.

This section first discusses several features of plant life histories that are relevant to the potential for evolutionary responses to air pollution stress. Where appropriate, traits that enable populations to avoid pollution stress are contrasted with those that enhance exposure, thereby intensifying selective pressures for the evolution of resistance within populations. The section ends with a discussion of plant responses to air pollution in the context of Grime's theory of adaptive strategies. The response of trees to pollution stress is selected for special consideration since certain features of the population biology of trees pose unique problems for the assessment of evolutionary change.

Seed Banks

Seed banks are a feature of many plants, particularly those that occupy environments which are frequently disturbed or exhibit a high degree of environmental uncertainty. Seeds of most species experience transient dormancy, however of primary concern here are those that are capable of developing a persistent seed bank, which by definition is one in which seeds remain viable for a period of a least 1 year in the soil. From an evolutionary perspective seed banks can be considered as sources of migration from the past. Templeton and Levin (1979) have modelled the effect of a seed bank in an annual population on selection for a single locus trait with two alleles. In their model, mating is assumed to be random in a population of infinite size, with overdominant (heterozygote advantage) fitness and a selection coefficient of 0.01 (weak selection). The main conclusions of Templeton and Levin's study are (1) seed banks decrease the rate of approach to equilibrium allele frequencies and hence the rate of evolutionary change. The degree to which the rate of allele frequency change is reduced increases with the average number of generations that germinating seeds spend in the seed pool. (2) Differences among species in seed viabilities, germination probabilities and total seed production result in different rates of evolutionary change. From their work it is evident that an increase in the mean number of generations in the seed bank from 1 to 2 years reduces the rate of evolution by half. Since the presence of a seed bank does not influence equilibrium allele frequencies, similar conclusions can be drawn for the case of directional selection, a situation more likely to occur with air pollution resistance (see below). In this case selection should lead to fixation of the selected allele, with rates to fixation decreasing with increasing time in the seed bank. One caveat worth considering is that resistance to air pollutants is more often a quantitative trait (Roose et al. 1982) than one controlled by a single major gene (although see Engle and Gabelman (1966) for a possible example of single gene control). With quantitive inheritance of resistance, rates of evolution are expected to be generally slower with persistent seed pools enhancing this effect.

With this basic model in mind other factors relevant to rates of evolution of resistance to air pollutants can be superimposed, such as whether or not the population is declining in number versus growing. The effective memory of the seed pool increases in populations that are in decline. Alternatively, the effect of a growing population is to reduce the memory of the seed pool, as seeds from past generations are swamped by the numbers of seeds produced by the larger population of recent years. Plant populations may experience an initial decline with the onset of pollution stress as susceptible individuals are eliminated. In this initial phase the contributions of the prepollution population to the seed bank will retard the evolution of a resistant population. However, if a second phase of rapid growth and expansion of the population occurs, particularly as more sensitive competitors are eliminated, the seed pool may no longer retard the rate of evolution of resistance to the same extent.

It is also possible that different contributions to the seed pool by plants of different size could have a similar effect. It is well established that the correlation between plant size and reproductive output is in the order of 0.90 for annuals (Heywood 1986). If individuals more resistant to pollution have significantly greater growth and subsequent seed production than sensitive individuals, then the tendency for seed pools to delay the evolution of resistance would be ameliorated. This would be the case if the reported differences in foliar injury and yield reductions of tolerant and sensitive genotypes under short-term pollutant exposures (reviewed in Roose et al. 1982) are translated into significant size differences over greater lengths of time. The significance of seed banks to rates of evolutionary response is likely to be less in perennial plants, particularly those of considerable longevity (e.g., forest trees). Dormancy periods tend to be relatively short in many tree species and in some tropical species may be absent altogether. However, perenniality itself may retard evolutionary change in much the same way as seed banks, since both create opportunities for matings between generations.

Phenotypic Plasticity

One of the primary distinctions between most animals and plants is the role that apical meristems play in permitting continual embryological development of the plant (White 1984). The plastic responses of the growth and morphology of plants to external environmental conditions is derived from this meristem potential. There is considerable evidence indicating the importance of plasticity in plant responses to air pollution. Pollution stress may elicit both qualitative and quantitative changes in the allocation of rsources to plant tissues and organs (reviewed by Lechowicz 1987). Both SO_2 and O_3 tend to suppress root growth more than shoot growth, thereby reducing rootshoot ratios. Allocation of resources to leaves tends to increase under SO_2 pollution whereas that to stems decreases. Reproductive investment is often reduced under pollution stress leading to a variety of responses (see below). Many other examples of plastic responses are reported throughout the air pollution literature, some of which are clearly under genetic control.

It has been suggested that phenotypic plasticity, developmental homeostasis, and genetic polymorphism represent alternative strategies for dealing with environmental unpredictability and heterogeneous environments (Jain 1979). Polymorphism has been thought to be adaptive where heterogeneity is manifested strongly in spatial aspects of the environment (e.g., mosaic soil conditions; see Snaydon and Davies 1972). Environmental heterogeneity is relatively predictable in this case, permitting adaptive gene frequency changes to occur over time. Plasticity and homeostasis, on the other hand, have been thought of as adaptive responses in environments characterized by unpredictable short-term changes that are manifested within the life cycle of individuals and which require immediate responses. Although it is often suggested that an inverse relationship between plasticity and heterozygosity may exist, a recent review of phenotypic plasticity in plants casts doubt upon this dichotomy (Schlichting 1986). Some studies have found an inverse relationship between heterozygosity and phenotypic plasticity, others have failed to corroborate this relationship. As Schlichting (1986) points out there is no apparent reason why the presence of genetic variability in a population should oppose the evolution of appropriate plastic responses unless there is an extra cost to an organism that is both plastic and heterozygous. As yet there is little empirical evidence for such a cost.

Plastic responses may be adaptive in the short term, but it is questionable whether plasticity alone could sustain growth and reproduction in the long term in populations subjected to chronic pollution stress. Plasticity could be a costly stress response involving increased expenditure to compensate for the decreased efficiency of photosynthesis through foliar injury. If pollution stress is persistent and severe enough, the plant may overextend its ability to maintain itself, resulting in greater susceptibility to pest and disease pressures (Lechowicz 1987). Furthermore, although plasticity can be selected, there are developmental and architectural constraints on the evolution of plasticity (Watson and Casper 1984). Therefore, although a high level of plasticity may help compensate for short-term pollution stress, it is unlikely to provide a mechanism of resistance in evolutionary terms. Populations with a high degree of phenotypic plasticity may, however, be just as likely to exhibit genetic variation for resistance as those that are less plastic.

Life Forms

The classification of Raunkiaer (1934) provides us with a means of evaluating life forms according to their potential susceptibilities to abiotic stresses (Hutchinson and Harwell 1985). With respect to air pollution, we can rank plants based on the degree to which phenology, architecture, and growth involve protection of sensitive meristems from gaseous air pollutants. Life forms whose perennating buds or shoot apices are borne well above ground (phanerophytes) are afforded little protection from exposure to air pollution. In addition, these life forms are often canopy species which form the first zone of pollutant capture (a characteristic exemplified by trees). The chamaephytes and hemicryptophytes are more capable of avoiding pollution stress since their sensitive meristems are borne either close to, or at ground level, and their above ground shoots senesce at the onset of unfavorable growing seasons. Perennials whose meristems are either protected underground as rhizomes, bulbs, or tubers (geophytes) or underwater (helophytes and hydrophytes) are probably best able to avoid direct meristem damage from atmospheric pollutants, however, the dissolution of air-borne pollutants into aquatic environments may constitute another form of pollutant stress.

The importance of meristem position is clearer if we consider the interaction of phenology with seasonal cycles of pollution. In parts of Europe pollution episodes can be most intense during the winter months (e.g., SO_2; see Venne et al. 1989). Life forms with subterranean organs will therefore avoid pollution stress. Dormancy of above ground buds and the absence of a canopy in deciduous trees may also restrict injury to growing plant organs. In parts of North America, some air pollution is greatest during the summer months (e.g., ozone pollution, acid advection fog). Where this occurs it can coincide with periods of peak plant growth. Therophytes (annuals) that complete their entire life cycle in a short time span may be under intense selection pressures from pollution stress at each developmental stage of the life cycle, from germination to reproduction. It is also worth considering the possibility that tropical plants, which often lack a dormant growth phase, may suffer from continual exposure to atmospheric pollutants, and that this susceptibility is compounded by the predominance of the phanerophyte life form in the tropics (Raunkiaer 1934).

It follows from the discussion above that clonal populations may be more resistant to air pollution stress than populations which rely exclusively on seed reproduction. The lower risk of mortality to vegetative offspring, whose growth can be sustained under adverse conditions by mobilization of resources from older ramets, may enable clonal growth and regeneration under conditions too stressful for establishment from seed. The storage reserves available to seeds at germination are usually much less than those available to their vegetative counterparts and, furthermore, the more rapid growth rates of seedlings may result in greater rates of pollutant absorption. However, if pollution levels vary greatly within plant communities, according to their structural complexity, it is possible that larger vegetative offspring (e.g., suckers and offshoots in trees) represent more extensive targets. In this situation seedlings may be afforded more protection by surrounding plants because of their small size. The relationship between plant size and susceptibility to pollution stress is probably not a straightforward association, but may depend more on the spatial and temporal location of growing points in relation to the timing and nature of pollution episodes.

Plant Strategies and Pollution Stress

Earlier in this section Grime's (1979) concept of adaptive plant strategies was introduced as a possible framework for predicting the stress responses of different plants to air pollution. This approach may be of value since the potential for evolution of pollution resistance is unlikely to be independent of other selective forces in the environment. The most important issue is whether life forms already adapted to surviving in stressful habitats will be "preadapted" to tolerating air pollution stress. This could be achieved through stress avoidance by means of slow growth rates, infrequent flowering, and the capacity to shutdown photosynthesis under adverse conditions. Other traits characteristic of stress-tolerators include perenniality, longevity of leaves and roots, and the ability to sequester resources. These may also provide plants with the resilience to endure periods of stress. However, several traits (e.g., longevity, infrequent reproduction) that may enhance resistance also delay rates of evolutionary change in populations and hence stress tolerators as a group may be slow in responding through genetic changes at the population level to air pollution stress.

Competitors are adapted to habitats with high productivity. Accordingly, they exhibit high rates of growth and resource acquisition, traits that enable them to compete effectively in dense stands of vegetation. As vegetation closes during the growing season they respond to increasing competition by rapid and continual readjustment of their absorptive surfaces. This involves large changes in reinvestment of their captured resources. Whether this kind of growth response can be maintained under chronic pollution stress is questionable. It seems likely that resource reserves would be rapidly exhausted, thus decreasing competitive ability and possibly increasing susceptibility to pest or pathogen attack. Since flowering is usually delayed until after periods of maximum growth, it may be suppressed if resource acquisition is compromised. This factor, and restricted opportunities for seedling establishment in productive habitats, both suggest that rates of evolutionary change in populations of competitors may be slow. Under stressed conditions allocation of resources to storage organs and to vegetative expansion may be at the expense of sexual reproduction.

The response of ruderals to stress conditions is markedly different from that of either competitors or stress tolerators. Many ruderals are annuals, and one adaptation to disturbed habitats is rapid completion of the life cycle. Onset of flowering occurs early in the growing season and there is usually a large investment of resources into seed production. Response to unfavorable conditions often includes the capacity to sustain limited seed production even under conditions of severe stress. It is here then that one might expect to find fertile ground for evolutionary change. The annual production of great quantities of seed in combination with good opportunities for seedling establishment are likely to provide the raw material on which natural selection can act. If the growing season coincides with periods of high pollution stress, this will serve to intensify selection pressures. It is also worth considering that the open structure of disturbed habitats may render ruderal species particularly vulnerable to exposure to air pollution. Furthrmore, if large areas of woody vegetation are destroyed through chronic air pollution and are replaced by open environments largely devoid of plant cover, it is likely to be ruderal species that rapidly colonize these new environments because of their high dispersability and rapid reproductive rates.

Colonizing populations may initially suffer considerable pollution induced mortality, however, the high selection pressures that this involves are likely to lead to the rapid evolution of resistant populations, particularly in species with short generation times. Although levels of genetic variation in many annual colonizers of disturbed environments are low in comparison with longer-lived plants (reviewed in Barrett and Shore 1989), sufficient genetic variation is likely to occur to enable evolutionary response to air pollutants, particularly where tolerance is polygenically controlled (see below). Two of the case studies on the evolution of populations resistant to air pollution involve ruderal species (*Geranium carolinianum*, see Taylor and Murdy 1975; *Lepidium virginicum*, see Murdy 1979).

Trees and Pollution Stress

Much attention has been focused on the impact of pollution stress on trees because they are highly visible components of ecosystems and because of their economic importance. Among the great diversity of tree life histories can be found species that fall within two of Grime's three primary strategies (see Fig. 18d in Grime 1979). Therefore, rather than discuss attributes of woody plants in the context of his theoretical framework, it is simpler to consider what particular genetic and ecological features of tree populations are likely to influence evolutionary responses to air pollutants.

Trees as a group contain higher levels of genetic variability than most other plant groups (Hamrick 1979; Ledig 1986). Therefore, it seems unlikely that the evolution of pollution resistance will be constrained by lack of appropriate genetic variation. Furthermore, unlike some annuals and short-lived perennials population sizes in many tree species, particularly the dominant species of temperate forests, can be very large, thus reducing effects of genetic drift. Large effective population sizes provide greater opportunities for selection response. Thus, it seems unlikely that stochastic forces will have the same influence in trees as in many annual colonizing species, unless population sizes are notably small or reduced dramatically by pollution-induced mortality. While the genetic potential for evolutionary response is likely to be present in populations of many tree species, the time scale and the dynamics of change are likely to be much slower than in herbaceous plants. Since trees are largely outcrossing the persistence of resistant genotypes, particularly when rare, may be short-lived since recombination will reshuffle tolerance genes into different genetic backgrounds during each mating cycle. The tempo of evolutionary change will be retarded if gene complexes conferring resistance need to be assembled from different individuals in the population and then refined and improved by selection. This constraint will be less severe, however, where many individuals contain resistance genes and levels of selective mortality are high.

Another potential constraint on evolutionary response to air pollution concerns the nature of recruitment patterns and the degree of spatial and temporal environmental variation in many forest environments. "Seedling banks" are a common feature of late successional tree species with growth rates of seedlings often extremely slow and mortality levels high. If conditions determining seedling mortality are different or more varied than those determining adult survival, then seedling populations will not necessarily be "preadapted" to the selection pressures they encounter as adults. For example, if pollution stress is only one of many factors contributing to seedling mortality, then surviving seedlings will not necessarily be the most resistant to pollution. Likewise, seedlings carrying resistance genes may succumb to other stresses. These factors will retard the evolution of resistance since the fate of resistance genes in the population will depend on the many factors influencing seedling survival.

The major factor influencing the nature of evolutionary change to air pollutants in tree populations is their great longevity. Since generation times of many tree species involve decades or centuries rather than years, it seems likely that most of the genetic effects that are likely to occur within populations will result from viability selection against sensitive genotypes. This process gives rise to populations composed of adults, that through possession of superior genotypes, are able to tolerate pollution stress. Although recombination and repeated mating cycles can potentially lead to improved resistance this is likely to occur over very long time scales in most tree species. Because of this our ability to measure evolutionary changes, resulting in increased adaptedness to pollution stress within tree populations, will be severely limited in comparison with most herbaceous plant populations. This difficulty may account for the recent interest in attempts to reveal fertility selection and gametophytic fitness differences in forest trees in response to pollution stress (Cox 1989; Venne et al. 1989).

Reproductive Systems

Sexual reproduction in flowering plants consists of flowering, pollination, fertilization, seed maturation, and dispersal. These events occur consecutively and the imposition of stress during any or ' can result in loss of reproductive potential with consequences for plant fitness. Most studies on the effects of air pollutants on plants have investigated vegetative parts with a particular emphasis on the relationships between foliar injury, photosynthetic rates, growth, and productivity (reviewed in Heath 1980). Far less work has been conducted on the influence of pollution stress on sexual reproductive processes despite the importance of this stage in the life cycle in relation to genetic transmission and population genetic structure. One of the problems in studying the effects of pollution on sexual reproduction, in contrast to vegetative growth, is that the events involved are complex, sensitive processes that often occur over limited time periods. As a consequence it can be extremely difficult to pinpoint which stages in the reproductive cycle are responsible for losses in reproductive potential, particularly where experiments are conducted under field conditions.

As discussed earlier a common reproductive response to stress factors is a curtailment of resources allocated to reproduction. This plastic response is usually manifested by either the inhibition of flowering or the abortion of buds, flowers, fruits, or seeds. The particular response will depend on the timing, duration, and intensity of stress (Lloyd 1980; Stephenson 1981). Where air pollution levels increase gradually, causing inhibitory effects on growth, the most likely effect on reproduction will be a reduction in levels of flowering, particulary in perennial plants (Bonte 1982; Ernst et al. 1985;

Lechowicz 1987; Taylor and Bell 1988). Although there are many reports of the stimulation of flowering by stress conditions, this often presages early mortality as a consequence of the cost of reproduction in individuals already severely limited by lack of resources. Where pollution is episodic in nature and populations become exposed to toxic levels during reproductive activity, responses are likely to be considerably more complex. They may involve effects on gametogenesis causing sterility of pollen and ovules as well as impaired reproductive function causing reductions in pollen viability, pollen germination, fruit and seed set (Houston and Dochinger 1977; Murdy 1979; DuBay and Murdy 1983; Cox 1988a,b). While there are a growing number of reports of the detrimental effects of air pollution on plant reproduction under field conditions, in many cases it is unclear which stage(s) in the reproductive process are responsible for observed reductions in fertility.

Pollination Systems

The transfer of compatible pollen between conspecific individuals is a critical stage in the mating cycle of outcrossing plants. The specific pollen vector(s) involved (e.g., wind, animals, water) depends on the pollination system of the species (reviewed in Faegri and van der Pijl 1971). In plants pollinated by animals (e.g., bees, flies, butterflies, birds, bats), there is the possibility that in heavily polluted areas the acute toxicity of pollutants will have a direct effect on the composition of the pollinator fauna. This could result in a reduced frequency of visits or loss of particular pollinator species. Such an effect seems unlikely for low-level regional air pollution but could be significant in highly polluted urban and industrial areas or at sites in close proximity to point sources. Tropical forests would be especially sensitive to such effects since most tree species are animal pollinated and outcrossing. Air contaminants are unlikely to directly influence pollen transport in wind-pollinated species (Smith 1981), but have been reported to be noxious to some pollinating insects causing reductions in their number (Bonte 1982). Reduced frequencies of pollinator visits may not necessarily lower fruit and seed set, since in many zoophilous plants reproductive output is apparently not pollen limited (Willson and Burley 1983). However, qualitative and quantitative changes in the pollinator fauna may have more subtle influences on mating through alterations in foraging behavior and its influence on male fertility and patterns of gene flow. Such effects may not be detectable by monitoring patterns of seed production in populations and may require the use of genetic markers (Brown et al. 1985).

Flower orientation may also be of significance in affording protection against certain types of air pollutants, particularly those that dissolve in rain. Tubular or bowl-shaped flowers that are held in an upright position may fill with rain, whereas those that are pendulous are not effected in this way. Related species of *Primula* with contrasting flower orientations possess pollen that respond differently to immersion in water (Eisikowitch and Woodell 1974). This suggests that chemical adaptations preventing germination may occur in species with exposed flowers. Other features of floral biology such as the phenology of flowering, the timing of stigma receptivity, and anther dehiscence, and the shape, structure, and orientation of reproductive parts are likely to influence the degree of sensitivity to air pollutants displayed by individual species (Cox 1984). In addition, the physical location of species in a community (e.g., understory, overstory) and hence the degree of protection they obtain from neighboring species may be important in modifying the impact of air pollutants on reproductive processes. Field studies that take into account the ecological heterogeneity of natural communities are needed to assess the variations in reproductive responses that are likely under natural conditions.

Pollen-Pistil Interactions

Following the deposition of pollen on stigmas of conspecific plants, a series of complex interactions occur between the male gametophyte and tissues of the maternal plant. The interactions are susceptible to a variety of stress factors, and it is known that pollen germination and pollen tube growth are among the more sensitive indicators of atmospheric pollution (Feder 1968; Stanley and Linskens 1974; Cox 1987). Wolters and Marten (1987) have reviewed in detail the effects of air pollutants on pollen biology and document numerous studies involving different pollutants (e.g., SO_2, O_3, CO_2, NO_2) and plant taxa. From these studies it is clear that pollen viability, pollen germination, and pollen tube growth can all be negatively affected by air pollutants both in vitro and in vivo. Although observed patterns vary with environmental conditions, species, and pollutant, in vivo pollen germination and pollen tube growth is usually more tolerant to air pollutants than under in vitro conditions. This difference is thought to reflect the extra buffering capacity of the stigma surface and the protection given by maternal tissue once pollen tubes have entered the style.

The extent to which pollen-pistil interactions are influenced by air pollutants in particular species is likely to depend on specific features of their compatibility systems, the size, structure, and cytochemistry of stigmatic surfaces, and the time over which pollen-pistil interactions normally take place. In species in which individual flowers last for several weeks, stigmatic receptivity and pollen tube growth are often of extended duration. By contrast, in flowers in which the anthesis periods last for less than a day, pollen germination, pollen tube growth, and fertilization often occur within hours of pollen deposition on the stigma. Primack (1985) has reviewed the literature on floral longevity and provides ecological and evolutionary explanations for the patterns observed. From the perspective of air pollution it would seem likely that the longer pollen remains on the stigma before germination and penetration of stigmatic tissues, the more vulnerable it may be to the effects of chemical modifications of the stigmatic secretion by air pollutants. Species with stigmas of large surface area (e.g., many wind-pollinated plants and also *Oenothera parviflora*; see Cox 1984) may be more susceptible to pollution-induced changes in pollen-stigma interactions than taxa with small stigmas. Comparative studies of the effects of air pollutants on the pollination process in closely related species with contrasting floral traits would be required to evaluate these ideas.

A particularly provocative suggestion concerning pollen-pistil function is the possibility that pollution-induced stresses may alter microgametophytic selection leading to fitness effects in the sporophyte generation (Mulcahy 1979; Searcy and Mulcahy 1985; Wolters and Martens 1987). Searcy and Mulcahy (1985) found that there was parallel expression of metal tolerance in pollen and sporophytes of two *Silene* spp. and in *Mimulus guttatus*. Pollen from metal-tolerant plants was able to germinate and grow in vitro at concentrations of metals which markedly inhibited the pollen from nontolerant individuals. They suggested that if tolerance is due to genes expressed in both the diploid sporophyte and the haploid microgametophyte, it could result in the rapid development of populations resistant to heavy metals. A different evolutionary scenario was envisioned by Cox (1984), who suggested that microgametophytic selection favoring pollen tolerant to the low pH conditions associated with acid precipitation might be selectively advantagous if sporophytes producing the pollen were locally adapted to acidic soils, but disadvantageous where they occur on calcareous soils.

Evolutionary changes through selection at the gametophytic stage of the life cycle of plant populations require a number of conditions to be met. First, pollen loads on stigmas following pollination need to be sufficiently high to enable microgametophyte competition to occur. The most intense competition will occur where ovule number per flower is low and pollen loads on stigmas are high. This condition occurs in many angiosperms but probably rules out the conifers since the number of pollen grains per ovule is normally less than five in most groups examined (Venne et al. 1989). Second, pollen genes must be expressed both post-meiotically and in the sporophyte generation resulting in a positive correlation between gametophytic and sporophytic vigor. There is some evidence for this overlap in gene expression (Tanksley et al. 1981; Willing and Mascarenhas 1984). Finally, for evolutionary changes to occur, heritable differences in pollen performance must occur *within* natural populations. Although many studies have demonstrated selective fertilization on the basis of pollen genotype (reviewed in Marshall and Ellstrand 1986; Snow 1986), few have examined the heritability of pollen performance. In one of the only studies of plants obtained from within a natural population, Snow and Mazer (1988) found no evidence in the wild radish (*Raphanus raphanistrum*) of heritable variation in pollen competitive ability. If a gene which greatly accelerated pollen tube growth arose in a population it would rapidly spread to fixation (Haldane 1932). However, genetic variation for pollen tube growth rate could exist within natural populations if there was a negative genetic correlation between the performance of the pollen and that of the zygotes which it produced. Thus, variation could be maintained if selection occurred in opposite directions in the haploid and diploid phases of the life cycle. Unfortunately, it is not yet clear whether evolutionarily significant amounts of gametophytic selection occur within natural populations, because as yet there is little evidence for heritable variation in pollen competitive abilities. Although it seems reasonable to assume that pollen competition occurs frequently in nature, it is not yet known whether this has led to fixation of genes that influence the fitness of sporophytes (Charlesworth et al. 1987).

Mating Systems and Gene Flow

Mating patterns in plant populations are largely governed by their breeding systems and the nature and magnitude of gene flow within and between populations. This stage in the reproductive cycle is of major importance in determining the genetic structure of populations and their potential for evolutionary change (Richards 1986). Roose et al. (1982) have suggested that outbreeding species are more likely than inreeding species to evolve resistance to air pollution stress. This suggestion follows from the observation that populations of outbreeders usually contain greater stores of genetic variation than inbreeders (Stebbins 1957; Jain 1976). Thus, it would be anticipated that greater opportunities for selection responses occur in outcrossing populations if they contain greater genetic variability for air pollution resistance. Although this suggestion seems reasonable on the basis of experience with the evolution of heavy metal tolerance, there are a number of reasons why we should not discount the possibility that inbreeding species may also rapidly evolve resistance to air pollutants where appropriate environmental conditions occur.

A larger proportion of the total genetic variation in inbreeding species is distributed among populations rather than within populations as is found in outbreeding species (Loveless and Hamrick 1984). However, the total amounts of genetic variation in species with contrasting levels of inbreeding versus outbreeding tend to be broadly similar. Furthermore, models of quantitative genetic variation, under different systems of mating, indicate that even in species which practice considerable inbreeding, high rates of mutation at loci controlling polygenic traits generate considerable amounts of quantitative genetic variation (Lande 1977). This variation is sufficient to allow rapid adaptive responses and considerable phenotypic divergence (Lande 1980).

Since air pollution effects are often regional in nature, with diffuse boundaries, large numbers of populations of a given species are likely to be exposed to altered selection pressures. This form of mass selection in inbreeding species may result in a greater likelihood that some populations will contain the necessary variation in comparison with cases of localized pollution stress. Where heavy metal contaminated soils or emissions from a single point source occur, relatively small numbers of individuals are usually exposed and, in these circumstances, absence of genetic variation within populations may frequently limit evolutionary responses. Under these circumstances the population genetic structure associated with outbreeding species would be more likely to maintain the necessary variation for selective responses. However, with regional air pollution selection responses may be extensive rather than localized, providing opportunities for interdemic selection in inbreeding species.

Another difference between the evolution of resistance to air pollution in comparison with heavy metals concerns the selective regime and type of population processes that are likely to occur. Because of the localized nature of mine wastes, a major force retarding the evolution of heavy metal resistance is the extent of gene flow from neighboring environments containing nonresistant plants. Indeed, it has been argued that resistant populations of some grass species have evolved reproductive isolating mechanisms in the form of altered flowering times and self-fertilization. These are believed to have developed in response to the disruptive effect of gene exchange and its effect in breaking down adaptive gene combinations responsible for resistance (Antonovics 1968). In the case of regional air pollutants, however, gene flow from sensitive or "unselected" populations is unlikely to be of major importance unless pollution sources are highly localized. In this case, wind-pollinated populations, particularly of tree species, are most likely to exchange genes over long distances and selfing populations are the least likely to do so. Since gene flow in flowering plants, irrespective of breeding system, is most often rather localized (Levin and Kerster 1974) it seems unlikely that genes for resistance to air pollution will be prevented from spreading within local populations owing to high levels of gene exchange with populations from unpolluted areas.

One of the likely population-level responses to severe air pollution stress is an alteration in the size and density of plant populations. Concomitant with increasing levels of stress is the selective mortality of individuals and species. Some cases of forest decline in Europe, for example, have resulted in dramatic reductions in population numbers of many forest tree species, resulting in open stands composed of small numbers of individuals. Similar patterns are evident in various parts of North America where high levels of pollution-induced mortality have resulted in sparsely vegetated areas containing few surviving plants. In these situations, resistant genotypes of short-lived species may soon multiply and population sizes have the potential of returning to their original levels, or even increasing in size, if few species can tolerate the polluted environment (e.g., *Deschampsia cespitosa* at Sudbury, Ontario; Cox and Hutchinson 1981). However, in longer-lived plants this process is likely to be considerably slower, giving rise to small populations composed of scattered individuals. In addition to increasing opportunities for genetic drift, this type of population structure could have potentially important consequences for the mating systems of survivors.

In small populations, particularly those at low density, levels of inbreeding may increase through self-fertilization or because of matings between related individuals. Where inbreeding occurs in normally outcrossing taxa, this usually leads to reduced fitness of progeny (Charlesworth and Charlesworth 1987). Under stress conditions, differences between offspring that result from outcrossing and selfing are usually magnified. In polluted environments, fitness decline may occur when survivors have descended from a small number of resistant individuals. In species incapable of self-fertilization, owing to self-incompatibility or dioecism, low density can potentially result in significant reductions in the reproductive output of individuals. These conditions may favor the evolution of self-fertilization, particularly in herbaceous groups (Baker 1955; Jain 1976; Lloyd 1979). This adaptive shift in mating system seems unlikely for most long-lived organisms, such as trees, because of their high genetic loads and the complex environments which they occupy (Ledig 1986). However in short-lived species, selfing may develop as a consequence of low density conditions in polluted environments and act secondarily as a mechanism for maintaining gene complexes that confer resistance to pollution stress.

Population Genetic Considerations

The evolution of resistance to air pollutants in plants is influenced by features of the life history and reproductive biology of individual species. This is because the ecological and demographic characteristics of populations have an important influence on population genetic structure, and, in addition, mating systems regulate patterns of genetic transmission and levels of recombination. Whether resistance develops in plant populations ultimately depends, however, on the presence of heritable variation for the ability to grow and reproduce in environments affected by air pollution. Genetic studies of the inheritance of resistance to air pollution are rudimentary but what little data is available indicate that resistance usually behaves as a quantitative trait governed by many genes with additive effects (Taylor 1978; Roose et al. 1982). If this turns out to be the case in most plants, then the evolution of resistance reduces to the problem of directional selction on a quantitative trait and the major empirical issues concern whether or not experiments can be devised to measure selection intensities and selection responses accurately.

Selection Responses

Selection pressures on sensitive genotypes will vary with the kinds and amount of atmospheric pollution that occur. In cases of severe pollution, for example, at sites in close proximity to smelters, power stations, and refineries, high levels of mortality would result in extremely high selection intensities. These conditions are similar to those involving heavy metal contaminated mine waste, and, in common with these situations, we may anticipate rapid selection responses if appropriate genetic variation is present within populations (Roose et al. 1982). A more complex situation prevails, however, where low levels of air pollution occur on a regional level or where pollution is episodic in nature. Unfortunately, we know relatively little about the effects on fitness of these types of pollution stress, but unless reductions in growth or fecundity are large, selection responses are likely to be considerably slower, particularly in long-lived, outbreeding plants with seed banks. At present the most convincing evidence for the evolution of resistance to air pollution involves short-lived herbaceous species where populations have been exposed to relatively high levels of pollution stress. In the case of the evolution of SO_2 resistance in the annual *Geranium carolinianum*, this has apparently occurred in approximately 30 generations (Taylor and Murdy 1975; Taylor 1978).

The amount and rate of response to directional selection on a quantitative trait is affected by the number and average effect of genes controlling the trait (additive effects) and the occurrence of dominance, epistasis, and pleiotropy (nonadditive effects). Other factors that limit or constrain the response to selection include (1) negative genetic correlations among fitness components, (2) the occurrence of favorable alleles at certain loci in gametic disequilibrium with alleles at other loci with negative effects on fitness, (3) additive genetic variance exhausted by selection, and (4) finite population size effects or inbreeding leading to a loss of heritable variation. Although it is often difficult to isolate which of these factors are important, phenotypic responses following relaxation of selection can help distinguish between constraints that result from exhaustion of variability versus negative genetic correlations among fitness components. No phenotypic change would occur if quantitative genetic variation is limiting response, whereas a change would be expected if negative genetic correlations among fitness components are involved (Falconer 1981; Hedrick 1985).

These considerations may be important for the evolution of air pollution resistance for several reasons. First, mechanisms of resistance may involve traits that are of adaptive significance in unpolluted environments (e.g., growth rates, patterns of stomatal opening and closure, and cuticle thickness), and, as a result, we may expect fitness constraints to be associated with the evolution of resistance (Roose et al. 1982). Fitness costs could be manifested by negative genetic correlations between traits conferring greater resistance and features of plant growth and reproduction that contribute to fitness. Similarly, where plant populations are exposed to several atmospheric pollutants simultaneously, the resistance mechanisms required to effectively combat each one may involve different sets of genes. In some cases the genes may be negatively correlated with one another, so that the selection response would be constrained.

Unfortunately, little is known about the genetic architecture of resistance to air pollutants or whether cotolerances, of the type reported for heavy metals (e.g., Cox and Hutchinson 1979), are likely to occur for different air pollutants. (Preliminary evidence does suggest that cotolerances to air pollutants can evolve; see Taylor and Bell 1988). In addition, where genetic studies have been conducted they have usually involved plants of similar age and developmental status screened for a single pollutant under a small number of doses. This approach is chosen to maximize the differences in response among genotypes. Under field conditions, however, selection pressures are a good deal more complex because of heterogeneous age structures, selection intensities that vary in space and time, and exposure to several pollutants at one time. This complexity makes it extremely difficult to predict with any certainty the types of selection responses that are likely to occur, particularly for regional air pollution involving an array of low-level pollutants.

Allelic Variation and Pollution Stress

Abundant evidence exists for the occurrence of genetic variation in resistance to different air pollutants at the species, cultivar, population, and genotype level (reviewed in Roose et al. 1982; Karnosky et al. 1989). Far less is known, however, about how pollution stress influences patterns of genetic variation in natural populations and whether selective mortality causes significant reductions in the kinds and amounts of genetic variation. It seems reasonable to assume that in environments that are exposed to severe pollution, and in which mortality levels are very high, loss of genetic diversity will occur. This will be particularly likely for alleles that occur at low frequency. However, to what extent this loss of allelic diversity may handicap the evolutionary potential of resistant populations is by no means clear. Moreover, how low-level pollution on a regional scale may influence population genetic structure, through effects on viability and fertility, is complicated because of the diverse historical, ecological, and genetic factors that also regulate patterns of genetic diversity.

Recently a number of workers have investigated the relationships between variation in resistance to air pollution stress and patterns of genetic diversity by the use of isozyme techniques (Mejnartowicz 1983;

Scholz and Bergmann 1984; Bergmann and Scholz 1985; Müller-Starck 1985; Geburek et al. 1987). In these studies two basic approaches have been employed using European tree species (Scots pine, Norway spruce, and European beech). In the first (e.g., Scholz and Bergmann 1984), plants originating from different provenances and maternal families were subjected to controlled fumigations (SO_2), their responses were monitored, and individuals were genotyped at several isozyme loci. The second approach (e.g., Müller-Starck 1985) involved comparisons under field conditions of the patterns of isozyme variability in trees which displayed contrasting symptoms to air pollution damage (i.e., no apparent damage or irreversible injury). In both types of study, significant differences were detected between the two groupings in the number and frequency of alleles at polymorphic loci as well as in levels of heterozygosity (see below). In most comparisons, trees in the tolerant group exhibited higher levels of genetic diversity than the sensitive group. These findings led several investigators to suggest that variation at some isozyme loci—e.g., ACP, Mejnartowicz (1983); G6PDH, Bergmann and Scholz (1985); and GDH and AAT, Geburek et al. (1987)—may be of adaptive significance in contributing towards resistance to pollution-induced stresses (and see Bergman and Scholz, 1989) and that heterozygote superiority (overdominance) may account for the apparently higher levels of heterozygosity in resistant genotypes (Geburek et al. 1987; Karnosky et al. 1989).

Although isozymes provide the most convenient means of assaying the amounts of genetic variation within plant populations, they are of less value in determining the nature of selection pressures operating on the variation. Despite two decades of detailed experimental investigations of enzyme polymorphisms, there is still relatively little evidence that most allelic variation at isozyme loci is maintained by selection in natural populations (Lewontin 1974, 1985). Many studies have attempted to establish a causal link between a particular electromorph and a selective factor, but in the vast majority of cases it has not been possible to distinguish whether observed fitness differences are the result of the alleles being maintained or the genetic background of the individuals carrying them. Because of these difficulties, as well as a large body of theoretical work, most workers in evolutionary genetics have abandoned attempts to establish measurable fitness differences that accompany electrophoretic differences. Instead, it is usually assumed that most variation is either selectively neutral or under such weak selection that it is impractical to measure. As a final note of caution, it is important to recognize that although studies may elucidate the functional significance of a particular enzyme to pollution or other forms of stress, this does not necessarily indicate that fitness differences between allelic variants of the enzyme occur.

Several of the studies in which different classes of allelic variants at isozyme loci were thought to be causally related to pollution stress-involved samples of trees obtained from different populations. This type of sampling scheme confounds many sources of genetic variation making it difficult to ascribe the observed differences to a single factor, e.g., pollution stress. Samples of resistant and susceptible genotypes of Norway spruce (Bergmann and Scholz 1985) differed principally in the presence of a higher number of low frequency alleles in the resistant plants. In this case it seems more likely that these alleles simply mark unique genetic backgrounds that possess other characteristics that confer tolerance. Even where more controlled sampling has been undertaken within a local area, e.g., a five-hectare plot of *Fagus sylvatica* (Müller-Starck 1985), the observed differences in allelic diversity and heterozygosity may result from sampling effects. Sample sizes were quite small ($n = 44$ trees per tolerant and sensitive group), and no particular attempt appears to have been made to control for age, size, or the presence of population substructure. Other studies of *Fagus sylvatica* have demonstrated spatial and temporal genetic differentiation at isozyme loci on a local scale indicating considerable differentiation among subpopulations (Gregorius et al. 1986). The sampling problems associated with these pollution studies confound many factors that are likely to influence patterns of genetic variation in forest trees. As a result it may be premature to conclude that there is any biological significance to the observed differences in allelic variation at isozyme loci that have been reported between resistant and sensitive trees.

At present there is too little empirical data to assess the genetic consequences of various types of pollution-induced stresses. In theory, strong directional selection or the occurrence of an adaptive bottleneck associated with the evolution of resistance may lead to an erosion or rapid loss of genetic variation within populations. However, the critical issues depend on the number of individuals carrying resistance genes, the strength of selection against susceptible individuals, and the population recovery rate following the initial decline. Clearly, in cases where only a few variants capable of tolerating anthropogenic stresses are responsible for founding new resistant populations, bottlenecks in allelic variation may be anticipated (e.g., *Deschampsia cespitosa* at Sudbury; see Bush and Barrett 1989). This may occur in heavy-metal tolerant populations that invade mine wastes because of the low frequency of resistant individuals (e.g., Gartside and McNeilly 1974; Bradshaw 1984) that usually occur in unpolluted areas (although see Verkleij et al. (1989) for an example where no differences in genetic diversity were detected between populations on polluted mine sites and nonpolluted areas). However, this process seems less likely to prevail where low-level pollution effects are manifested over large areas and mortality levels are not sufficient to eliminate large numbers of individuals within local populations. Loss of some allelic variation is an inevitable consequence of pollution-induced stress. Whether such losses are biologically relevant to the evolutionary potential of natural populations is another matter.

Heterozygosity and Pollution Stress

The relationship between fitness and heterozygosity has been a major theme in evolutionary genetics since Lerner (1954) argued that heterozygosity gave organisms greater developmental and genetic homeostasis. These features are believed to enhance the ability to maintain high fitness over a broad range of environmental stresses. Although some evidence exists for a relationship between heterozygosity and fitness traits in some animal groups (e.g., Zourous et al. 1980; Leary et al. 1983), studies on plants have provided a more complex picture. For example, in forest trees heterozygosity and fitness components show a variety of relationships depending on the scales of measurement, genetic background, environment, and age (Mitton et al. 1981; Strauss 1987). Where positive relationships have been detected these have usually been in older forest stands and may reflect the greater homeostasis of heterozygotes, in the face of year to year variation, and/or the accentuation of differences by competition between heterozygotes and homozygotes as the canopy closes (Ledig et al. 1983).

The observation that heterozygosity at isozyme loci is higher in resistant individuals than sensitive ones in field trials where tree populations were exposed to multiple stress factors (several pollutants and natural stress conditions) has been interpreted as evidence in support of the classical hypothesis of heterozygote superiority (Karnosky et al. 1989). In addition, it has been claimed that high heterozygosity results in a "high adaptive potential for populations exposed to varying stress factors" (Geburek et al. 1987). There are several reasons why these conclusions may not be fully justified. Mitton and Pierce (1980) and Chakraborty (1981) have shown that heterozygosity measured at a small number of loci is a poor indicator of total genomic heterozygosity for an individual. Relatively small sample sizes for individuals and isozyme loci were employed for comparing heterozygosity in resistant and nonresistant trees (Mejnartowicz 1983; Müller-Starck 1985; Geburek 1987), and problems concerned with the confounding of genetic, environmental, and developmental variables already eluded to above are also evident in these analyses. However, even if concerns over sampling error are cast aside, their are still problems concerning the interpretation of the causes of the apparent heterozygote superiority that was observed. Many years of detailed biometrical studies of plants (e.g., see Jinks 1983) indicate that true overdominance either does not exist or is very rare. In addition, studies of isozyme variation in forest trees have usually failed to detect any heterosis associated with single isozyme loci (Ledig et al. 1983). Because of these results it seems more reasonable to interpret the apparent heterozygote superiority found in some studies as the result of inbreeding depression in homozygotes as a result of coancestry of descent (Ledig et al. 1983). Despite their outcrossed mating systems, populations of many tree species practice some inbreeding as a result of limited pollen and seed dispersal. With this mating pattern and high genetic loads, offspring carrying varying numbers of deleterious genes are produced at every generation giving rise to wide fitness variation (Ledig 1986). It seems more likely that it is this source of variation that accounts for the association between fitness components and heterozygosity rather than overdominance at isozyme loci or groups of linked genes that they mark.

Conclusions

In this chapter an attemp has been made to evaluate some of the features of plant life histories and reproductive systems that may facilitate or constrain the evolution of resistance to air pollutants in natural populations. Much of the discussion has been speculative since there is still insufficient empirical data to allow a comparative survey of a wide variety of plant species. Although there is a growing literature documenting effects of various atmospheric pollutants on plants, most studies have measured a small number of response variables under controlled conditions in plants of economic importance. This work is of importance in determining which aspects of plant growth and reproduction are sensitive to pollution stress, however, it is of less value in assessing how pollution in natural communities affects the life time fitness of genotypes within populations. Yet, a complete demonstration of natural selection will require this information in combination with studies of the heritability of traits conferring resistance (Endler 1986).

The best evidence for the evolution of resistance to air pollutants involves comparative data obtained from populations exposed to different levels of pollution (Taylor and Murdy 1975; Bell and Mudd 1976; Murdy 1979; Horsman et al. 1979; Ayazloo and Bell 1981; Ernst et al. 1985; Taylor and Bell 1988). In these studies genotypes from populations growing in highly polluted areas were found to be more tolerant to controlled exposures than genotypes obtained from unpolluted sites. In all cases the species involved in these studies are relatively short-lived herbs. The vast majority of examples of natural selection in wild populations of plants also involve herbaceous plants (see Table 5 in Endler 1986). This pattern highlights the difficulties that are associated with providing convincing experimental evidence of the direct effects of natural selection in long-lived organisms. Karnosky et al. (1989) recently stated that "air pollution has served and continues to serve as a strong natural selection factor in the evolution of forest ecosystems (Scholz 1981; Sinclair 1969)." A major future challenge will be to devise experiments to confirm this assertion. Although population biologists have recently devised techniques for the measurement of the direction and intensity of natural selection on phenotypic traits in natural populations (Arnold and Wade 1984a,b; Lande and Arnold 1983; Endler 1986), it remains to be seen whether these approaches can be usefully employed to study evolutionary processes in forest trees.

Acknowledgments. We thank Roger Cox for comments on the manuscript, Tom Hutchinson for valuable discussions, and the Natural Sciences and Engineering Research Council of Canada for financial support. The manuscript was prepared while the senior author was a recipient of an E.W.R. Steacie Memorial Fellowship funded by NSERC, Canada.

References

Antonovics J (1968) Evolution in closely adjacent plant populations. V. Evolution of self fertility. Heredity 23:219–238

Antonovics J, Bradshaw AD, Turner RG (1971) Heavy metal tolerance in plants. Advance in Ecological Research 7:1–85

Arnold SJ, Wade MJ (1984a) On the measurement of natural and sexual selection: theory. Evolution 38:709–719

Arnold SJ, Wade MJ (1984b) On the measurement of natural and sexual selection: applications. Evolution 38:720–734

Ayazloo M, Bell JNB (1981) Studies on the tolerance to sulphur dioxide of grass populations in polluted areas. I. Identification of tolerant populations. New Phytologist 88:203–222

Baker HG (1955) Self-compatibility and establishment after "long distance" dispersal. Evolution 9:347–348

Barrett SCH, Shore JS (1989) Isozyme variation in colonizing plants. In: Soltis DE., Soltis PS. (eds) Isozymes in plant biology. Dioscorides Press, Washington, pp 106–126

Bell JNB, Mudd CH (1976) Sulphur dioxide resistance in plants: a case study of *Lolium perenne.* In: Mansfield TA (ed) Effects of air pollutants on plants. Cambridge University Press, Cambridge, pp 87–114

Bergmann F, Scholz F (1985) Effects of selection pressure by SO_2 pollution on genetic structure of Norway Spruce (*Picea abies*). In: Gregorius HR (ed) Population genetics in forestry. Springer-Verlag, Berlin, pp 267–275

Bonte J (1982) Effects of air pollutants on flowering and fruiting. In: Unsworth MH, Ormrod DP (eds) Effects of gaseous air pollution in agriculture and horticulture. Butterworth Scientific, London, pp 207–223

Bradshaw AD (1984) Adaptation of plants to soils containing toxic metals—a test for conceit. In: Ciba Foundation Symposium 102 (eds) Origins and development of adaptation. Pitman, London, pp 4–19

Bradshaw AD, McNeilly T (1981) Evolution and pollution. Arnold, London

Brown AHD, Barrett SCH, Moran GF (1985) Mating system estimation in forest trees: models, methods and meanings. In: Gregorius HR (ed) Population genetics in forestry. Springer-Verlag, Berlin, pp 32–49

Bush EJ, Barrett SCH (1989) Colonization genetics of the invasion of metal contaminated areas in Ontario by *Deschampsia cespitosa.* American Journal of Botany (Supplement) 76:145

Chakraborty R (1981) The distribution of the number of heterozygous loci in an individual in natural populations. Genetics 98:461–466

Charlesworth D, Charlesworth B (1987) Inbreeding depression and its evolutionary consequences. Annual Review of Ecology and Systematics 18:237–268

Charlesworth D, Schemske DW, Sork VL (1987) The evolution of plant reproductive characters: sexual vs natural selection. In: Stearns SC (ed) The evolution of sex and its consequences. Birkhauser-Verlag, Basel, pp 317–335

Cox RM (1983) Sensitivity of forest plant reproduction to long range transported air pollutants: in vitro sensitivity of pollen to simulated acid rain. New Phytologist 95:269–276

Cox RM (1984) Sensitivity of forest plant reproduction to long range transported air pollutants: in vitro sensitivity of *Oenothera parviflora* L. pollen to simulated acid rain. New Phytologist 97:63–70

Cox RM (1987) The response of plant reproductive processes to acidic rain and other air pollutants. In: Hutchinson TC, Meema KM (eds) Effects of atmospheric pollutants on forests, wetlands and agricultural ecosystems. Springer-Verlag, Berlin, pp 155–170

Cox RM (1988a) The sensitivity of pollen from various coniferous and broad leaved trees to combinations of acidity and trace metals. New Phytologist 109:193–201

Cox RM (1988b) The sensitivity of forest plant reproduction to long range transported air pollutants: the effects of net deposited acidity and copper on reproduction of *Populus tremuloides*. New Phytologist 110:33–38

Cox RM (1989) Natural variation in sensitivity of reproductive processes in some boreal forest trees to acidity. In: Scholz F, Gregorius HR, Rudin D (eds) Genetic effects of air pollutants in forest tree populations. Springer-Verlag, Berlin, (in press) 1, pp 77–88

Cox RM, Hutchinson TC (1979) Metal co-tolerances in the grass *Deschampsia cespitosa*. Nature 279:231–233

Cox RM, Hutchinson TC (1981) Environmental factors influencing the rate of spread of the grass *Deschampsia cespitosa* invading areas around the Sudbury, nickel-copper smelters. Water, Air and Soil Pollution 16:83–106

Crawford TJ (1984) What is a population? In: Shorrocks B (ed) Evolutionary ecology. Blackwell Scientific Publications, Oxford, pp 135–173

Dubay DT, Murdy WH (1983) Direct adverse effects of SO_2 on seed set in *Geranium carolinianum* L.: a consequence of reduced pollen germination on the stigma. Botanical Gazette 144:376–381

Eisikowitch D, Woodell SRJ (1974) The effect of water on pollen germination in two species of *Primula*. Evolution 28:692–694

Endler J (1986) Natural selection in the wild. Princeton University Press, Princeton, New Jersey

Engle RL, Gabelman WH (1966) Inheritance and mechanism for resistance to ozone damage in onion, *Allium cepa* L. Journal of the American Society for Horticultural Science 89:423–429

Ernst WHO, Tonneijck AEC, Pasman FJM (1985) Ecotypic response of *Silene cucubalus* to air pollutants (SO_2, O_3). Journal of Plant Physiology 118:439–450

Fäegri K, Van der Pijl L (1971) The principles of pollination ecology, 2nd ed. Pergamon, Oxford

Falconer DS (1981) Introduction to quantitative genetics, 2nd ed. Longman, New York

Feder WA (1968) Reduction in tobacco pollen germination and tube elongation, induced by low levels of ozone. Science 160:1122

Gartside DW, McNeilly T (1974) The potential for evolution of heavy metal tolerance in plants. II. Copper tolerance in normal populations of different plant species. Heredity 32:335–348

Geburek TH, Scholz F, Knabe W, Vornweg A (1987) Genetic studies by isozyme gene loci on tolerance and sensitivity in an air polluted *Pinus sylvestris* field trial. Silvae Genetica 36:49–53

Gregorius HR, Krauhausen J, Müller-Stark G (1986) Spatial and temporal genetic differentiation among the seed in a stand of *Fagus sylvatica* L. Heredity 57:255–262

Grime JP (1979) Plant strategies and vegetation processes. Wiley, Chichester

Haldane JBS (1932) The causes of evolution. Harper, New York

Hamrick JL (1979) Genetic variation and longevity. In: Solbrig OT, Jain S, Johnson GB, Raven PH (eds) Topics in plant population biology. Columbia University Press, New York, pp 84–114

Harper JL (1977) Population biology of plants. Academic Press, London

Heath RL (1980) Initial events in injury to plants by air pollutants. Annual Review of Plant Physiology 31:395–431

Hedrick PW (1985) Genetics of populations. Jones and Bartlett Publishers, Boston

Heywood JS (1986) The effect of plant size variation on genetic drift in populations of annuals. American Naturalist 127:851–861

Horsman DC, Roberts TM, Bradshaw AD (1979) Studies on the effect of sulphur dioxide on perennial ryegrass (*Lolium perenne* L.). Journal of Experimental Botany 30:495–501

Houston DB, Dochinger LS (1977) Effects of ambient air pollution on cone, seed and pollen characteristics in eastern white and red pines. Environmental Pollution 12:1–5

Hutchinson TC (1984) Adaptation of plants to atmospheric pollutants. In: Ciba Foundation Symposium 102 (eds) Origins and development of adaptation. Pitman, London

Hutchinson TC, Harwell MA (1985) Environmental consequences of nuclear war, vol 2. Ecological and agricultural effects. Wiley, Chichester

Hutchinson TC, Meema KM (1987) Effects of atmospheric pollutants on forests, wetlands and agricultural ecosystems. Springer-Verlag, Berlin

Jain S (1979) Adaptive strategies: polymorphism, plasticity and homeostasis. In: Solbrig OT, Jain S, Johnson GB, Raven PH (eds) Topics in plant population biology. Columbia University Press, New York, pp 160–187

Jain SK (1976) The evolution of inbreeding in plants. Annual Review of Ecology and Systematics 7:469–495

Jinks JL (1983) Biometrical genetics of heterosis. In: Frael R (ed) Heterosis—reappraisal of theory and practice. Springer-Verlag, Berlin, pp 1–46

Karnosky DF, Scholz F, Geburek TH, Rudin D (1989) Implications of genetic effects of air pollution on forest ecosystems—knowledge gaps. In: Scholz F, Gregorius HR, Rudin D (eds) Genetic effects of air pollutants in forest tree populations. Springer-Verlag, Berlin, pp 199–201

Lande R (1977) The influence of the mating system on the maintenance of genetic variability in polygenic characters. Genetics 86:485–498

Lande R (1980) Genetic variation and phenotypic evolution during allopatric speciation. American Naturalist 116:463–479

Lande R, Arnold SJ (1983) The measurement selection on correlated characters. Evolution 37:1210–1226

Leary RF, Allendorf FW, Knudsen KL (1983) Developmental stability and enzyme heterozygosity in rainbow trout. Nature 301:71–72

LeBaron HM, Gressel J (1982) Herbicide resistance in plants. Wiley, New York

Lechowicz MJ (1987) Resource allocation by plants under air pollution stress: implications for plant-pest-pathogen interactions. Botanical Reviews 53:281–300

Ledig FT (1986) Heterozygosity, heterosis, and fitness in outbreeding plants. In: Soulé ME (ed) Conservation biology: the science of scarcity and diversity. Sinaeur Association, Sunderland, pp 77–104

Ledig FT, Guries RP, Bonefield BA (1983) The relation of growth to heterozygosity in pitch pine. Evolution 37:1227–1238

Lerner IM (1954) Genetic homeostasis. Wiley, New York

Levin DA (1978) Some genetic consequences of being a plant. In: Brussard, P (ed) Ecological genetics: the interface. Springer-Verlag, Berlin, pp 189–912

Levin DA, Kerster HW (1974) Gene flow in seed plants. Evolutionary Biology 7:139–220

Lewontin RC (1974) The genetic basis of evolutionary change. Columbia University Press, New York

Lewontin RC (1985) Population genetics. Annual Review of Genetics 19:81–103

Lloyd DG (1979) Some reproductive factors affecting the selection of self-fertilization in plants. American Naturalist 113:67–79

Lloyd DG (1980) Sexual strategies in plants. I. An hypothesis of serial adjustment of maternal investment during one reproductive session. New Phytologist 86:69–79

Loveless MD, Hamrick JL (1984) Ecological determinants of genetic structure in plant populations. Annual Review of Ecology and Systematics 15:65–95

Marshall DL, Ellstrand NC (1986) Sexual selection in *Raphanus sativus*: experimental data on non-random fertilization, maternal choice, and consequences of multiple paternity. American Naturalist 127:446–461

Mejnartowicz LE (1983) Changes in genetic structure of Scots Pine (*Pinus silvestris* L.) population affected by industrial emission of fluoride and sulphur dioxide. Genetica Polonica 24:41–50

Mitton JB, Knowles P, Sturgeon KB, Linhart YB, Davis M (1981) Associations between heterozygosity and growth rate variables in three western forest trees. In: Conkle MT (ed) Proceedings of the symposium on isozymes of North American forest trees and forest insects. USDA. Forest Service, General Technical Report PSW-48, pp 27–34

Mitton JB, Pierce BA (1980) The distribution of individual heterozygosity in natural populations. Genetics 95:1043–1054

Mulcahy DL (1979) Rise of the angiosperms. Science 206:20–23

Müller-Starck G (1985) Genetic differences between "tolerant" and "sensitive" beeches (*Fagus sylvatica* L.) in an environmentally stressed adult forest stand. Silvae Genetica 34:241–247

Murdy WH (1979) Effect of SO_2 on sexual reproduction in *Lepidium virginicum* L. originating from regions with different SO_2 concentrations. Botanical Gazette 140:299–303

Pitelka LF (1988) Evolutionary response of plants to anthropogenic pollutants. Trends in Ecology and Evolution 39:233–236

Primack RB (1985) Longevity of individual flowers. Annual Review of Ecology and Systematics 16:15–38

Raunkiaer C (1934) The life forms of plants and statistical plant geography. Oxford University Press, Oxford

Richards AJ (1986) Plant breeding systems. George Allen and Unwin, Boston

Roose ML, Bradshaw AD, Roberts TM (1982) Evolution of resistance to gaseous air pollutants. In: Unsworth MH, Ormrod DP (eds) Effects of gaseous air pollution in agriculture and horticulture. Butterworth Scientific, London, pp 379–409

Schlichting CD (1986) The evolution of phenotypic plasticity in plants. Annual Review of Ecology and Systematics 17:667–693

Scholz F (1981) Genecological aspects of air pollution effects on northern forests. Silva Fennica 15:384–391

Scholz F, Bergmann F (1984) Selection pressure by air pollution as studied by isozyme-gene systems in Norway spruce exposed to sulphur dioxide. Silvae Genetica 33:238–241

Scholz F, Gregorius HR, Rudin D (eds) (1989) Genetic effects of air pollutants in forest tree populations. Springer-Verlag, Berlin, pp 201

Searcy KB, Mulcahy DL (1985) The parallel expression of metal tolerance in pollen and sporophytes of *Silene dioica* L. *Clairv.*, *S. alba* (*Mill.*) *Krause* and *Mimulus guttatus DC.* Theoretical and Applied Genetics 69:597–602

Sinclair WA (1969) Polluted air: potent new selective force in forests. Journal of Forestry 69:305–309

Smith WH (1981) Air pollution and forests: interactions between air contaminants and forest ecosystems. Springer-Verlag, New York

Snaydon RW, Davies MS (1972) Rapid population differentiation in a mosaic environment. I. The response of *Anthoxanthum odoratum* populations to soils. Evolution 24:257–269

Snow AA (1986) Evidence for and against pollen tube competition in natural populations. In: Mulcahy DL, Mulcahy GB, Ottaviano E (eds) Biotechnology and ecology of pollen. Springer-Verlag, New York

Snow AA, Mazer SJ (1988) Gametophytic selection in *Raphanus raphanistrum*: a test for heritable variation in pollen competitive ability. Evolution 42:1065–1075

Solbrig OT (1980) Demography and evolution in plant populations. Blackwell Scientific Publications, Oxford

Stanley RG, Linskens HF (1974) Pollen: biology, biochemistry and management. Springer-Verlag, Berlin

Stebbins GL (1957) Self-fertilization and population variability in the higher plants. American Naturalist 91:337–354

Stephenson AG (1981) Flower and fruit abortion: proximate causes and ultimate functions. Annual Review of Ecology and Systematics 12:253–279

Strauss SH (1987) Heterozygosity and developmental stability under inbreeding and crossbreeding in *Pinus attenuata*. Evolution 41:331–339

Tanksley SD, Zamir D, Rick CM (1981) Evidence for extensive overlap of sporophytic and gametophytic gene expression in tomato. Science 213: 453–455

Taylor GE (1978) Genetic analysis of ecotypic differentiation within an annual plant species, *Geranium carolinianum* L., in response to sulphur dioxide. Botanical Gazette 139:362–368

Taylor GE, Bell JNB (1988) Studies on the tolerance to SO_2 of grass populations in polluted areas. V. Investigations into the development of tolerance to SO_2 and NO_2 in combination and NO_2 alone. New Phytologist 110:327–338

Taylor GE, Murdy WH (1975) Population differentiation of an annual plant species, *Geranium carolinianum*, in response to sulfur dioxide. Botanical Gazette 136:212–215

Templeton AR, Levin DA (1979) Evolutionary consequences of seed pools. American Naturalist 114:232–249

Venne H, Scholz F, Vornweg A (1989) Effects of air pollutants on reproductive processes of poplar (*Populus* spp) and Scots pine (*Pinus sylvestris* L.). In: Scholz F, Gregorius HR, Rudin D (eds) Genetic effects of air pollutants in forest tree populations. Springer-Verlag, Berlin, pp 89–106

Verkleij JAC, Bast-Cramer B, Koevoets P (1989) Genetic studies in populations of *Silene cucubalus* occurring on various polluted and unpolluted areas. In: Scholz F, Gregorius HR, Rudin D (eds) Genetic effects of air pollutants in forest tree populations. Springer-Verlag, Berlin, pp 107–114

Watson MA, Casper BB (1984) Morphogenetic constraints on patterns of carbon distribution in plants. Annual Review of Ecology and Systematics 15:233–258

White J (1984) Plant metamerism. In: Dirzo R, Sarukhan J (eds) Perspectives in plant population ecology. Sinaeur Association, Sunderland, pp 15–47

Willing RP, Mascarenhas JP (1984) Analysis of the complexity and diversity of mRNAs from pollen and shoots of *Tradescantia*. Plant Physiology 75:865–868

Willson MF, Burley N (1983) Mate choice in plants: tactics, mechanisms and consequences. Princeton University Press, Princeton

Wolters JHB, Marten MJM (1987) Effects of air pollutants on pollen. Botanical Review 53:372–414

Zourous ES, Singh SM, Miles HE (1980) Growth rate in oysters: an overdominant phenotype and its possible explanations. Evolution 34:856–867

Population-Level Processes and Their Relevance to the Evolution in Plants Under Gaseous Air Pollutants

FLORIAN SCHOLZ

Introduction

It was Darwin's dilemma that he had derived a new and revolutionary theory that could explain the numerous observations of his journeys and was free from obvious contradictions but was not verified by controlled studies.

Today we are in a different situation. We have a rich body of results from genetic experiments which we can use for discussing hypotheses on evolutionary processes belonging to the past. Furthermore, the byproducts of industrial processes produce environmental regimes, by which we may become witnesses of evolutionary changes in relatively small temporal and large spatial scales.

This can be used for extending the knowledge of the evolutionary processes, but it should be done in regard to our responsibility for the further evolutionary development under anthropogenically altered conditions.

Essential Questions in Evolution Research

In ecological genetic studies of plants under air pollution stress, investigators more often study the development of resistance than the costs associated with the adaptation. Therefore, this Commentary points out some essential questions which are important concerning the loss of genetic diversity and its possible consequences for ecosystems.

First, one should consider limits to the process of adaptation. Some aspects will be discussed here. Along with genetic diversity, heritability of characters, life history characteristics, reproductive systems, and selection pressure, time is the most important feature in evolution. This limiting factor becomes the more important the faster the environment is changing and the longer the reproductive cycles are. Furthermore, the balance of ecosystems may be endangered by the differences in the quickness of evolutionary progress of the involved species. An obvious example is forest ecosystems with their evolutionarily slow tree species and the evolutionarily quick parasite species. The advantage of the latter may impair the host-parasite equilibria.

Second, accepting that in several species there is considerable potential for adaptation to recently occurred pollutants and that this widespread environmental change will be continued, it is necessary to accept that losses of genetic diversity cannot be excluded. The necessary proof of this hypothesis is even more complicated than testing other hypotheses on evolution, since one must prove concrete allelic losses, i.e., the reduction or even the absence of something. This is a challenge for evolutionary biologists and geneticists. However, the rich body of literature on evolution of resistance may be utilized for considerations of losses in genetic diversity as both processes are related.

Last but not least, one must consider ways in which genetic losses affect evolution. Such effects are difficult to investigate since some aspects are stochastic processes (e.g., drift), and it will never be possible to predict the exact path of future evolution. But the boundaries of possible developments are determined by the present genetic structure of the species. Hence, and because of the unpredictable temporal variation of pollutant components and regional climatic changes due to the global greenhouse effect, maintenance of genetic diversity in order to maintain genetic adaptability is the primary requirement.

To date the interest of scientists and foresters has mainly been in resistance to pollutants, its evolution, and its breeding. The more pollution becomes a large-scale problem, the more species and ecosystems are involved, and the more global consequences such as climate change occur, the less is breeding for resistance a solution to detrimental effects on ecosystems. Reading Barret and Bush's contribution (Chapter 6) entitled "Population processes in plants and the evolution of resistance to gaseous air pollutants" and replacing the term "resistance" by "response" will give the piece the more general importance it really has. And, where it applies, replacing the term "evolution of resistance" by "possible losses of genetic diversity" will facilitate focuse on the (often unconsciously) hidden aspect of the costs of evolution of resistance and point out the importance of the respective biological features for possible losses of genetic diversity.

Evolution Potential in Plants

As air pollutants become less severe locally but more common over large regions and influence increasing parts of the ranges of many species, we become increasingly obliged to investigate selective processes within exposed populations. Hence, we have to ask for the relationships between adaptedness, adaptability, and genetic diversity (Gregorius 1989a) and the limits of the processes of adaptation. Barret and Bush (Chapter 6; this volume) discuss limiting factors of adaptability in life history characteristics and reproductive systems which themselves are the result and subject of evolution.

The importance of genetic diversity within species for ecosystems in changing environments becomes obvious when Bradshaw (1984a) summarizes his comprehensive work on adaptation of plants to contaminated soils:

Darwin, and many biologists afterwards, have seen few, if any, limits to the processes of adaptation by evolutionary change. Perhaps we have been conceited. A study of heavy-metal tolerance, and other conditions to which evolutionary adaptation has occurred, should overwhelm us with evidence for limits to the evolutionary process and limits to the adaptation it achieves. These limits clearly arise from restrictions in the supply of genetic variability. Nearly all species are in a condition of genostasis, in which there is a lack of appropiate variability for further evolutionary change.

Although the extent to which genostasis holds for different species may remain open, the importance of sufficient genetic diversity for adaptation to anthropogeneous environmental change is obvious. And many species reveal insufficient genetic diversity in this respect (Bradshaw 1984a).

Along with genetic diversity, time is an important feature in evolution. Environmental conditions change in time and so does the genetic structure of the species during evolution. But just as sufficient genetic diversity is indispensable, sufficient time is a crucial point in evolution. Compared with natural environmental changes during the history of the earth, changes by air pollutants through their toxicity or their influence on climate are rapid. This is of particular importance for long-lived species with long reproductive cycles such as forest trees. Even if the given genetic structure of a species and the inherent potential for evolution would allow adaptation to a single qualitatively or quantitatively new environmental condition, the adaptability to various cooccurring new factors may be limited.

For each species of an ecosystem the other species are environmental factors which themselves will adapt to the changing environmental conditions. Adaptation may affect the relationship between the cooccurring species, whether these relationships be neutral, competitive, symbiotic, or parasitic. The coevolution of species in the various ecosystems is influenced by and results in life cycles and reproductive systems that correspond in a way that enables survival and evolution of all participants. However, if selective forces of new anthropogenic environmental factors or short-term climatic changes require a quicker evolutionary response than some species can afford, the equilibria in the ecosystem may be endangered (e.g., the host-parasite equilibria).

Our knowledge of such coevolutionary aspects of air pollutants is very poor. Life history and reproductive systems represent key information for understanding the ecological genetics of species interacting in ecosystems. This knowledge will help us in managing ecosystems under threat of environmental pollution. Nevertheless, our influence on life history characteristics, e.g., by artificially shortening the reproductive cycle, is limited, and our first requirement is to keep ecosystems stable by ensuring the adaptability potential of the involved species.

Genetic Diversity and Adaptedness

Genetic diversity is not only the basis for evolution in the future, it is also the result of evolution in the past. During that evolution and still under present environmental conditions, adaptedness, at least to a certain extent, is a prerequisite for further evolution. Without sufficient adaptedness, survival and/or reproduction would not occur. Does this coincidence correspond with a relation between adaptedness and genetic diversity? This relation was comprehensively reviewed and discussed by Ledig (1986) for outbreeding plants.

Barret and Bush (Chapter 6, this volume) discuss results of genetic investigations in polluted forest tree populations. Some more results shall be added here. The results are based on investigations of genetic structures in three types of experiments. Type A compares genetic structures of subpopulations with different sensitivities to air pollutants which were scored after controlled fumigation or growth in polluted stands (Bergmann and Scholz 1989; Geburek et al. 1986, 1987; Mueller-Starck and Hattemer 1989). Type B compares the genetic structure of initial populations with that of the residual population after reduction of plant number by selection under detrimental conditions (Mueller-Starck and Hattemer 1989). Type C involves reproduction, thus including feritility selection (Bergmann and Scholz 1989).

In type A investigations with various species, at certain gene loci heterozygous genotypes are more often to be found in the tolerant than in the sensitive population (Mejnartowicz 1983; Geburek et al. 1987; Bergmann and Scholz 1989; Mueller-Starck and Hattemer 1989). According to these results, individual genetic diversity at certain gene loci (i.e., heterozygosity) is advantageous.

Further investigations show that also multilocus diversity was higher in tolerant subpopulations than in sensitive ones; see the examples in Scholz et al. (1989) and Mueller-Starck and Hattemer (1989).

Type B investigations are in accordance with type A results. In different type B populations, during exposure of beech seedlings for 2 years in contaminated soils, the average conditionnal heterozygosity (according to Gregorius et al. 1986) increased by between 4.6% and 21.3%, depending on the tested provenances.

For several reasons, type C investigations have to be interpreted very carefully. First results, however, indicate that heterozygosity at certain loci is advantageous (Bergmann and Scholz 1989).

Theoretically, the advantage of genetic diversity should be revealed especially in heterogeneous environments. Any toxic or stimulating pollutant influences the spatial and temporal environmental heterogeneity. At low levels it will increase the heterogeneity. But with high concentrations it will decrease environmental heterogeneity the more it becomes the prevailing environmental factor. Under such conditions, genetic diversity may be of no advantage or even a load. Such relations between heterozygosity and tolerance may not only depend on pollutant concentration by also on life history characters of the respective plant species. This limits the generalization of results.

On the Evidence of Genetic Losses

One individual is not capable of carrying all alleles of its species. Each individual carries a different sample of alleles, forming their genetic identity and causing the gentic source of phenotypic variation in response to the environment. The alleles are not randomly distributed over the range of the species, thus forming provenance variation. It is obvious that with the extinction of a species its gene pool is lost. It is also obvious that if certain alleles are only present in one provenance, they will be lost if that provenance goes extinct as a result of the effects of air pollutants.

However, this applies only to those few regions where entire populations of the respective species go extinct. Beyond these more obvious cases, a loss of genetic diversity within surviving populations is of more general interest. It may even be of greater importance since medium or low level concentrations over great regions comprise far larger areas than highly polluted regions. Adaptation of surviving populations can be accompanied by gene loss when a pollution regime causes viability and/or fertility selection.

Adaptation to natural and to pollutant stress factors is similar in many respects. In one respect, however, natural stress factors are quite different from anthropogeneous toxic air pollutants. Natural stress factors change only in their occurrence of intensity. Their character, however, has been constant for long periods of earth history. Adaptation to them may have resulted in genetic loss, but, it was a gain which could be utilized in the future with in principle the same factors. This applies also for those anthropogenic pollutants which already naturally existed for ages, but only in those biotopes where they occurred.

For various new pollutants that the respective populations hitherto never experienced, adaptability cannot be assumed without further study. Each situation has to be regarded separately and this applies for the pollutants as well as for the species. Generalization is not possible. Furthermore, each new anthropogenic toxic pollutant has to be regarded individually, each of them creating a new stress and adaptation of its own, if resistant factors are available.

For copper, Bradshaw (1984a) pointed out that sensitive species which had no information in their gene pool for evolving tolerance were excluded form survival. Subsequent stress by different pollutants would probably endanger the whole species community because it is likely that from the species remaining after exposure to one factor, only few can evolve tolerance to a new pollutant while the already excluded species are no longer available. For simultaneous impact of different pollutants the situation is more complex.

The situation within a species may resemble that example with respect to its gene pool. Subsequently acting pollutants may in turn eliminate different parts of the gene pool, thus, further reducing genetic diversity and limiting adaptability.

Forest tree species in Central Europe are an example for selective processes where pollutant conditions are not severe enough to cause extinction. They show decline over a great range of distribution and great individual differences in viability. Selective processes should be expected.

Genecology studies show that adaptation to different environments does not only include changes in allele frequencies but also differences in their presence. In the case of pollutants, losses of alleles cannot be excluded whereever adaptive genes are present. From the investigations on relations between genetic diversity and adaptedness mentioned above, conclusions on selective processes were drawn. The type A experiments showed that certain alleles only or with higher frequency occur in the sensitive subpopulations, thus indicating the danger of gene loss by loss of sensitive trees (Bergmann and Scholz 1989; Geburek et al. 1987; Mejnartowicz 1983; Mueller-Starck and Hattemer 1989). In the type B experiment Mueller-Starck and Hattemer (1989) found that the average number of alleles per locus was reduced from 3.0 in the initial population to 2.69 alleles in the residual population. This means a loss of alleles by 16% within 2 years of growth in that experiment. The consequence is the reduction of the potential to generate genotypic diversity in the next generation. The hypothetical gametic multilocus diversity was reduced by 25%.

Such examples reveal considerable evidence for genetic loss by air pollutants. Loss of genetic multiplicity, however, is a threat to the adaptability of species. The examples reported here are supported by genecological studies at the provenance level in silver fir. Provenances from northern Europe show decline symptoms whereas southern provenances (especially from Calabria) are healthy in their original growing area as is the case with old provenance trials in northern Europe (Larsen 1981). They are also more tolerant to frost and pollutants (Larsen 1986; Larsen et al. 1988). And, in accordance with the results reported above, they have higher genetic diversity (Bergmann et al. 1990). According to Larsen the northern provenances have lost parts of the initial genetic information during their reimmigration after the ice age.

The estimation of the amount and importance of such losses in various species and the consequences for the adaptability of ecosystems are still open questions. Facing the prognosed global climatic change, however, preserving genetic diversity, presumably by controlling air pollutants deserves high priority.

Outlook and Research Needs

For various species the essential prerequisites for selection by pollutants, genetic variability in response, viability, and fertility are fulfilled. Experimental results support the hypothesis that selection really takes place—even under moderate pollution regimes. Also, changes in the genetic structure not only may occur for allele frequency but also include loss of alleles.

Hitherto, most investigations on genetic aspects were carried out to understand the process of evolution of tolerance and to breed for tolerance. Now, when pollution is no longer a local problem, the research interest shifts to the fundamental prerequisite of evolution, the genetic diversity and its importance for adaptability of species and ecosystems. In this urgent field of research few results on few species are available. At present, the most appropriate tool for such investigations are isozymes as gene markers (Bergmann et al. 1989). Considerations on the sampling procedure are given by Gregorius (1989b), and most knowledge gaps on genetic effects of air pollutants on forest ecosystems as reported by Karnosky et al. (1989) are also valid for other ecosystems.

If our knowledge of ecological genetics can be expanded and appropriately used, it may contribute to the protection of ecosystems. This also means that air pollution control should consider protection of genetic systems, thus taking into account the maxim of "the importance of evolutionary ideas in ecology—and vice versa" (Bradshaw 1984b). For this purpose the genetic systems of the endangered species should be investigated, i.e., the generation, storage, modification, and inheritance of genetic diversity. And it is necessary to investigate causes and mechanisms of threats to these systems, to estimate the potential genetic loss and its consequences in order to conceive measures for genetic management in biological conservation.

Acknowledgments. The author gratefully appreciates the suggestions of T. Geburek, H-R Gregorius, L. Pitelka, H. Venne, and an anonymous reviewer for improvement of the manuscript.

References

Bergmann F, Gregorius HR, Larsen JB (1990) Levels of genetic variation in European silver fir (*Abies alba*). Are they related to the species' decline? Genetica 82:1–10

Bergmann F, Gregorius HR, Scholz F (1989) Isoenzymes, indicators of environmental impacts on plants or environmentally stable gene markers? In: Scholz F, Gregorius HR, Rudin D (eds) Genetic effects of air pollutants in forest tree populations. Springer-Verlag, Berlin, pp 17–25

Bergmann F, Scholz F (1989) Selection effects of air pollution in Norway spruce (*Picea abies*) populations. In: Scholz F, Gregorius HR, Rudin D (eds) Genetic effects of air pollutants in forest tree populations. Springer-Verlag, Berlin, pp 141–160

Bradshaw AD (1984a) Adaptation of plants to soils containing toxic metals—a test for conceit. Origins and development of adaptation. Libu Foundation Symposium 102. Pitman, London, pp 4–19

Bradshaw AD (1984b) The importance of evolutionary ideas in ecology—and vice versa. In: Shorrocks B (ed) Evolutionary ecology. 23rd Symposium of the British Ecology Society Leeds 1982. Blackwell, London, pp 1–25

Geburek T, Scholz F, Bergmann F, (1986) Variation in aluminum sensitivity among *Picea abies* (L.) Karst. seedlings and genetic differences between their mother trees as studied by isozyme-gene-markers. Angew Bot 60:451–460

Geburek T, Scholz F, Knabe W, Vornweg A (1987) Genetic studies by isozyme gene loci on tolerance and sensitivity in an air polluted *Pinus sylvestris* field trial. Silvae Genetica 36:49–53

Gregorius H-R (1989a) The importance of genetic multiplicity for tolerance of atmospheric pollution. In: Scholz F, Gregorius H-R, Rudin D (eds) Genetic effects of air pollutants in forest tree populations. Springer-Verlag, Berlin, pp 163–172

Gregorius H-R (1989b) The attribution of phenotypic variation to genetic or environmental variation in ecological studies. In: Scholz F, Gregorius H-R, Rudin D (eds) Genetic effects of air pollutants in forest tree populations. Springer-Verlag, Berlin, pp 3–15

Gregorius H-R, Krauhausen J, Mueller-Starck G (1986) Spatial and temporal genetic differentiation among the seed in a stand of *Fagus sylvatica* L. Heridity 57:255–262

Karnosky DF, Scholz F, Geburek T, Rudin D (1989) Implications of genetic effects of air pollution on forest ecosystems—knowledge gaps. In: Scholz F, Gregorius H-R, Rudin D (eds) Genetic effects of air pollutants in forest tree populations. Springer-Verlag, Berlin, pp 199–201

Larsen JB (1981) Waldbauliche und ertragskundliche Erfahrungen mit verschiedenen Provenienzen der Weißtanne (*Abies alba* Mill.) in Dänemark. Forstw Cbl 100:274–286

Larsen JB (1986) Die geographische Variation der Weißtanne (*Abies alba* Mill.). Wachstumsentwicklung und Frostresistenz. Forstw Cbl 105:396–406

Larsen JB, Quian XM, Scholz F, Wagner I (1988) Ecophysiological reactions of different provenances of European silver fir (*Abies alba* Mill.) to SO_2 exposure during winter. European Journal of Forest Pathology 18:44–50

Ledig FT (1986) Heterozygosity, heterosis, and fitness in outbreeding plants. In:

Soule (ed) Conservation biology: the science of scarcity and diversity. Sinauer Association, Sunderland, pp 77–104

Mejnartowicz LE (1983) Changes in genetic structure of Scots pine (*Pinus sylvestris* L.) population affected by industrial emission of fluoride and sulphur dioxide. Genetica Polonica 24: 41–50

Mueller-Starck G, Hattemer H-H (1989) Genetische Auswirkungen von Umweltstreß auf Altbestände und Jungwuchs der Buche (*Fagus sylvatica* L.). Forstarchiv 60:17–22

Scholz F, Gregorius HR, Rudin D (1989) Genetic effects of air pollutants in forest tree populations. Springer-Verlag Berlin, pp 201

7
Consequences of Evolving Resistance to Air Pollutants

WILLIAM E. WINNER, JAMES S. COLEMAN,
CHRISTOPHER GILLESPIE, HAROLD A. MOONEY, and
EVA J. PELL

Introduction

The prospects are high that plant populations are evolving in response to air pollutants. Evidence for this (Table 1) exists in a number of isolated observations (e.g., Houston and Stairs 1973; Horsman and Wellburn 1977; Ayazloo and Bell 1982; Ayazloo et al. 1982; Berang et al. 1986; Pitelka 1988). For example, wild populations of *Geranium* sp. growing either proximal to, or distant from, an SO_2 point source were found to differ in sensitivity to this pollutant. As expected growth of individuals from the population growing close to the pollution source was less sensitive to SO_2 than the growth of individuals from the population growing in clean air (Taylor and Murdy 1975; Taylor 1978, 1981). Examples also exist in which agronomists have inadvertently developed crop lines which are progressively less sensitive to air pollutants. *Lolium perenne* L. cultivars exist in England which are less sensitive to SO_2 than ancestral cultivars, with this change in SO_2 response largely attributed to agronomic breeding programs (Horsman et al. 1979). From these, and other examples discussed in other chapters (Bell et al. Chapter 3, this volume), the genetic basis for plant responses to air pollution has been established, along with the idea that air pollution resistance is heritable (Table 2).

The idea that plants may evolve in response to air pollution stress is provocative and raises a number of fundamental questions of importance to foresters, agronomists, botanists, environmental scientists, and those charged with managing natural resources. This chapter focuses on an important question which must be included when discussing the possibility that plants may evolve resistance to air pollutants: Will the process of evolution necessarily result in vegetation that is similar to that of today but which is also unaffected by ambient air pollutants? The significance of this question is obvious. Some may suggest that since the process of evolution will produce vegetation that is resistant to air pollutants, it is unnecessary to either regulate air pollution emissions or study plant responses to this stress. Such work would be unnecessary because the pro-

TABLE 1. Examples of intraspecific variation in resistance to O_3, SO_2, and other pollutants.

Pollutant	Life form	Species
Ozone	Trees	*Populus tremuloides* MichX.
		Pinus taeda L.
		Picea abies (L.) Karst.
	Herbs	*Lolium perenne* L.
		Glycine max (L.) Merr.
		Zea mays L.
Sulfur dioxide	Trees	*Picea abies* L.
		Pinus sylvestris L.
		Populus tremuloides MichX.
	Herbs	*Loluim perenne* L.
		Glycine max (L.) Merr.
		Zea mays L.
Hydrogen flouride	Trees	*Pinus sylvestris* L.
	Herbs	*Zea mays* L.

From Pitelka 1988; Roose et al. 1982.

TABLE 2. Inheritance of air pollution resistance of various species.

Species	Air pollutant
Allium cepa L.	O_3
Geranium carolinianum L.	SO_2
Nicotiana tabacum L.	O_3
Petunia hybrida Vilm.	O_3, PAN
Phaseolus vulgaris L.	O_3
Pinus strobus L.	SO_2, O_3
Pinus taeda L.	O_3
Populus tremuloides MichX.	O_3, SO_2
Zea mays L.	O_3

From Roose et al. 1982.

cess of natural selection will act as a tool which brings mitigation of air pollution damage.

Although the process of natural selection will result in the selection of species, or genotypes within species, that are resistant to air pollution, this is likely to occur with some consequences to individual plants, plant populations, and ecosystems. For example, natural selection resulting in plant genotypes resistant to heavy metals in mine spoils occurs within a few generations (Bradshaw and McNcilly, Chapter 2, this volume; Antonvics et al. 1971). However, there are at least two severe consequences of this selection. First, heavy metal resistant plants are not able to grow as well or compete with nonresistant plants under conditions where heavy metals were not present, suggesting a physiological cost of resistance. Second, the strong selective force of heavy metal pollution removes a great deal of the biological and genetic diversity that previously existed in populations and communities. Thus, although natural selection resulted in resistance to heavy metal pollution, it leaves a small amount of genetic variation for natural or even agricultural selection to act upon in the future.

Since it is clear that air pollution can act as a powerful selective force on plants, it is important to consider consequences that may arise from the natural selection of air pollution resistance. In some cases these "costs" may be measured in terms of lost metabolites or productivity. In other cases, we may infer costs in a more general sense in that the evolution of resistance to one stress may in some way reduce the capacity of plants to resist or evolve resistance to other stresses. A reduced ability to evolve resistance to other stresses is related to reductions in the amount of genetic variation that occurs as a result of intense natural selection (see Schmalhausen 1949; Simpson et al. 1960).

We have chosen to focus on three specific areas in which a cost of air pollution resistance may be important:

1. At the leaf level, evolution of air pollution resistance is apt to have physiological consequences. What are the resistance mechanisms and what are the consequences?
2. At the whole plant level, air pollution resistance is apt to compromise the capacity of the plant to tolerate other stresses that co-occur with air pollution or to change the interactions of plants with other organisms (i.e., pathogen, herbivores, symbionts, other plants). How will these types of interactions affect our capacity to predict plant responses to combinations of evironmental stresses?
3. At the population level, evolution of air pollution resistance is apt to occur at the expense of those genotypes most sensitive to air pollutants resulting in unpredictable shifts in the gene pools and a severe loss of genetic variation in the population of affected species. Can we develop conceptual models of expected shifts in gene pools and the consequences of altered genetic variation?

The objectives of this chapter will be to focus on these issues that pertain to leaves, whole plants, and populations. Consequently, the content will involve discussions of stomatal and photosynthetic responses to air pollutants, effects of air pollutants on carbon allocation and partitioning patterns of plants, and potential for air pollutants to alter gene pools and genetic variation within populations.

We have also chosen to emphasize ideas built around plant responses to either SO_2 or O_3. Sulfur dioxide was selected because this air pollutant has unique attributes as a selective force (Roose et al. 1982; Karnosky 1985; Coleman et al. 1990). Ozone is discussed because this pollutant is viewed as the one which poses the greatest threat to plant productivity in the

United States (US EPA 1986). However, many other gaseous air pollutants such as HF, NO_2, H_2S, and NH_3, have the potential to act as selective forces. In addition, human activity is resulting in other environmental
changes, such as the deposition of heavy metals, increases in UVB radiation as a result of decreased stratospheric O_3, and increases in global
atmospheric CO_2 concentration. However, with the exception of studies
on plants that invade metal-laden mine tailings (e.g., Bradshaw 1952;
Antonovics et al. 1971), little is known about the effects of many of these
anthropogenic stresses on the genetics of plant populations. Our intent is
to develop material around the analysis of plant responses to SO_2 and O_3
in a general way so that many of the ideas are relevant to any air pollutant.

Finally, the approach of this paper is to develop the idea that there may
be costs to plants for the evolution of air pollution resistance. Although the
focus is on costs, it is important to keep in mind that selection is impartial
with respect to the cause of environmental change. More specifically,
selection acts equally upon variation in plant responses to environmental
change regardless of whether human activity is influencing the habitat.
The process of natural selection is not to be judged as either a positive or
negative force, but rather regarded as simply inescapable.

Costs of Leaf Level Increases in Air Pollution Resistance

The Process of Leaf Level Resistance

Increases in air pollution resistance of single leaves are recognized both
in the short term, as products of acclimation, and in the long term, as
products of evolution. For example, plants that rapidly acclimate to SO_2
and O_3, perhaps through the induction of a specific enzyme or by rapidly
closing stomata, may show the greatest fitness in a population of plants
that experience high concentrations of O_3 or SO_2. Consequently, their
genotype would be selected for under those conditions. However, it is
also possible that the evolution of leaf-level resistance to air pollution
will not involve acclimation. For example, plants with leaves containing
a constitutive enzyme that detoxifies an air pollutant might be selected
over plants whose leaves do not contain this compound in a polluted
environment.

One problem in studying the evolution of leaf level resistance to air
pollution is defining the criteria for assessing resistance. Natural selection
is based on variation in fitness between individuals, and leaf level characters may or may not relate easily to the reproductive performance of
plants (Lechowicz 1984). Photosynthesis is often used as the criteria for
assessing resistance because measurements of this character are nondestructive, reveal dynamic patterns of physiological response to exposures,

and quantify effects of exposures on rates of carbon gain, which is fundamentally important to plant function (see Winner 1987).

Although links between photosynthesis, growth, and reproduction are complex and involve many aspects of carbon allocation (e.g., Mooney 1972; Lechowicz 1984; Mooney et al. 1988; Coleman et al. 1989), photosynthesis is clearly related to the processes of growth, onset of reproduction, and fitness. Another criterion that has been used to assess leaf-level responses to SO_2 and O_3 has been visible foliar injury (see Berang et al. 1986). However, this measure is often unrelated to changes in plant growth or to any plant process thought to be important to reproductive fitness (Winner et al. 1987). Without a direct connection to reproductive performance, it is almost impossible to connect changes in visible leaf injury to the natural selection of air pollution resistance in plant populations. Consequently, this chapter focuses on changes, in leaf-level resistance of photosynthesis because there is at least a partial link between net carbon gain and reproductive performance (Mooney 1972).

Two general mechanisms are known to account for both short-and long-term changes in photosynthesis as a result of air pollution. One mechanism involves stomata (Tingey and Andersen, Chapter 8, this volume). For example, leaves which have either low stomatal conductance or close stomata in response to exposure to air pollutants often have less sensitivity to SO_2 or O_3, with regard to long-term carbon gain, than leaves with continuously high stomatal conductance (see Winner 1989 and references therein). A second mechanism involves processes which take place in the leaf mesophyll: once air pollutants are absorbed through stomata and enter the mesophyll, their impacts on biochemical processes can be analyzed. Factors which might influence the sensitivity of the mesophyll to air pollutants include enzymes that detoxify pollutants or the sequestering of absorbed pollutants into vacuoles (see Horsman and Wellburn 1977; Pell 1979; Garsed 1985; Heath 1987). Thus analysis of costs of leaf-level changes in air pollution resistance must include discussions of both stomatal and mesophyll factors.

Stomatal Factors

Many examples exist in which leaves with high conductance are more sensitive to air pollutants than are leaves with low conductance (reviewed in Black 1985; Winner 1989). High conductance likely evolved to maximize carbon gain and plant performance in environments without water limitations or air pollution (see Larcher 1980). Yet, there is clearly a cost of high conductance when plants are forced to respond to air pollutants; plants with high conductance absorb a greater amount of air pollutants than plants with lower conductance and subsequently receive more damage. For example, low stomatal conductance results in low rates of air pollution absorption and is often associated with high air pollution resistance. This was shown to be true for deciduous and evergreen Californian shrubs as well as for C_3 and C_4 *Atriplex* species, even though the intrinsic properties of stomata differ within these groups of plants (Winner and Mooney 1980a, 1980b, 1980c).

Species also differ in stomatal responses to air pollutants. For example, SO_2 can cause increases in stomatal conductance for some species, such as *Vicia faba* L. (Black 1985), and decreases in conductance for other species, including crops shrubs and trees (Winner 1989). Furthermore, the response of stomata to SO_2 may be dependent on other environmental factors. Stomata of *Vicia faba* L. open when plants are exposed to SO_2 and high relative humidity but close when exposed to SO_2 and low relative humidity (Black 1985). Stomatal responses to O_3 are less well studied than those for SO_2, but a survey of studies shows that O_3 almost always causes stomatal closure in a number of crop and forest species (Winner 1989).

One way to gauge the importance of stomatal responses to SO_2 in the evolution of plant resistance to this pollutant is to examine stomatal responses of plant species that have evolved in environments with naturally high levels of SO_2. These environments include areas that are proximal to active volcanoes (Winner and Mooney 1980d, 1985). *Meterosiderous collina* Rock is a tree that occupies a habitat in Hawaii where periodic exposures to high concentrations of volcanic SO_2 have occurred with some frequency for hundreds of years. Older leaves of this species closed their stomata when exposed to SO_2-containing volcanic fumes and subsequently survived an eruption (Winner and Mooney 1980d). Young leaves, however, maintained high conductances in the presence of volcanic fumes and subsequently died. Consequently, leaf survival, and perhaps plant survival, was likely related to reduced uptake of fumes by old leaves (Winner and Mooney 1980d). It is likely that natural selection in this environment has acted to produce *Meterosiderous* plants that have some leaves which are capable of closing their stomata in response to the presence of volcanic fumes.

Consequences of Resistance Due to Stomatal Factors

Selection by natural processes or by agronomists could lead to increased numbers of forest and crop plants which close their stomata when exposed to air pollutants and are therefore resistant to noxious gases. Unfortunately, the evolution of mechanisms assuring stomatal closure during air pollution exposure will create new problems in the relationship between plants and their environment.

Plants continuously adjust stomata in response to many factors, such as light, water, nutrients, and temperature in order to maximize carbon gain (photosynthesis) while minimizing water loss (transpiration) (see Larcher 1980). Changes in conductance caused by O_3 would have the effect of moving conductance from this optimal value and would therefore compromise either rates of carbon gain or water loss. Since stomatal closure results in greater relative decreases in photosynthesis than in transpiration for C_3 plants (Fig. 1), O_3-caused reductions in conductance are apt to have the largest detrimental impact on carbon metabolism. More specifically, the benefit of O_3-caused reductions in stomatal conductance would be in terms of reduced O_3 absorption rates, and the consequence of this response would often be in terms of reduced photosynthetic rates.

Nevertheless, reductions in transpiration as a result of stomatal closure could also have serious effects on leaves through changes in leaf energy balance. Transpiration is one mechanism that reduces the heat load on leaves, so severe reductions in transpiration during extended periods of high levels of incoming solar radiation could cause leaf temperature to increase to the point where enzyme degradation begins to occur. This possibility is of concern in environments with O_3 pollution, because high levels of O_3 are often associated with bright sunny days with large amounts of incoming solar radiation (SU EPA 1986). Consequently, another cost of air pollution-caused stomatal closure could be a reduced capacity for leaves to cool themselves via transpiration. This would subsequently reduce carbon gain or even leaf longevity.

An interesting possibility exists that plants could evolve the capacity to close stomata only when O_3 concentrations exceeded dangerous, threshold values or when leaves were at a physiological state where they were most

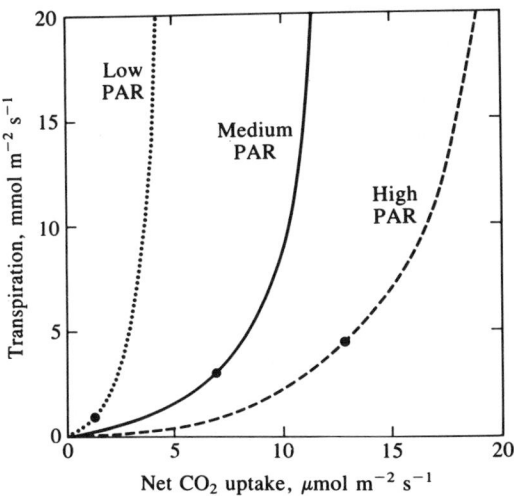

FIGURE 1. Relation between net photosynthesis (J_{CO_2}) and transpiration (J_{WV}) as stomatal conductance is varied (Nolbel 1983). The three *curves* depict various PAR levels, indicated as "low," "medium," and "high." The *circles* indicate where the slope $\delta J_{WV}/\delta J_{CO_2}$ is 1,000 H_2O/CO_2 ($\delta J_{CO_2}\, \delta J_{WV} = 0.0010\ CO_2/H_2O$). Cuticular transpiration is ignored.

vulnerable to O_3. It is also possible that evolution could result in leaves which did not completely close their stomata in the presence of O_3, but rather closed stomata only to the degree necessary to reduce O_3 absorption rates to the maximum acceptable value with regard to tissue damage. Even in these situtations, there would be a consequence (e.g., reduced carbon gain; nonoptimal ratio of carbon gain to water loss) of resistance to air pollution conferred by the mechanisms of O_3-caused stomatal closure in comparison with plants grown in clean air.

Nonstomatal Factors

Most of the effects of air pollution on plants occur after pollutants are absorbed into the leaf mesophyll. As a pollutant passes through the cell wall and then cell membrane, it may undergo oxidation or reduction depending upon the chemical character of the compound (see Heath 1987). Often free radicals are generated as a result of these primary reactions (Heath 1987; Richardson et al. 1989; Pell and Dan 1990). Thus, significant toxicity is attributed to reactions between cellular constituents and secondary reaction products. Resultant effects include altered membrane integrity, enzyme function, and hormone balance (Pell 1979; Heath 1987). These effects are manifest as physiological effects such as reduced photosynthesis, changes in carbon partitioning, and enhanced rates of senescence (Pell 1979).

Evolving resistance to absorbed air pollutants will clearly involve processes that reduce the toxicity of absorbed pollutants, resulting in plants with a number of cellular and molecular mechanisms for reducing toxic effects of O_3 and SO_2. These mechanisms can be constitutive (i.e., expressed irrespective of the presence of air pollution) or induced (i.e., are only present after exposure to pollutants). For example, *Solanum tuberosum* cultivars that are susceptible to O_3 have more tritratable sulfhydryl groups in rubisco than the native rubisco protein from O_3-resistant cultivars. These constitutive differences in the structure of rubisco molecules may account for the differences in vulnerability to oxidation and subsequent proteolysis that cells from these different cultivars exhibit (Pell and Dan 1990). There are also examples where inherent differences in the metabolic rate of plants influence their response to air pollution. For example, drought-deciduous shrubs that are native to California have a higher metabolic rate and intrinsic photosynthetic capacity than co-occurring evergreen shrubs. Consequently, the evergreen shrubs were more resistant to SO_2 than the drought-deciduous shrubs (Winner and Mooney 1980a, 1980b, 1980c).

Evolution may produce mechanisms by which plants can resist or minimize the toxic effects of absorbed air pollutants, and the mechanisms that exist can differ widely across plant species (e.g., Tingey et al. 1976; Guderian et al. 1985; Tingey and Andersen, Chapter 8, this volume).

Cells can increase levels of enzymes like peroxidase, superoxide dismutase, or glutathione reductase in response to elevated levels of certain stresses like O_3 and SO_2 (e.g., Horsman and Wellburn 1977; Alscher and Amthor 1987; Richardson et al. 1989). Nonenzymatic antioxidant compounds including glutathione and ascorbate also are known to increase in the presence of air pollution stress and their increase may be under genetic control (e.g., Alscher and Amthor 1987). Alternatively, plant cells may minimize air pollution effects by rendering the toxins unavailable to the cell by sequestering absorbed pollutants in vacuoles and isolating them from sites of metabolism. For example, some plant species exposed to SO_2 accumulate large quantities of various oxidized sulfur ions in vacuoles (Garsed 1985).

Although it is clear that differences in the sensitivity of different species or cultivars to air pollution can be explained by some of the mechanisms described above, there can be substantial variation in the resistance of different aged leaves from a single plant to air pollution. For example, leaves of plants with highest photosynthetic capacities are often the most sensitive to air pollution absorption (Reich 1983; Coleman 1986). Furthermore, compounds that can detoxify pollutants (e.g., polyamines and superoxide dismutase) are present in higher concentrations in young leaves than in old foliage (Smith 1984 and references therein), and very young tissue is more O_3 resistant than older foliage on the same plant (see Pell 1979; Reich 1983). Since air pollution resistance mechanisms may differ from leaf to leaf, or since plants may only employ a given mechanism on leaves of a specific age, cost estimates of air pollution resistance mechanisms might involve assessment of individual leaves.

Costs of Resistance Due to Nonstomatal Factors

Costs of constitutive characteristics of air pollution resistance will differ for several reasons. For example, synthesizing a larger amount of a constitutive antioxidant might enable a leaf to survive an O_3 episode but also cause the plant to invest carbon in that compound at the expense of other plant tissues or even plant growth and reproduction. This investment, however, may confer plant resistance to other types of environmental stress, so the actual cost should be weighed against the benefits that an air pollution resistance mechanism confers in given environments. Other constitutive characteristics may be one-time investments in a genetic change in molecular structure which provides lasting air pollution resistance. Such costs are minimal for maximum gain if the molecular changes do not carry reductions in other critical functions. The induction of biochemicals or processes responsible for detoxification of air pollutants also carry a cost. Thus, creating vacuoles to store toxic ions and energetically moving toxins into those vacuoles, producing enzymes to detoxify toxic ions, or even cellular repair mechanisms require increased investment of metabolic resources that are diverted from growth and reproduction. However, in order for these diversions of resources to be of selective advantage, they must result in plants which ultimately have greater fitness in the presence of air pollution than plants that do not exhibit these mechanisms. Thus, a cost (e.g., reduced growth) resulting from this diversion of resources leading to air pollution resistance might only be evident in environments without air pollution.

Resistance to air pollutants that result from changes in plant chemistry could have a number of ecological "costs." In one case, a number of pollinating insects use specific chemical cues to find suitable host flowers, while a number of herbivorous insects are either stimulated or deterred from feeding by the presence or absence of specific biochemicals (see Jermy 1966, 1984). Consequently, a slight change in the biochemistry of plant foliage or any plant tissue could have a large effect on pollinators and herbivores which would then immediately feed back to affect plant fitness, and these changes can be quite complicated. In another case, a short-term O_3 fumigation caused a small reduction in the production of phenol glycosides in cottonwood leaves with a concomitant increase in the deposition of phenolic compounds on to cell wall and membrane material (Jones and Coleman 1989, 1990). Decreased phenol glycosides apparently stimulated the feeding of a leaf beetle; the deposition of phenolics reduced the ability of beetles to digest leaves and reduced beetle growth rates, and a yet unknown change reduced the acceptability of foliage for oviposition by female beetles (Coleman and Jones 1988a; Jones and Coleman 1988).

The effects of air pollution on foliar chemistry and herbivory could continue on through an ecosystem so that the interactions within insect communities and interactions between insects and their predators could change. For example, although a leaf beetle showed drastic responses to changes in cottonwood leaf chemistry as a result of O_3 fumigation, a leaf aphid did not (Coleman and Jones 1988a, 1988b). Thus, it is likely that the relative abundance of beetle versus aphid populations could show large changes as a result of O_3 fumigation of cottonwood and subsequent changes in leaf chemistry. Additionally, the chemical quality of leaves is a major determinant of their decomposition rate (Mellilo et al. 1982). Conceivably, air pollution-induced changes in biochemistry could alter the rate of leaf decomposition as well as the rate at which nutrients cycle in an ecosystem.

Changes in leaf-level resistance to air pollutants may even scale up to affect the structure and productivity of ecosystems. For example, it is clear that leaves from the most productive plants, those with high photosynthetic capacities and metabolic rates, are the most susceptible to air pollution. Thus, the implications for selection are straight forward. Air pollutants, like many severe abiotic stresses (see Grime 1979), will select against plants with high photosynthetic capacities and metabolic rates and for plants with lower metabolic rates. In the California shrub community, this might mean that the rapidly metabolizing drought deciduous shrubs would be replaced by the less productive broad leafed evergreen species (Winner and Mooney 1980a, 1980b, 1980c). Thus, one major cost of evolving resistance to air pollution absorption would be a reduction in intrinsic metabolic capacity and a likely reduction in growth rates.

Calculating Construction Costs of Tissues

The evolution of mechanisms necessary to resist biochemical damage from absorbed air pollutants will bring biochemical changes to plants which reflect both the detrimental effects of the stress itself as well as the compensatory response of the plant. The cost of these responses can be calculated and expressed in glucose equivalent units. Glucose equivalent units are simply the grams of glucose needed to construct a gram of a specific constituent. This cost has been worked out for most plant constituents such as carbohydrates, organic acids, fatty acids, proteins, and amino acids (Table 3). Thus, as air pollutants cause changes in plant metabolism and therefore changes in the constituents of tissue, the cost of change with regard to constructing compounds can be calculated from biochemical analysis of specific plant compounds. However, it is still

TABLE 3. Glucose equivalents, construction costs, and biosynthesis efficiencies for carbohydrates.

Compound	Molecular mass (g/mol)	GE' (g glu/g)	Cost (g glu/g)	Biosynthetic efficiency (E_B)
Carbohydrates				
Glucose	180	1.000	1.000	1.000
Fructose	180	1.000	1.028	0.973
Mannose	180	1.000	1.028	0.973
Galactose	180	1.000	1.056	0.947
Lactose	342	1.053	1.082	0.973
Cellulose	(162)[c]	1.111	1.173	0.947
Hemicellulose[a]	—	1.132	1.296	0.874
Hemicellulose[b]	—	1.132	1.205	0.939
Sucrose	342	1.053	1.096	0.960
Starch	(162)[c]	1.111	1.173	0.947
Ribose 5-P	230	0.652	0.671	0.973

Adapted from Williams et al. 1987.
Values are also determined for other plant constituents such as organic acids, fatty acids and triglycerides, lignins, monoterpines, tannins, amino acids, proteins, nucleic acids, and other nitrogenous compounds.
[a] Hemicellulose with residue composition reported by Bauer et al. 1973. Myoinositol pathway used.
[b] As above but with pathways employing dehydrogenases.
[c] The molecular mass of the monomer in the polymer.

difficult to calculate the costs associated with the maintenance of the cellular machinery necessary to produce new compounds, as well as the opportunity costs associated with investing resources in one function as opposed to another (Chapin 1989).

The task of calculating energetic costs of individual plant tissues is burdensome, fraught with technical problems, and prone to errors. Consequently, other approaches have been developed. One approach is to analyze the elemental composition of plant tissue (Williams et al. 1987). If done for the elements of carbon, hydrogen, nitrogen, and oxygen, then equations can be used to calculate total costs of plant tissues in glucose equivalents. Table 4 shows an example of a elemental analysis of three species of chaparral shrub leaves where the effects of leaf aging are evident.

The equations used to calculate costs of plant tissues on the basis of elemental analysis have been tested against measurements of the heat of combustion for the same tissues (Fig. 2). The close regression between calculated tissue costs and heat of combustion values indicates the usefulness of analyzing elemental analysis. Thus, elemental analysis can be used to document costs of constructing plant tissues, provide baseline information about the chemical changes in tissues accounting for changes in construction costs, and show that heat of combustion measurements can

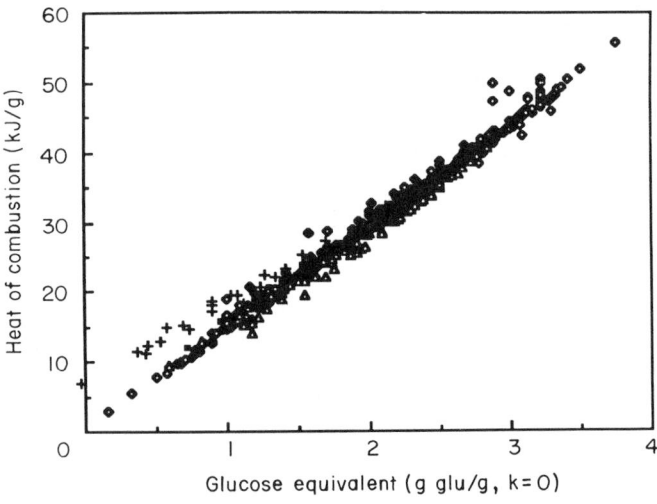

FIGURE 2. The relationship of heat of combustion (ΔH_c) to GE' for 545 organic compounds (Williams et al. 1987). Data for compounds containing nitrogen bonded directly to oxygen ($n = 22$), other forms of nitrogen ($n = 102$), or sulphur ($n = 134$) are displayed separately from those containing only carbon, hydrogen and/or oxygen ($n = 287$), as indicated. CH(O) = $\diamondsuit$; Ch(O)N = $\blacklozenge$; CH(ON)-NO$_x$ = +; CH(ON)S ($m = 4$) = $\triangle$, CH(ON)S ($m = 6$) = $\blacksquare$.

TABLE 4. Heat of combustion and elemental composition of chaparral shrub leaves. Five leaf age classes are represented.

	ΔH_c (ash-free) (kJ/g)	Kjeldahl N (%)	Percent composition					Percent recovery
			C	H	N	O	Ash	
Lepechinia calcina	(Benthe.) Epl.							
I[a]	22.44	2.65	48.66	6.65	2.76	35.83	6.51	100.41
II	21.48	2.92	48.46	5.99	3.11	35.64	6.88	100.08
III	—	—	45.00	6.00	2.41	—	11.30	—
IV	20.51	2.71	44.08	5.40	2.55	35.94	8.51	96.48
V	20.36	2.39	43.72	5.85	2.30	36.50	10.29	98.66
Diplacus aurantiacus	(Curtis) Jeps.							
I	23.67	1.93	51.24	6.93	1.70	33.95	3.65	97.47
II	22.68	1.90	51.06	6.34	1.99	35.45	4.26	99.10
III	22.04	1.91	49.20	6.88	1.59	35.88	4.29	97.84
IV	21.47	1.85	48.35	6.89	1.63	36.54	5.49	98.90
V	20.84	1.57	46.26	6.39	1.32	37.41	6.33	97.71
Heteromeles arbutifolia	(Ait.) Roem.							
I	20.28	2.33	46.48	6.43	2.03	38.05	4.90	97.89
II	20.52	1.84	46.45	6.60	1.49	38.10	4.80	97.44
III	20.64	1.51	46.39	6.64	1.37	38.65	5.65	98.70
IV	20.49	1.13	46.34	6.84	1.01	39.39	5.15	98.73
V	20.44	0.85	47.23	6.84	0.73	39.22	4.35	98.37

[a] I = youngest; V = oldest.

also be used to calculate tissue construction costs. The relationship between percentages of carbon, hydrogen, and oxygen in tissues can help provide clues as to changes in specific biochemical compounds in tissues due to stresses such as O_3 and SO_2. These techniques for assessing tissue construction costs have yet to be applied to plants stressed with air pollutants and may represent a promising approach for defining the true energetic and biochemical costs of resistance to air pollution.

Costs of Evolution of Whole Plant Resistance to Air Pollution

The Process of Whole Plant Resistance

Whole plant responses to air pollutants are the product of responses of various plant parts, including leaves, roots, shoots, and reproductive structures. Shifts in plant resources between tissues may have adaptive values, and plants that have evolved air pollution resistance mechanisms might be expected to have a greater capacity for partitioning biomass between tissues than do nonadapted plants. For example, it is generally thought that species with greater inherent plasticity in their allocation of resources between plant parts are better able to grow in a wide range of environments (see Grime 1979).

The single response of whole plants which has received the greatest attention is that of carbon partitioning between plant roots and shoots (see Lechowicz 1987; Miller 1987; Mooney and Winner 1990). Partitioning of carbon at the whole plant level is of interest because partitioning is the fundamental link between photosynthesis, growth, and reproduction (Mooney 1972; Mooney and Chiariello 1984), it defines plant architecture, and is known to be sensitive to many environmental factors (Mooney and Winner 1991), including air pollutants.

Links between stress-caused changes in photosynthesis, growth, and reproduction are complex. For example, compared with control plants, air pollution may cause a 10% reduction in photosynthesis throughout the life of the plant without causing a 10% decrease in final plant dry weight (Lechowicz 1987; Mooney et al. 1988). This may be because plants can compensate for reduced carbon exchange capacity by producing a greater proportion of their biomass as leaves thereby reducing the loss of growth potential (see Mooney et al. 1988). A reduction in photosynthesis and the compensatory response of decreased root/shoot ratios (i.e., the plant allocates more to shoots) are common responses of plants to above ground stresses such as low light or air pollution (Mooney and Winner 1990). Alternatively, below ground stresses (such as limiting water or nutrient availability) usually cause increases in root/shoot ratios (Szaniawski 1987; Mooney and Winner 1991) allowing the plant to have a greater root surface area for absorption of water and nutrients. Consequently, since these stress responses may have adaptive value (see Bloom et al. 1985), it is possible that air pollution might select for greater plasticity in the partitioning of resources between plant parts.

Plants may also be able to change allocation of their resources to various physiological processes, such as growth, maintenance respiration, production of secondary compounds, and luxury consumption of nutrients, in response to air pollution stress (Bloom et al. 1985). For example, radish plants exposed to SO_2 had reduced rates of night-time root respiration and luxury uptake of nitrate. This response apparently compensated for reduced carbon gain because these plants grew as well as control plants as long as nitrate availability remained high (Coleman et al. 1989). Furthermore, cottonwood plants that were exposed to O_3 showed reduced photosynthesis and production of phenol glycosides without a reduction in total growth. This suggests that reduced production of the secondary compound might have compensated to some degree for reduced carbon gain (Jones and Coleman 1989, 1990). Thus, plasticity in resource allocation may also have adaptive value for plants exposed to air pollution, and selection could act to favor plant genotypes that exhibit the ability to easily shift resource allocation between different physiological processes.

Consequences of Whole Plant Resistance

Air pollution-caused changes in the partitioning and allocation of carbon represent compensatory mechanisms that plants can use to minimize the effect of air pollution on resource acquisition, plant growth, and reproduction. These forms of compensation may be beneficial in the short term but result in some large consequences over the long term.

Patterns of carbon allocation and partitioning for plants in clean air are genetically defined and may be optimal for plant growth under existing conditions (Mooney and Chiariello 1984; Bloom et al. 1985). Thus, decreases in plant root/shoot ratios or changes in carbon allocation in response to air pollution may enable plants to compensate for the effect of air pollutants, but these responses may have negative effects on plant growth or reproduction when other environmental stresses occur simultaneously or in sequence with air pollution. For example, a reduction in nitrate uptake and storage apparently enabled SO_2-fumigated radish plants to grow as well as control plants when nitrate availability was high. However, when nitrate availability was reduced, the growth rates of SO_2-fumigated plants did not recover as quickly or completely as control plants (Coleman et al. 1989). Additionally, O_3 fumigated cottonwood leaves produced smaller concentrations of phenol glycosides than control leaves and this apparently helped plants compensate for reduced carbon gain. However, the reduced phenol glycosides stimulated the feeding of a leaf beetle on O_3-treated leaves resulting in increased amounts of short-term leaf damage by herbivores (Jones and Coleman 1988, 1989, 1990). Lastly, if drought follows an air pollution episode, then plants which showed decreases in root/shoot ratios in response to air pollution may be at greater disadvantage than plants which had not responded because a smaller relative root biomass may not be sufficient for absorbing enough water to meet the needs of growing shoots.

Other possibilities for drought-O_3 interactions exist. Drought may coincide with O_3 stress rather than follow air pollution exposures. If so, the two stresses could cause antagonistic shifts in root/shoot partitioning. In this case, O_3-caused decreases in root/shoot ratios might be less than in the case of the sequential stress. Regardless of whether O_3 and drought stress occur simultaneously or in series, an O_3-caused decrease in root/shoot ratios will differ in metabolic costs during ensuing drought. The size of the cost in water use efficiency will depend upon whether stomata close during drought. If stomata do not close, the cost of excess water loss could be high. If stomata close during drought, water loss during the stress period will be low regardless of root/shoot ratio. In addition, drought may stimulate root growth as the plant successfully compensates for both O_3 and water stress.

A great challenge is to estimate costs of plant responses to combinations of stresses that might co-occur or follow air pollution stress. This would be less difficult if the predictability of air pollution and other stresses in the environment were equal. In this case, we might expect for selection to produce genotypes with evolved mechanisms for minimizing multiple stress interactions on plant fitness. However, many stresses occur at much more stochastic frequencies than air pollution (e.g., drought, herbivory, disease, temperature). Thus, the consequences of evolving whole plant mechanisms to resist air pollution stress might be great in environments where air pollution is the strongest constant selective force, but other strong selective forces may occur at stochastic frequencies (i.e., the drought-O_3 interaction discussed above). Here the cost of evolution of the whole plant resistance to air pollution stress, via shifts in carbon partitioning and/or allocation, might be increased vulnerability to other more infrequent stresses such as drought, limiting nutrients, changes in light, excessively warm or cold temperatures, and biotic factors.

Costs of Population Level Resistance to Air Pollution

The Process of Changes in Resistance Within a Species Population

Natural selection works by acting on the genetic variability between individuals in a population to select genotypes with the highest fitness (Fisher 1930; Schmalhausen 1949; Bradshaw and McNeilly, Chapter 2, this volume). Intraspecific variability in air pollution resistance is known to exist for many plant species, to be genetically based, to be a strong selective force, and to reflect variable degrees of resistance attributable to the leaf level and whole plant resistance mechanisms previously discussed (Houston and Stairs 1973; Karnosky 1977, 1985; Roose, Chapter 5, this volume). Consequently, natural selection for air pollution resistance can work by increasing the frequency of resistant genotypes in a population at the expense of sensitive genotypes and perhaps even the amount of genetic variability within the population. As chronic levels of air pollution persist, this kind of selection within a species will probably occur (Bell et al. Chapter 3, this volume).

A clear example of the capacity for intraspecific variability to result in the development of populations more resistant to air pollution can be seen from recent greenhouse/laboratory experiments with a population of radish plants (Gillespie and Winner 1989). Initially, a large population of plants was screened to identify those individuals most sensitive and most resistant to either O_3 or SO_2. The sensitive and resistant individuals were raised to flowering, cross pollinated within respective groups, and allowed to set seed. Seeds from the parents represented the F_1 generation and represented genetic lines selected for sensitivity to O_3 and SO_2 and resistance to the air pollutants. These F_1 seeds were germinated and exposed to both O_3 and SO_2. This experiment showed several important results:

1. Even though seeds of a single genetic cultivar were selected for uniform size and weight, there was a large range of SO_2 and O_3 resistance within the population of radishes. Thus, it seems that among both native vegetation and agricultural species, there can be large intraspecific differences in resistance to O_3 and SO_2.
2. The F_1 plants selected for O_3 resistance were significantly more resistant to O_3 than were plants from the original seed stock (Fig. 3). These results indicate that O_3 might be a powerful selective force and be capable of influencing intraspecific resistance to O_3 damage within a single generation.
3. The SO_2-resistant F_1 plants tended to be larger than nonselected plants when both plant types were raised in SO_2, however, these differences between SO_2 sensitivity groups were not as large as those for O_3 (Fig. 3). Therefore, intraspecific variability in SO_2 resistance may take more generations than for O_3.
4. Oddly, the genetic lines selected to be sensitive to O_3 and SO_2 were not more sensitive to these pollutants than nonselected plants (Fig. 3).

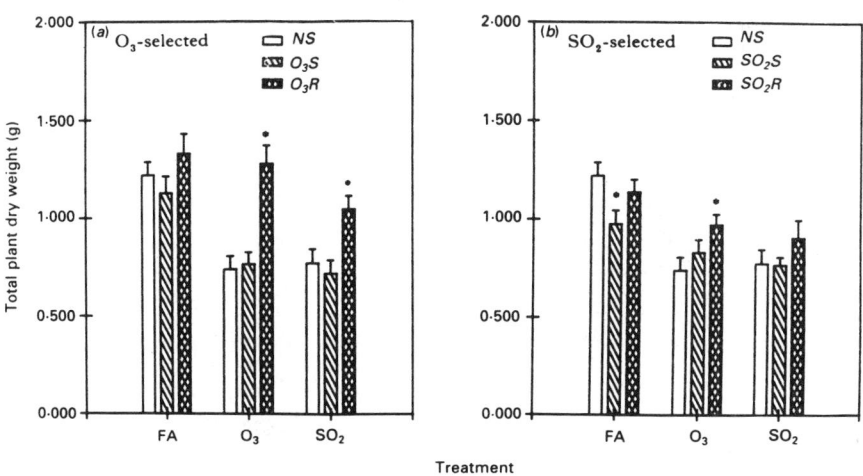

FIGURE 3. Total plant dry weight of O_3-selected (a) and SO_2-selected (b) radish plants to 100 ppb O_3, 500 ppb, or filtered air (Gillespie and Winner 1989). Values are means ± SE, $n = 15$, * = means are significantly different than nonselected plants at the $P \leq 0.05$ level. $O_3S = O_3$-selected; $O_3R = O_3$-resistant; $SO_2S = SO_2$-selected SO_2S SO_2-resistant; NS = nonselected.

Why resistance to the pollutants is heritable, but sensitivity is not, remains unclear.

5. The F_1 plants selected for O_3 resistance were not only of greater resistance to O_3 than the original population, but they were also of greater resistance to SO_2 (Fig. 3). Since the mechanism of O_3 resistance for F_1 plants was at least partly related to stomatal closure during O_3 exposure (Table 5), O_3-resistant plants apparently also closed stomata in response to SO_2. Thus, selection for resistance to O_3 also conferred resistance to SO_2.

Although this experiment clearly shows that selection for air pollution resistant genotypes can occur in a greenhouse study, how might the process of natural selection develop air pollution resistance in field populations? First, since natural selection acts on the genetic variability between in-

TABLE 5. Stomatal conductance (mol m^{-2} s^{-1}) of leaves from second leaf pair of plants at end of 4-h fumigation.

| | Treatment | | |
Plant type	O_3	FA[a]	SO_2
O_3R	0.195 ± 0.026	0.467 ± 0.081	—
O_3S	0.446 ± 0.046	0.670 ± 0.078	—
NS	0.404 ± 0.087	0.334 ± 0.052	0.382 ± 0.052
SO_2R	—	0.420 ± 0.046	0.366 ± 0.063
SO_2S	—	0.347 ± 0.044	0.403 ± 0.048

From Gillespie and Winner 1989; values are means $\pm$ SE, $n = 10$.
[a] R = resistant; S = sensitive; NS = not selected; FA = filtered air.

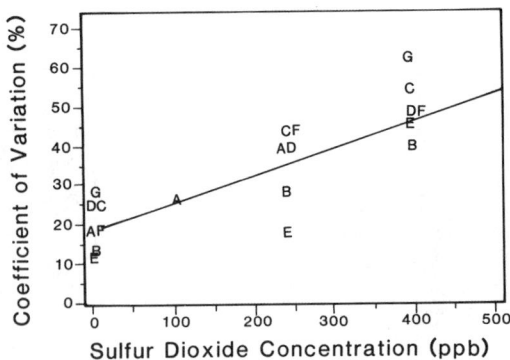

FIGURE 4. Scatter diagram of coefficient of variation of radish biomass production vs concentration of SO_2 from several different experiments (CV $= 19.8 + (69.3 \cdot SO_2$ dose (ppm)); $r = 0.6$ (Coleman et al. 1990). Experiments involved both wild type (A, B, D, and G) and "cherry belle" (C, E, F) radishes raised at different temperatures, levels of nitrogen availability, and SO_2 exposures.

dividuals in a population, the first step for the natural selection of air pollution resistance is to expose the genetic variation in a population for this resistance. Thus, variability in growth and fitness should show a large increase with an initial air pollution stress. For example, Coleman et al. (1990a) showed that the degree of variability in biomass accumulation between radish plants was linearly related to the degree of stress (defined as SO_2 dose) that they received (Fig. 4). When variation in fitness of a population increases as a result of air pollution stress, those plants with the greatest fitness will contribute a larger amount of genetic material to the next generation than they would in the absence of stress. Consequently, the proportion of air pollution resistant genotypes would increase exponentially within the population resulting in severely reduced genetic variation between individuals in subsequent generations. That is why the amount of variability present for a given character in a population is usually inversely proportional to the intensity of previous selection pressures for that character (Fisher 1930; Simpson et al. 1960). For example, plant populations that evolved on mine spoils laden with heavy metals now have extremely little genetic variability in their ability to tolerate, grow, and reproduce on heavy metals. We might expect the same to occur for plant populations evolving in highly O_3 or SO_2 polluted areas. However, it must also be understood that different species will undergo this process at different rates, i.e., annual plants might be selected for at much faster rates than perennial plants (Bazzaz and Sultan 1987; Barrett and Bush, Chapter 6, this volume).

There was little selective pressure by air pollution on plant populations prior to the past 50 years. So where could the variability within a population for air pollution resistance come from? One possibility is that "hidden mutations" build up in a plant population over many generations. These genetic mutations are not expressed by individuals but they are passed on from generation to generation because they do not negatively effect plant fitness (Schmalhausen 1949; Hall 1988). However, under conditions of novel environmental conditions or environmental extremes, these mutations might be expressed resulting in differential success of individuals within a population (Schmalhausen 1949). Since air pollution is a novel environmental condition for many plants, it is easy to see how the expression of "hidden mutations" might result in increased variability in the population leading to selection for air pollution resistance.

Another interesting possibility is that plants within a population may experience spontaneous or directed mutations in their meristems as a result of air pollution stress. These kinds of mutations have been shown to occur in bacteria grown in stressful environments (Hall 1988). In this case, a bacteria cell placed in a stressful environment might exhibit increased mutation rates in a specific area of the genome relating to resistance of the stress, ultimately producing a stress-resistant cell. Consequently, stress-resistant bacteria strains evolve much quicker than they would given random mutation rates (Hall 1988; Hall et al. 1989). Although this may seem a little far-fetched for plants, it is known that plants exhibit somatic mutations in their meristems (Whitham and Slobodchikoff 1981) and that these mutations can be expressed by subsequent generations. Consequently examining the ability of plants to "direct their own mutations" may result in important information regarding how plant populations survive in stressful environments.

Consequences of Population Resistance

The consequences of air pollution resistant populations can be viewed in two ways. The selection for a given resistance mechanism to air pollution might preclude the ability of plants in a population to respond to other changes in the environment. Secondly, the drastic changes in the variability within a population that occurs as a population becomes resistant to air pollution might have consequences to the ecological interactions of that population with other organisms or to human uses of that plant population.

One measure of consequences for increased air pollution resistance to plant growth in changing environments might be made by evaluating growth of F_1 plants differing in air pollution resistance, but which are the result of selections from common stock. Growing F_1 plants in clean air which are either resistant to air pollution or from original stock should show whether the increase in air pollution resistance compromises growth of the resistant genotypes when air pollution is not present (Fig. 5). In this environment air pollution resistance might be considered as "excess baggage" and its cost most apparent.

The F_1 generation O_3 resistant radish plants from the selection experiment above were grown in O_3 free air alongside plants from the parental

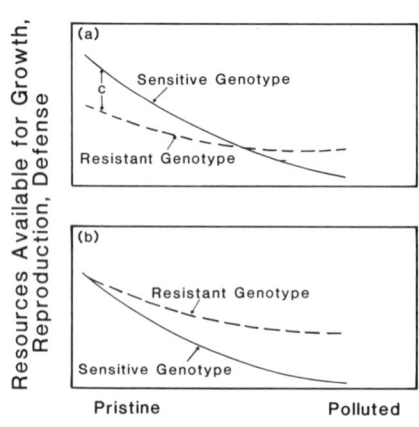

FIGURE 5. Analysis of fitness costs for air pollution resistance. Resistance may have a cost (a from Pitelka 1988) or may be independent of costs (b from Gillespie and Winner 1989).

seed supply. Results show that plants from the original seed supply and F_1 plants which had increased O_3 resistance had similar growth in O_3 free air. Thus, in this analysis the evolution of O_3 resistance did not come with a detectable cost with regard to growth of the resistant genotype in an environment without O_3.

The analysis of the growth of these radishes selected for O_3 resistance is a first attempt to assess the possible cost of air pollution resistance within a population. It is important to recognize that even though no growth costs of air pollution resistance were recognized in this experiment, such costs might exist but, for several reasons, were not detected. For example, the radish plants were raised in O_3 free air and in a close to optimal environment. Thus, in addition to clean air, each pot contained only one plant and was supplied with liberal amounts of water, nutrients, and light. However, if other resources had been limiting, the air pollution resistant genotypes may not have fared well in the clean air environment, or perhaps, even in an O_3 polluted environment.

Perhaps the greatest consequences of selection for air pollution resistance are related to changes in the amount of genetic variation present in the population (see Bazzaz and Sultan 1987). It is clear that intense forces of selection can eventually remove a large portion of the variability within a population (Antonovics et al. 1971). The consequences of this are manifold. First, variability between individual plants is an important mechanism for keeping herbivore and pathogen populations from reaching epidemic levels (see Segal et al. 1980; Whitham 1983). Thus, a reduction in variability between individuals might result in the creation of plant populations that are severely at risk to predators and pathogens. Second, a reduction in the genetic diversity of a population that occurs with the selection of air pollution resistance might result in a population containing plants that do not have sufficient genetic diversity to evolve resistance to novel stresses. Consequently, the probability of an unusual environmental event causing extinction of the population could increase. Third, changes in the variation of a population may occur over many generations. As previously mentioned, the first response of a population to an intense selective force is to exhibit increased variability and this may occur for several generations before the nonresistant genotypes are completely removed. This initial increase in variability may have substantial consequences for humans with regard to our ability to predict plant growth in natural or even agricultural environments.

Conclusions

There are a number of mechanisms, at the leaf level, for resisting damage from air pollutants. These mechanisms involve both stomata and metabolic processes within the leaf. The consequences of leaf responses to air pollution may differ with species and leaf age. These responses, if evolved and conferring resistance to air pollution damage, may protect foliage from air pollution stress. The cost of this protection may be in terms of reduced carbon gain, altered patterns of carbon use, increased water loss, change in leaf energy balance, change in foliar nutrient content, and shortened period of leaf longevity. These costs can be quantified by analysis of foliar elements and calorimetery.

Whole plants respond in systematic ways to naturally occurring stresses, such as drought, shading, and insufficient nutrients. One well-studied response is the partitioning of carbon and other plant resources between tissues such as leaves, stems, and roots. These responses are integrated across multiple stresses, since these stresses do not occur singly, but rather in combinations. Imposition of anthropogenic stresses, such as tropospheric O_3, SO_2, NO_x, heavy metals, and acid deposition constitute yet additional stress factors with which plants must cope. Such additional stresses may be most taxing for plants already pushed to physiological limits because they are growing in marginal habitats.

Air pollution, as any environmental stress, has the capacity to act as a selective force within a species. It does so by reducing fitness of air pollution sensitive individuals. Air pollution has the capacity to increase variability of growth within a species and to bring differences in air pollution resistance within a single generation of some species. Loss of individuals that are sensitive to air pollutants does not necessarily mean that genetic diversity within the species will decrease.

References

Alscher RG, Amthor JS (1987) The physiology of free-radical scavenging: maintenance and repair processes. In: Schulte-Hostede S, Dorrall NM, Blank LW, Wellburn AR (eds) Air pollution and plant metabolism. Elsevier Applied Science Publishers, New York, pp 94–115

Antonovics J, Bradshaw AD, Turner RC (1971) Heavy metal tolerance in plants. Advances in Ecology Research 7:1–85

Ayazloo M, Bell JNB (1982) Studies on the tolerance to sulphur dioxide of grass populations in polluted areas. I. Identification of tolerant populations. New Phytologist 88:203–222

Ayazloo M, Garsed SG, Bell JNB (1982) Studies on the tolerance to sulphur dioxide of grass populations in polluted areas. II. Morphological and physiological investigations. New Phytologist 90:109–126

Bazzaz FA, Sultan SE (1987) Ecological variation and the maintenance of plant diversity. In: Urbanska K (ed) Differentiation in higher plants. Academic Press, London, pp 69–93

Berang PD, Karnosky DF, Mickler RA, Bennett JP (1986) Natural selection for ozone tolerance in *Populus tremuloides*. Canadian Journal of Forestry Research 16:1214–1216

Black VJ (1985) SO$_2$ effects on stomatal behavior. In: Winner WE, Mooney HA, Golstein RA (eds) Sulfur dioxide and vegetation: physiology, ecology and policy issues. Stanford University Press, Stanford, CA, pp 96–117

Bloom AJ, Chapin FS III, Mooney HA (1985) Resource limitation in plants—an economic analogy. Annual Review of Ecological Systems 16:363–392

Bradshaw AD (1952) Populations of *Agrostis tenuis* resistant to lead and zinc poisoning. Nature 109:1098

Chapin FS III (1989) The cost of tundra plant structures: evaluations of concepts and currencies. American Naturalist 133:1–19

Coleman JS (1986) Leaf development and leaf stress: increased susceptibility associated with sink-source transition. Tree Physiology 2:289–299

Coleman JS, Jones CG (1988a) Plant stress and insect performance: cottonwood, ozone and a leaf beetle. Oecologia 76:57–61

Coleman JS, Jones CG (1988b) Acute ozone stress on eastern cottonwood (*Populus deltoides* Bartr.) and the pest potential of the aphid, *Chaitophorus populicola* Thomas (Homopter: Aphididae). Environmental Entomology 17:207–212

Coleman JS, Mooney HA, Gorham JN (1989) The effect of multiple stresses on radish growth and resource allocation. I. The response of wild radish plants to a combination of SO$_2$ exposure and decreasing nitrate availability. Oecologia, 81:124–131

Coleman JS, Mooney HA, Winner WE (1990) Anthropogenic stress and natural selection: variability in radish biomass accumulation increases with increasing SO$_2$ dose. Canadian Journal of Botany 68:102–106

Fisher RA (1930) The genetical theory of natural selection. Clarendon Press, Oxford, UK

Garsed SG (1985) SO$_2$ uptake and transport. In: Winner WE, Mooney HA, Golstein RA (eds) Sulfur dioxide and vegetation: physiology, ecology and policy issues. Stanford University Press, Stanford, CA, pp 75–95

Gillespie CT, Winner WE, (1989) Development of radish lines differing in resistance to O$_3$ and SO$_2$. New Phytologist 112:353–361

Grime JP (1979) Plant strategies and vegetation processes. Wiley, New York

Guderian R, Tingey DT, Rabe R (1985) Effects of photochemical oxidants on plants. In: Guderian R (ed) Air pollution by photochemical oxidants. Springer-Verlag, Berlin, pp 129–333

Hall BG (1988) Adaptive evolution that requires multiple spontaneous mutations. I. Mutations involving an insertion sequence. Genetics 120:887–897

Hall BG, Parker LL, Betts PW, DuBose RF, Sawyer SA, Hartl DL (1989) IS103, a new insertion element in *E. coli*: characterization, and distribution in natural populations. Genetics 121:423–431

Heath RL (1987) The biochemistry of O$_3$ attack on the plasma membrane of plant cells. Recent Advances in Phytochemistry 21:29–54

Horsman CC, Roberts TM, Bradshaw AD (1979) Studies on the effect of sulphur dioxide on perennial ryegrass (*Lolium perenne* L.) Journal of Experimental Biology 30:495–501

Horsman DC, Wellburn AR (1977) Effect of SO$_2$ polluted air upon enzyme activity in plants originating from areas with different annual mean atmospheric SO$_2$ concentrations. Environmental Pollution 13:33–39

Houston DB, Stairs GR (1973) Genetic control of sulfur dioxide and ozone tolerance in Eastern White Pine. Forestry Science 19:267–271

Jermy T (1966) Feeding inhibitors and food preference in chewing phytophagous insects. Entomology Experimental and Applied 9:1–12

Jermy T (1984) Evolution of the insect/host plant relationships. American Naturalist 124:609–630

Jones CG, Coleman JS (1988) Plant stress and insect behavior: cottonwood, ozone and the feeding and oviposition preference of a beetle. Oecologia 76:51–56

Jones CG, Coleman JS (1989) Biochemical indicators of air pollution effects in trees: unambiguous signals based on secondary metabolites and nitrogen in fast-growing species? In: National Research Council, Biologic markers of air pollution stress and damage in forests. National Academy Press, Washington, DC, pp 261–274

Jones CG, Coleman JS (1990) Plant stress and insect herbivory: toward an integrated perspective. In: Mooney HA, Winner WE, Pell EJ (eds) Integrated responses of plants to stress. Academic Press, New York (in press)

Karnosky DF (1977) Evidence for genetic control of response to sulfur dioxide and ozone in *Populus tremuloides*. Canadian Journal of Forestry Research 7:437–440

Karnosky DF (1985) Genetic variability in growth responses to SO_2. In: Winner WE, Mooney HA, Golstein RA (eds) Sulfur dioxide and vegetation: physiology, ecology and policy issues. Stanford University Press, Stanford, CA, pp 346–356

Larcher W (1980) Physiological plant ecology. Springer-Verlag, Berlin

Lechowicz MJ (1984) The effects of individual variation in physiological and morphological traits on the reproductive capacity of the common cocklebur, *Xanthium strumarium* L. Evolution 38:833–844

Lechowicz MJ (1987) Resource allocation by plants under air pollution stress: implications for plant-pest-pathogen interactions. Botany Review 53:281–300

Melillo JM, Aber JD, Muratore JF (1982) Nitrogen and lignin control of leaf litter decomposition dynamics. Ecology 63:621–626

Miller JE (1987) Effects of ozone and sulfur dioxide stress on growth and carbon allocation in plants. Recent Advances in Phytochemistry 21:55–100

Mooney HA (1972) The carbon balance of plants. Annual Review of Ecological Systems 3:315–346

Mooney HA, Chiariello N (1984) The study of plant function—the plant as a balanced system. In: Dirzo R, Sarukhan S (eds) Perspectives on plant population ecology. Sinauer Associates, Sunderland, MA, pp 305–321

Mooney HA, Winner WE (1991) Effect of environmental stress on the partitioning of resources between plant roots and shoots. In: Mooney HA, Winner WE, Pell EJ (eds) Integrated responses of plants to stress. Academic Press, New York (in press)

Mooney HA, Kuppers M, Koch G, Gorham J, Chu C, Winner WE (1988) Compensating effects to growth of carbon partitioning changes in response to SO_2-induced photosynthetic reduction in radish. Oecologia 75:502–506

Nobel PS (1983) Biophysical plant physiology and ecology. W.H. Freeman, San Francisco, CA

Pell EJ (1979) How air pollutants induce disease. In: Horsfall JG, Cowling EB (eds) Plant disease, vol 4. Academic Press, New York, pp 273–292

Pell EJ, Dan ME (1990) Multiple stress and plant senescence. In: Mooney HA, Winner WE, Pell EJ (eds) Integrated responses of plants to stress. Academic Press, New York (in press)

Pitelka LF (1988) Evolutionary responses of plants to anthropogenic pollutants. Trends in Ecology and Evolution 3:233–236

Reich PB (1983) Effects of low concentrations of O_3 on net photosynthesis, dark respiration, and chlorophyll contents in aging hybrid poplar leaves. Plant Physiology 73:291–296

Richardson CJ, DiGuilio RT, Tandy NJ (1989) Free-radical mediated processes as markers of air pollution stress in trees. In: National Research Council, Biologic markers of air pollution stress and damage in forests. National Academy Press, Washington, DC, pp 251–260

Roose ML, Bradshaw AD, Roberts TM (1982) Evolution of resistance to gaseous air pollutants. In: Unsworth MH, Ourmrod DP (eds) Effects of gaseous air pollution in agriculture and horticulture. Butterworth, London, pp 379–409

Schmalhausen II (1949) Factors of evolution: the theory of stabilizing selection. University of Chicago Press, Chicago, IL (reprinted in 1986)

Segal A, Manisterski J, Fisschbeck G, Wahl I (1980) How plant populations defend themselves in natural ecosystems. In: Horsfall JG, Cowling EB (eds) Plant disease, vol 4. Academic Press, New York, pp 273–292

Simpson GG, Roe A, Lewontin RC (1960) Quantitative zoology. Harcourt, Brace and World, New York.

Smith TA (1984) Putrescine and inorganic ions. Recent Advances in Phytochemistry 18:7–54

Szaniawski RK (1987) Plant stress and homeostasis. Plant Physiology and Biochmemistry 25:63–72

Taylor GE Jr (1978) Genetic analysis of ecotypic differentiation within an annual plant species, *Geranium carolinianum* L., in response to sulfur dioxide. Botany Gazette 139:362–368.

Taylor GE Jr (1981) Physiology of ecotypic plant response to sulfur dioxide in *Geranium carolinianum* L. Oecologia 49:76–82

Taylor GE Jr, Murdy WH (1975) Population differentiation of an annual plant species, *Geranium carolinianum*, in response to sulfur dioxide. Botany Gazette 136:212–215

Tingey DT, Fites RC, Wickliff C (1976) Differential foliar sensitivity of soybean cultivars to ozone associated with differential enzyme activities. Physiology of Plants 37:69–72

US EPA (1986) Air quality criteria for ozone and other photochemical oxidants. Research Triangle, NC, US Government Doc. #EPA/600/8-84-020cf

Whitham TG (1983) Sources of interplant variation and consequences for herbivores. In: Denno RF, McClure MD (eds) Variable plants and herbivores in natural and managed systems. Academic Press, New York, pp 1–42

Whitham TG, Slobodchikoff CN (1981) Evolution by individuals, plant-herbivore interactions, and mosaics of genetic variability: the adaptive significance of somatic mutations in plants. Oecologia 49:287–292

Williams K, Percival F, Merino J, Mooney HA (1987) Estimation of tissue construction cost from heat of combustion and organic N content. Plant Cell Environments 10:725–734

Winner WE (1989) Photosynthesis and transpiration measurements as biomarkers of air pollution effects on forests. In: National Research Council, Biologic markers of air pollution stress and damage in forests. National Academy Press, Washington, DC, pp 303–316

Winner WE, Cotter IS, Powers HR JR, Skelly JM (1987) Screening loblolly pine seedling responses to SO_2 and O_3: analysis of families differing in resistance to Fusiform rust. Environmental Pollution 47:205–220

Winner WE, Mooney HA (1980a) Ecology of SO_2 resistance: I. Effects of fumigations on gas exchange of deciduous and evergreen shrubs. Oecologia, 44:290–295

Winner WE, Mooney HA (1980b) Ecology of SO_2 resistance: II. Photosynthetic changes of shrubs in relation to SO_2 absorption and stomatal behavior. Oecologia, 44:296–302

Winner WE, Mooney HA (1980c) Ecology of SO_2 resistance: III. Metabolic changes of C_3 and C_4 *Atriplex* species due to SO_2 fumigations. Oecologia, 46:49–54

Winner WE, Mooney HA (1980d) Responses of Hawaiian plants to volcanic SO_2: Stomatal behavior and foliar injury. Science, 210:789–791

Winner WE, Mooney HA (1985) Ecology of SO_2 on native Hawaiian plants. Oecologia, 86:387–393

Estimating Costs of Air Pollution Resistance

JAMES R. EHLERINGER

Introduction

When plants are exposed to an environmental stress such as limited soil water, enhanced ultraviolet radiation from the sun, or high metal concentrations in the soil, there are two possible responses: some plants will persist, while others cannot. Depending on the severity and duration of stress, fewer or more plants will fall into one category versus the other. Differences in responses to stress should be expected at the intra- and interspecific levels, and both have been described for plants exposed to atmospheric pollutants. Although several of the previous contributions in this book have focused on the ability of plants to resist air pollution stress, this commentary focuses instead on the concepts of costs associated with air pollution resistance and stems from the chapter by Winner et al. (Chapter 7, this volume).

Resistance and Its Measurement

For those plants resistant to an air pollutant stress, it is important to disinguish between stress tolerance versus stress avoidance. The term "tolerance" is often used loosely in the air pollution literature to indicate a resistance to a gaseous pollutant. Whereas the term resistance implies no mechanistic understanding, tolerance indicates a specific metabolic or enzymatic shift, which allows plants to maintain metabolic activity while tissues are exposed to air pollution stress (Levitt 1972).

A second major point in discussing resistance to air pollution stress is whether the stress is acute (short-term) or chronic (long-term). Studies cited by Winner et al. (Chapter 7, this volume), Roose et al. (1982), Darrall (1989) and Tingey and Andersen (Chapter 8, this volume) clearly point out that plant responses to these two kinds of exposure regimes are qualitatively different, resulting from physiologically dissimilar mechanisms of action. Clarification of these concepts and terms is important if we are to focus on the means by which plants are responding to an air pollution stress, and more importantly in providing an understanding of the costs associated with plant persistence when plants are exposed to stress.

A difficulty in evaluating costs associated with air pollution stress or in assessing resistance is knowing what parameter(s) to measure. Ultimately, all investigators are interested in estimating fitness, but this feature is not easily quantified, especially with just a single measure, because many parameters which are highly correlated with fitness may be elusive. Evaluation of integrative physiological parameters (such as growth rate) are likely to be more reliable than short-term measures (such as instantaneous photosynthetic rate). Yet the time constraints associated with reliably estimating many physiological activities have resulted in the majority of the studies focusing on an evaluation of a resistant versus a sensitive genotype rather than assessing variance in resistance levels within the population. Several of the more easily obtained characters, such as plant height or visible leaf injury, have been measured as estimators of fitness, but there is little evidence to indicate that these are the most appropriate features. Other promising approaches that measure integrated aspects of plant metabolism and that can be used to rapidly screen large numbers of individuals over a short time period have been proposed, but their utility is not well documented. These approaches include carbon isotope ratios to estimate stomatal limitations, fluorescence to estimate photosynthetic activity, thermal scans to estimate transpiration rates, and spectral analyses to estimate protein concentrations.

The Cost of Evolving Resistance

"What is a cost?" and "How is it measured?" are two major questions, which in the abstract are easy to discuss, but at the experimental level are difficult to determine. From an ecophysiological perspective, cost can be measured in metabolic terms. For example, costs associated with resistance to an air pollution stress can be measured in terms of ATP, NADPH, and organic carbon levels. On a dynamic basis, cost can be assessed in terms such as (1) reduced photosynthetic rates, (2) increased respiration rates, and (3) premature senescence. The cost of reduced photosynthesis is estimated as the actual rate of photosynthesis relative to that expected if there were no stress, and appears most often when stomata close to restrict gas diffusion rates. Similarly, costs may appear as increased respiration rates, such as those associated with repair processes. Again, cost can be assessed as the increased rate of respiration relative to that expected of a plant in an environment not exposed to stress. Presenescence is often neglected in cost estimates, where the focus is on the immediate reduction in metabolic activity. Yet a common plant response to air pollutants is not only a reduction in photosynthetic rates, but a change in tissue longevity. The presenescence cost should be viewed as the overall reduction in carbon gain by a leaf due to the combination an immediate reduction in photosynthesis compounded by the reduction in the life expectancy of that leaf.

Cost is occasionally used loosely in ways that either cannot be quantified or that actually represent a minor metabolic cost. For example, it is not appropriate to consider an "evolutionary cost" of air pollution resistance. That is, even though plants may in fact have a reduced photosynthetic rate when exposed to a stress, one can not quantify a "cost to evolve resistance." In these situations, what is often meant is a measure of the likelihood that a new DNA sequence will arise that produces a gene which confers resistance to the stress. Changes in the composition of specific enzyme systems are likely not to have significant costs associated with them unless there are significant changes in the specific activities of tolerant versus sensitive enzyme systems. In Chapter 7, Winner et al. showed that when costs are evaluated in terms of the metabolic cost to construct leaves of different biochemical compositions, these static costs are minor relative to the dynamic, continual costs associated with changes in metabolic rate.

Measuring Costs of Resistance

Costs can be evaluated at the physiological level as changes in metabolic rate and at the whole plant level as changes in both carbon allocation and phenology. Each approach is necessary to describe plant responses to an air pollutant stress and each spans a different time scale. To ascertain any cost of air pollution resistance at the metabolic level, it is essential to know in what ways the plant has responded. That is, resistance is based on two major processes—avoidance and tolerance, neither of which are mutually exclusive, but which potentially have different consequences.

Avoidance appears to be a common initial resistance mechanism among a broad diversity of plant species (Darrall, 1989; Winner et al., 1989; Tingey and Andersen, Chapter 8, this volume). Avoidance is based upon stomatal closure, which reduces a pollutant's uptake rate. Stomatal closure has significant primary and secondary effects. Carbon dioxide diffusion into the leaf, and therefore photosynthesis, is reduced and this will certainly constitute a cost to the plant's carbon balance. Secondarily, reducing transpiration alters the leaf energy balance, having secondary effects on both respiration and photosynthetic rates. In addition, exposure of leaves to high solar irradiances while stomata are closed could contribute to photoinhibitory damage and a reduction in photosynthetic capacity. Stomatal closure appears most likely to be a short-term, avoidance response, and probably contributes little to the resistance of plants subjected to a long-term chronic exposure to pollutants. However, stomatal closure does provide short-term resistance to a large array of atmospheric pollutants, all of which must penetrate to internal target tissues via the stomatal pores.

Tolerance of the internal leaf tissues to air pollutants seems essential in chronic exposure regimes. The extent to which tolerance has costs associated with it seems a much-debated issue with insufficient information to offer a conclusion. In some cases, increased respiration rates are common in tolerant genotypes. These respiration costs are definable in terms of the energetic costs associated with sequestering pollutants, detoxifying pollutants, or repairing/resynthesizing damaged proteins and membranes. Competition for photoreduction energy is also a cost that can be directly calculated. For example, within the chloroplast, both CO_2 and SO_2 compete for photosynthetic light reaction products (ATP and NADPH). The reduction of SO_2 to H_2S is associated with tolerance to SO_2, yet there is a reduction in CO_2 fixation rates (photosynthesis) since ATP and NADPH allocated to SO_2 reduction can not also be allocated to CO_2 reduction.

Kinds of Costs Exhibited Within a Population Undergoing Change in Response to an Air Pollutant Stress

In their introductory chapter, Bradshaw and McNeilly (Chapter 2, this volume) referred to stages in how plant populations change upon exposure to stress: an elimination of the least resistant genotypes, followed by selection for more resistant genotypes. Bell et al. (Chaper 3, this volume) noted that resistance to acute injury evolves faster than resistance to chronic injury. These observations would suggest that upon exposure to an air pollution stress, genotypes which initially avoid the stress by closing their stomata would constitute the majority of the surviving population. Yet avoidance indeed has a cost, since stomatal closure would substantially reduce photosynthesis and growth. Thus, it seems reasonable to expect that over time the population structure would shift to genotypes that contain both avoidance and tolerance mechanisms. Ultimately, if the stress persists and the cost of avoidance exceeds the cost of tolerance, the population structure would shift such that the majority of genotypes possessed tolerance mechanisms with relatively few individuals exhibiting avoidance mechanisms.

Summary: Evolving Resistance Without Cost?

Perhaps one of the most intriguing issues is the possibility that resistance could evolve without a cost. That is, is it possible that resistance mechanisms exist that do not reduce metabolic activity under nonstress conditions. Given the earlier discussions, it is likely that this may not be possible if resistance is based on an avoidance mechanism. Reducing pollutant uptake rates appears inextricably associated with avoidance, and, since photosynthetic gas exchange will always be reduced under these situations, the cost is evident in terms of reduced photosynthetic activity. However, if air pollution resistance were associated with a tolerance mechanism, it is possible that resistance could have no or, at best, a minimal cost. In the plant pathogen literature as well as in the plant-animal interaction literature, there are numerous examples of increased resistance resulting from small but significant changes in the chemical structure of a particular compound (protein, secondary metabolite, etc.) (Harborne 1982). The cost to produce the resistance versus tolerant component is trivial in metabolic terms, but the consequences for survival are significant. Perhaps such analogies might be appropriate for aspects of air pollution resistance.

This possibility is perhaps best visualized comparing two models of plant responses to an air pollution stress (Chapter 7, Fig. 6). Consider two genotypes and their performances under nonstress and stress conditions. Pitelka (1988) suggested that a resistant genotype must experience a greater metabolic cost under nonstress conditions, such that performance (and therefore fitness) is less than that of a sensitive genotype. As the stress level increases, the response functions (i.e., fitness) of sensitive and resistant genotypes cross, such that the resistant genotype would have the greater performance under stress conditions (Chapter 7, Fig. 6A). Alternatively, both resistant and sensitive genotypes could have the same performance under nonstress condition (Chapter 7, Fig. 6B). Such a situation could be possible if the tolerance mechanism required little or no energetic cost (for instance, proteins functional and stable under exposure to a pollutant). If this were the case, overall plant performance might not be expectd to decrease under exposure to stress as it would in the sensitive genotype. Few data are available to evaluate costs of resistant versus sensitive genotypes, but clearly these tradeoffs need to be considered before any conclusions are offered as to the costs of air pollution resistance.

References

Darrall NM (1989) The effect of air pollutants on physiological processes in plants. Plant Cell Evironment 12:1–30

Harborne JB (1982) Introduction to ecological biochemistry. Academic Press, London

Pitelka LF (1988) Evolutionary responses of plants to anthropogenic pollutants. Trends in Ecology and Evolution 3:233–236

Roose ML, Bradshaw AD, Roberts TM (1982) Evolution of resistance to gaseous air pollutants. In: Unswoth MH, Ourmrod DP (eds) Effects of gaseous air pollution in agriculture and horticulture. Butterworth, London, pp 379–409

Levitt J (1972) Responses of plants to environmental stresses. Academic Press, New York

8
The Physiological Basis of Differential Plant Sensitivity to Changes in Atmospheric Quality

DAVID T. TINGEY and CHRISTIAN P. ANDERSEN

Introduction

During the next several decades vegetation will continue to be exposed to a wide variety of pollutants, although the types of compounds, their concentrations, and their spatial patterns may change. Some are distributed globally, whereas others are distributed regionally or locally. These airborne pollutants can have either a direct impact on the plant foliage or act indirectly through deposition onto the soil and subsequent uptake by roots. These effects can range from subtle modifications of cellular biochemistry and whole-plant physiology (e.g., carbon allocation) to overt foliar injury and effects on plant growth, yield, and/or reproduction.

Numerous studies have established that plants exhibit differential sensitivity to air pollutants (e.g., Tingey and Taylor 1982; Roose et al. 1982; Tingey and Olszyk 1985; Scholz et al. 1989). Various hypotheses have been proposed to explain differential pollutant sensitivity, however, in most cases the mechanism(s) of the differential response is not known. Nevertheless, the underlying source of variation in plant response is disparate physiological states among the various individuals. These disparate states are important because they are due, at least in part, to genetic variation.

Pollutant resistance is often a quantitative trait involving numerous genes (Pitelka 1988; Roose, Chapter 5, this volume). Heritability for resistance tends to be high and the genes for resistance often work in an additive manner (Roose et al. 1982; Dragoescu et al. 1988; Houston and Stairs 1973; Karnosky 1977; Taylor 1978; Butler et al. 1979; Hanson et al. 1976). Resistance enables the plant to survive, grow, and reproduce in the presence of a particular pollutant (Pitelka 1988). Populations possessing latent variability for pollutant resistance can evolve resistant populations if exposed to high concentrations of pollutant for sufficient time (Pitelka 1988). In contrast, populations lacking such variation cannot become resistant and may decline under pollutant stress.

Pollutant stresses can have a profound impact on genetic diversity even though stressed individuals rarely die as a direct result of current ambient pollutant levels. Pollutant stresses can reduce growth, reproductive fecundity, and competitive status in sensitive individuals, without causing death. The population consequence of reduced fecundity would be a shift from the sensitive to resistant types in future generations. In this manner, pollutant stresses can impact evolutionary processes occurring in plant communities (Bradshaw and McNeilly, Chapter 2, this volume). The overall result is a reduction in genetic diversity with a concomitant reduction in tolerance to other stresses (Karnosky et al. 1989).

Other chapters in this book specifically address population level responses to changes in atmospheric quality (e.g., Tonsor and Kalisz, Chapter 10, this volume). This chapter addresses the physiological basis of differential plant sensitivity to pollutants. A generic model will be used to provide a conceptual framework for discussing differential plant response. After a brief description of the model, potential feedback control of physiological responses will be illustrated with examples from the literature. Plant responses to multiple pollutants, including the nonadditive nature of stress response are briefly addressed.

Factors Affecting Plant Response

The conceptual model (described below and shown in Fig. 1) is not specific for any pollutant or pollutant type (i.e., gaseous or wet deposition) but provides a conceptual framework to understand the influence of genetic

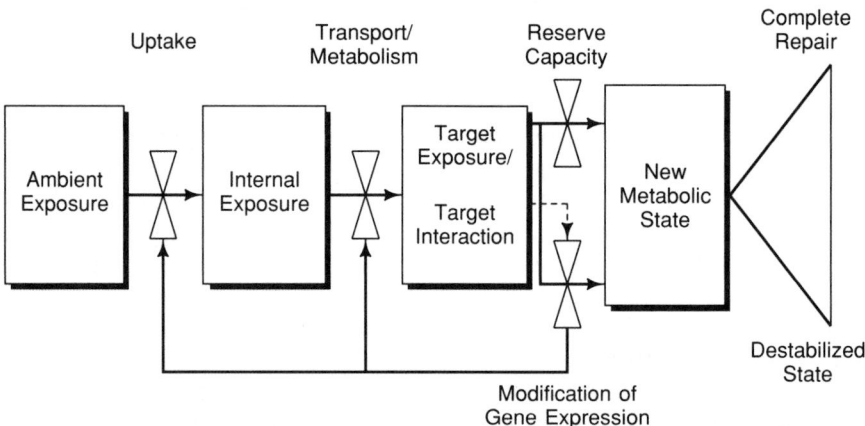

FIGURE 1. Conceptual model of the processes controlling differential sensitivity to pollutants. The pollutant concentration will decrease as it moves from the ambient exposure to the target site. The generalized avoidance and tolerance processes are represented as values. Avoidance processes control the amount of pollutant at the target site and tolerance processes control the plant response (repair/compensation) to the stress. The response may also be a feedback process that influences the avoidance mechanisms.

control on differential plant sensitivity. The model proposes a common progression of events, based on a multiplicity of interactive physiological mechanisms.

Plant response to a pollutant is the consequence of the interaction among various avoidance and tolerance mechanisms (Ariens et al. 1976; Tingey and Taylor 1982) (Fig. 1). Avoidance factors influence both the amount and the rate at which the pollutant or its metabolites reach the target site; tolerance factors operate through processes that repair or compensate for the stress. Various feedback mechanisms can also influence plant response. The specific mechanisms for avoidance and tolerance may be broad-based (i.e., responding to all stresses) or stress-specific. Although the target site for each stress is unique, it is the interactive balance among avoidance and tolerance processes that controls plant response to any stress.

Avoidance Factors

Avoidance factors influence the amount/rate of the pollutant reaching the target site and consist of two phases: pollutant uptake and pollutant transport/metabolism (Fig. 1).

POLLUTANT UPTAKE

A pollutant stress response can occur only after the stress factor enters the plant. The uptake of airborne pollutants is dependent on: (1) the concentration of the pollutant in the surrounding air, (2) the stomatal and cuticular conductances of the pollutant, and (3) the pollutant's solubility. Stomata are the principal plant barrier influencing the uptake of gaseous pollutants. Cuticular characteristics are important for controlling the uptake of wet and dry deposited pollutants.

During the day, stomatal factors are most important while at night cuticular characteristics predominate. However, for gases such as O_3, cuticular conductance is several orders of magnitude smaller than stomatal conductance (Kerstiens and Lendzian 1989).

POLLUTANT TRANSPORT/METABOLISM

The processes of pollutant transport/metabolism include (1) absorption and distribution over the tissues, (2) transport (via membranes) and binding to the tissue components, and (3) metabolic alteration of the toxicant including changes in its chemical properties/sequestering, and if possible, excretion.

Target Site Interaction

The pollutant target site is likely specific for a given class of compounds. If the pollutant reaches the target site in sufficient quantity, it will elicit a plant response. The concentration of the toxicant attained in the target tissue and the specific characteristics of the target site influence the magnitude of the resultant biological stress.

Tolerance Factors

Tolerance (repair and compensation) mechanisms lead to a new metabolic state following a target interaction and can be separated into two categories: reserve capacity and modification of gene expression (Queiroz 1988) (Fig. 1).

RESERVE CAPACITY

The plant utilizes only a fraction of its existing metabolic capacity under normal conditions, i.e., enzyme capacity generally exceeds demand (Queiroz 1988). Consequently the initial response to a pollutant episode or increased pollutant stress is achieved using existing enzyme capacity with little additional energy cost.

MODIFICATION OF GENE EXPRESSION

If there is insufficient enzyme capacity to achieve a new metabolic homeostasis, de novo enzyme synthesis may occur (Queiroz 1988). This includes increased enzyme concentrations or the synthesis of new iso-enzymes that may be more efficient. This pathway, which involves the synthesis of new protein, is energetically expensive. The pathway also includes the possibility of posttranslational enzyme modification such as covalent modifications of enzymes by phosphorylation/dephosphorylation and synthesis of secondary metabolites.

Consequently, the initial response to a pollutant episode or increased pollutant level is achieved by utilizing the currently available enzyme capacity, with further response resulting in de novo enzyme synthesis. If repair is complete, there will be no detectable effect except an increase in energy for repair and maintenance of the metabolic pools. The magnitude of injury reflects the extent to which the metabolic machinery is able to repair or compensate for the stress. It is probable that both paths act in concert when the plant is exposed to a pollutant. If the plant cannot reestablish metabolic homeostasis, it will be destabilized (Fig. 2).

Repair/compensation may not be achieved if the pollutant impact disrupts the metabolic process(es) before the tolerance mechanisms are operative. This inhibition of metabolic adjustments offers an explanation for the synergistic effects from combinations of NO_2 and SO_2. Plants are able to tolerate low levels of NO_2, which may even be beneficial (Wellburn 1982). This tolerance is accomplished by the enzymatic reduction of NO_3^- and NO_2^- to NH_4^+ (Zeevaart 1976). However, SO_2 inhibits the induction of additional nitrite reductase, and the bioenergetic levels are significantly lower in plants receiving the combined exposure. These factors combine to cause the injury.

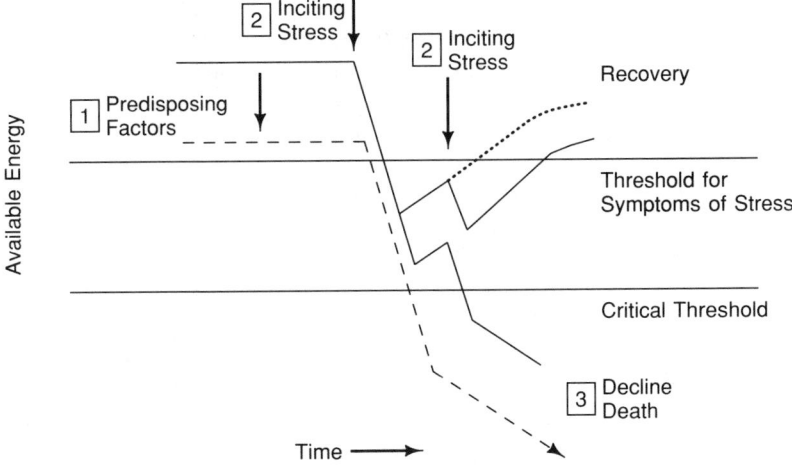

FIGURE 2. Conceptual model of energy (carbon) available to the plant for avoidance and repair/compensation processes (adapted from Johnson, 1989); see text for details.

The conceptual model (Fig. 1) proposes a progression of events, based on a multiplicity of interactive physiological mechanisms, with multiple points for environmental/metabolic control. Consequently, any one or a combination of several events may contribute to the observed differential response. This has led to the suggestion that different mechanisms of resistance may be involved among various species and pollutants. However, from a conceptual viewpoint this is unlikely. A more reliable hypothesis is that the relative importance of avoidance and tolerance mechanisms varies among species/genotypes. For example, pollutant exclusion (avoidance) may be the dominant mechanism in one genotype while a modification of gene expression (tolerance) is more important in another. In either case, the primary basis of the differential sensitivity conforms to the general model of plant response to pollutants. Examples will be drawn from the literature to illustrate the various components of the model (Fig. 1) and show how they influence the response to various pollutants.

Mechanisms of Response

Avoidance Mechanisms

POLLUTANT UPTAKE

Phytotoxicity results from biochemical processes that occur within the plant cell or on its surface. Consequently, the ease with which the pollutant moves from the ambient air into the plant is a key factor controlling plant response (Fig. 1). Gaseous pollutant uptake is affected by factors including patterns of stomatal opening and closing, stomatal frequency, morphological characteristics, and cuticle composition and structure. Using a combination of either physical or chemical processes, the pollutant stress can be either partially or completely excluded from the plant (Levitt 1972). Resistant cultivars of *Allium cepa* L. displayed rapid stomatal closure when the plants were exposed to O_3, which resulted in complete stress avoidance (Engle and Gabelman 1966). However, partial stress avoidance is more common (Tingey and Taylor 1982). For example, in *Petunia hybrida* Vilm. O_3 injury was positively correlated with leaf conductance (Thorne and Hanson 1976).

Ozone and SO_2 can alter normal patterns of stomatal opening and closing (Winner et al. 1988), thereby affecting further pollutant uptake. Pollutant-induced avoidance may result from a direct impact of the pollutant on the guard cells or may act indirectly through changes in the carboxylation capacity and a subsequent change in the internal CO_2 concentration. Direct stomatal closure in response to SO_2 may result from guard cell death and subsequent loss of turgor (Black and Black 1979a, 1979b). Indirect stomatal closure has been observed in *Raphanus sativus* L. and other species exposed to SO_2 (400 ppb) (Winner et al. 1988; Alscher et al. 1987). In these cases, photosynthesis decreased prior to a decrease in conductance, and stomatal closure may have resulted from an increase in internal CO_2 concentration. The mechanisms of stomatal closure may vary depending on pollutant concentration. Feedback mechanisms such as stomatal closure ultimately act to exclude further pollutant uptake, effectively increasing the ambient air threshold concentration necessary for response.

The stomatal control of gas exchange is under both intrinsic and environmental control (Larcher 1975; Nobel 1974). Some species have intrinsically higher O_3 uptake rates than others. For example, most herbaceous plants have a higher uptake rate than woody species; uptake is highly correlated with the transpiration rate (Thorne and Hanson 1972). Although broad-based generalizations are difficult to make, species exhibiting high stomatal conductances tend to be more susceptible to gaseous pollutant uptake (Reich 1987).

The intrinsic (gene) control of pollutant uptake can be overridden by environmental or edaphic factors (e.g., Tingey and Taylor 1982). For example, drought stress can override intrinsic plant sensitivity to O_3. Tingey et al. (1982) reported a decrease in O_3 sensitivity in water stressed *Phaseolus vulgaris* L. which was correlated with decreased leaf conductance. The response was related to decreased O_3 uptake in water-stressed plants and was reversible when water stress was removed. Other environmental factors such as humidity can also override intrinsic factors affecting pollutant uptake (Black and Unsworth 1980; Winner et al. 1988; McLaughlin and Taylor 1981). McLaughlin and Taylor (1981) found that foliar O_3 and SO_2 uptake was increased two- to fourfold by an increase in relative humidity from 35% to 75%. The mechanism underlying these differences was related to a change in internal leaf conductance rather than an effect on the stomata.

Genetic factors controlling cuticular composition and structure can also modify plant response to pollutants. Raddi and Rinallo (1989) correlated the pollutant sensitivity of two provenances of *Abies alba* Mill. to epistomatal wax characteristics. The resistant provenance maintained fibrillar waxes longer with age and pollutant exposure. Resistant trees growing in both polluted and nonpolluted areas maintained their fibrillar integrity, suggesting a genetic basis for their relative sensitivity. While other morphological factors could not be eliminated, the results suggest that pollutant sensitivity was related to cuticular properties and that amorphous plug formation may adversely effect gas exchange capabilities. In another study, O_3 exposure altered epicuticular wax structure in *Picea abies* (L.) Karst. (Barnes et al. 1988). These workers found a change from the crystalline arrangement of fine tubes to an amorphous layer that ultimately occluded the stomata in the trees exposed to O_3. This change is typically associated with needle aging, and O_3 was shown to accelerate the process. Stomatal occlusion may limit further uptake of pollutants, however, CO_2 diffusion may also be limited and could contribute to premature needle shed (Sauter and Voss 1986). In addition, stomatal occlusion may result in greater photooxidation and increased leaf temperatures due to a reduction in evaporative cooling. Barnes and Davison (1988) found that O_3 exposure increased cuticular transpiration in Norway spruce, which may predispose trees to winter desiccation. Genotypic variation in epicuticular wax degradation appears to be high, suggesting genetic control of composition may influence pollutant response.

Genotypic differences in cuticular composition or thickness may be part of an avoidance mechanism for UV-B. Enhanced UV-B radiation significantly increased cuticular waxes through accelerated wax biosynthesis (Steinmüller and Tevini 1986). Although most cuticular waxes do not absorb within the visible and near UV-wavelengths, the structural arrangement of the waxes on the leaf surfaces appears to enhance light reflection and scattering (Steinmüller and Tevini 1986). One could postulate O_3-induced changes in surface wax structure (Barnes et al. 1988) could also alter its UV reflectance and consequently influence the avoidance mechanism.

Intrinsic and environmental factors affecting leaf conductance (stomatal and cuticular) affect the rate at which pollutants diffuse into the leaf interior. This relationship has obvious consequences for the concentration of the pollutant or its metabolites in the cell, which influences the magnitude of plant response to the pollutant. However, in most cases, knowledge of pollutant uptake rate alone is insufficient to predict plant sensitivity (e.g., Tingey and Taylor 1982; Tingey and Olszyk 1985) because of the range of interactions which influence plant response. Pollutant uptake is only one of several important factors that control pollutant sensitivity (Fig. 1).

Transport to Target Site

Once a pollutant comes in contact with the leaf interior, factors such as solubility, absorption rate, transport, metabolism, and detoxification processes influence the pollutant concentration/amount that reaches the target site.

Foliar sensitivity to O_3 has been associated with ascorbic acid levels; plants with high levels are less sensitive than plants with low levels (e.g., Hanson et al. 1971; Lee et al. 1984; Tanaka et al. 1985). Recent calculations by Chameides (1989) have suggested that a major fraction of the O_3 diffusing through the stomata could be detoxified via reaction with ascorbic acid in the extracellular H_2O. This possible mechanism for reducing the O_3 concentration before it reaches the target site is supported by the observation that O_3 exposures initially reduce ascorbic acid concentrations (e.g., Tanaka et al. 1985). Subsequently, there is an increase in the ascorbic acid concentration (to $1 mM$) and cationic peroxidase activity in the extracellular water of leaves (e.g., Lee et al. 1984; Castillo and Greppin 1986).

When SO_2 dissolves into H_2O in and on the cell, it is rapidly hydrated, forming HSO_3^{-1} and SO_3^{-2} (Peiser and Yang 1985). These compounds are substantially more toxic than SO_4^{-2} and have both been implicated in SO_2 toxicity (Thomas 1961). In *Lycorpersicon esculentum* Mill. foliar SO_3^{-2} levels increased with SO_2 exposure, and tissue SO_3^{-2} concentrations were highly correlated with foliar injury (Howe and Woltz 1981).

The plasma membrane was found to be a barrier to the uptake and transport of both SO_2 and SO_3^{-2} in studies on the inheritance of SO_2 resistance in crosses and inbred lines of *Cucumis sativus* L. (Bressan et al. 1981). The authors suggested that this membrane barrier contributed to differential SO_2 sensitivity. They supported their conclusion by the observation that the relative resistance to SO_2 or HSO_3^{-1} injury was the same for a variety of other stresses likely to affect membrane properties.

Detoxification systems capable of protecting sensitive tissues from pollutants and their derivatives are common in plants and are subject to genetic control (avoidance mechanisms). The oxidation of SO_3^{-2} to SO_4^{-2} has been proposed as a potential detoxification mechanism for SO_2. Exogenous SO_3^{-2} was metabolized more rapidly in *Glycine max* (L.) Merr. cultivars resistant to foliar SO_2 injury than in sensitive cultivars (Miller and Xerikos 1979). Kondo et al. (1980) reported that SO_2 tolerance in various species was correlated with sulfite oxidase activity. Ayazloo et al. (1982) found that a sensitive genotype of *Lolium*, S23, had a higher proportion of inorganic S as SO_3^{-2} than the resistant genotypes after both light and dark incubations in SO_3^{-2} solutions.

The photoreduction of SO_3^{-2} to H_2S and subsequent volatilization is another possible detoxification mechanism (Rennenberg 1984). Dodrill (1976) suggested that the emission of H_2S was a method of detoxifying SO_2, since H_2S emissions were stimulated at lower SO_2 concentrations than would cause foliar injury. Plants exposed to SO_2 emit H_2S in the light (DeCormis 1968; Silvius et al. 1976; Hällgren and Fredriksson 1982). Hydrogen sulfide was detected within 15 to 30 min after the start of exposure, with emissions reaching a maximum within 45 min (DeCormis 1968; Taylor and Tingey 1983). An uptake rate of at least $48 \mu mol$ SO_2 m^{-2} over 15 min was required to stimulate emissions, suggesting that the cellular concentration had to exceed a threshold before emissions occurred (Taylor and Tingey 1983). However, differential emission rates ("detoxification") were not correlated with ecotypic differences in SO_2 sensitivity (Taylor et al. 1986).

Ozone and SO_2 give rise to free radicals and H_2O_2 within plant cells (e.g., Heath 1980; Tingey and Taylor 1982; Alscher and Amthor 1988). Biological systems have evolved protective scavenging or buffering systems such as superoxide dismutase (SOD), catalase (CAT), peroxidases, ascorbate, and α-tocopherol in response to free radicals (e.g., Heath 1980; Alscher and Amthor 1988). Superoxide dismutase converts the superoxide anion to H_2O_2 which is subsequently decomposed to H_2O and O_2 by catalase. The importance of free-radical scavenging mechanisms in aerobic metabolism (Fridovich 1978; Salin 1988) suggests that they are important processes controlling the amount of O_3 reaching the cellular target site. The rates of pollutant absorption appear to be more important than the total absorbed dose (Bennett et al. 1984). Healthy leaf mesophyll tissue can quantitatively absorb subinjurious concentrations of O_3. If the redox enzymes and other detoxifying processes are functioning, limited protection may be expected if saturation conditions are not exceeded.

The pools of SOD and CAT are under genetic and environmental control (Scandalios 1987). Ethylene diurea (EDU), a chemical that reduces O_3 injury, induces higher tissue levels of SOD and CAT in *Phaseolus vulgaris* L. leaves; these elevated levels have been correlated with reduced foliar injury (Lee and Bennett 1982; Bennett et al. 1984). However, this clear association between elevated levels of scavenging enzymes and foliar injury is not found in all *Phaseolus vulgaris* L. cultivars. For example, O_3 exposure increased SOD activity in *Phaseolus vulgaris* L. leaves and in some cultivars lipid-soluble antioxidants (McKersie et al. 1982). However, these increases in enzyme activity were not always the physiological basis of differential cultivar sensitivity. Chanway and Runeckles (1984a) found that EDU did induce protection from O_3 injury, but SOD levels were not increased in their cultivars. However, SOD levels did increase simultaneously with the appearance of visible foliar injury (Chanway and Runeckles 1984b).

The apparent differential importance of SOD and CAT among *Phaseolus vulgaris* L. cultivars discussed above does not invalidate the role of SOD as an avoidance mechanism. The conceptual model indicates that there are several control points (both avoidance and tolerance) that are involved in differential sensitivity. The various *Phaseolus vulgaris* L. studies failed to provide a broad-scale assessment of the various avoidance and tolerance factors that controlled plant response. It is likely that among various cultivars the several avoidance and tolerance mechanisms were of differential importance.

Tolerance Mechanisms

To understand the effects of a pollutant, it is more important to understand the processes driving the metabolic adjustments rather than the metabolic fate of specific pollutant molecules. Plant metabolism is functionally compartmentalized with interconnected metabolic pathways that are optimized by feedback loops. Balance between the metabolic pathways is continuously controlled and adjusted in response to environmental changes. The responses to environmental change involve coordinated readjustments in the plant's metabolic paths rather than a few "specific" reactions (Queiroz 1988).

Plant response to stress is a function of the interaction between the frequency or intensity of the stress and the rate at which the plant allocates resources for maintenance, repair, and defense. Plants allocate resources in a manner which overcomes the resource deficiency that most limits plant growth. Allocation occurs in a direction that allows the plant to deal with or compensate for the stress. Such a result has the obvious evolutionary advantage of bringing the plant into balance with its biotic and abiotic environment.

RESERVE CAPACITY

Repair and compensatory mechanisms that lead to a new metabolic state following the target interaction can utilize the existing capacity of the plant, without a modification of gene expression (Queiroz 1988). Wellburn et al. (1972) reported that the SO_2-induced swelling of thylakoids was reversible if polluted air was replaced by clean air. In *Chlorella*, O_3 increased potassium efflux, however, within 2–3 min after ozonation the normal efflux rate was reestablished suggesting an active repair process (Heath et al. 1974). The rate of recovery was too rapid to be attributed to new protein synthesis. In *Phaseolus vulgaris* L. leaves, the membrane repair rate was enhanced by continuous light or glucose supplements and retarded by low temperatures or darkness following exposure (Sutton and Ting 1977a, 1977b). One early symptom of acute pollutant injury is the appearance of water-soaked interveinal areas on leaves. If these areas are not subjected to extensive drying by either wind or light, H_2O may be reabsorbed into the cells, and thus result in little or no visual injury; the initial stages of injury are therefore reversible.

Ozone has been found to increase ATP and total adenylate contents (Pell 1974) and increase respiration rates (Amthor 1988; Amthor and Cumming 1988; Barnes 1972; Skärby et al. 1987). Skärby et al. (1987) found that accumulated respiration was 60% higher in O_3-exposed shoots than in controls of 20-year-old *Pinus sylvestris* L. Amthor (1988) and Amthor and Cumming (1988) mathematically partitioned respiration into growth and maintenance components in *Phaseolus vulgaris* L. exposed to O_3. They found that growth respiration was unaffected by O_3 while that due to maintenance was increased by as much as 25%. If sufficient carbon substrates are available, the reserve capacity may be sufficient to accommodate the increased maintenance and repair costs associated with O_3 stress.

Another example of homeostatic readjustment via reserve capacity can be found in the glutathione system, which prevents oxidative breakdown of unsaturated lipids and biomembranes, and maintains proteins, cysteine and homocysteine in a reduced state (i.e., metabolically active) (Leshem 1988; Rennenberg 1982). This protection is mediated by the oxidation of the thiol groups on the tripeptide to form the glutathione dimer conjugate (GSSG). In chloroplasts, this oxidation is mediated by the ascorbate-dehydroascorbate cycle (Leshem 1988; Salin 1988); glutathione can also be directly oxidized by oxidants such as O_3 (Mudd 1982). Glutathione reductase reduces GSSG to glutathione using NADPH for the reducing equivalents (Leshem 1988; Rennenberg 1982). The reduction of GSSG is important to maintain metabolism as GSSG can inactivate light-activated enzymes in the Calvin cycle (Rennenberg 1982). The reduction of GSSG to glutathione via glutathione reductase would be expected to stimulate the pentose phosphate pathway (Mudd 1982). Ozone has been shown to stimulate the activity of glucose-6-phosphate dehydrogenase, suggesting the pentose phosphate pathway was stimulated (Tingey et al. 1976). The increased enzyme occurred earlier in the sensitive cultivar than the resistant one, possibly as a result of greater glutathione oxidation.

Plant protective mechanisms to UV may be grouped into three main classes: (1) UV damage is repaired or its effects negated; (2) the amount of UV radiation reaching the target site is reduced; and (3) plant response minimizes the effects of the UV stress (Beggs et al. 1986). An example of a repair process is the photoreactivation of UV stress. The inhibitory effect of UV on leaf enlargement is less if the plants are placed in light immediately following UV exposure than if they are placed in the dark (Beggs et al. 1986). UV stress, as measured by the UV induction of coumestrol, was reversible if the plants were placed in white light but not blue light (Beggs et al. 1985). Sunlight counteracts the UV inhibition of phytochrome-mediated anthocyanin formation (Beggs et al. 1986). In *Glycine max* (L.) Merr., UV-B inhibited photosynthesis only in leaves that expanded and were irradiated under a low PPFD indicating that a high PPFD during leaf expansion or irradiation was effective in minimizing the detrimental effects of UV-B (Mirecki and Teramura 1984). Warner and Caldwell (1983) reported the relative magnitude of photosynthesis inhibition was similar under either light-limited or light-saturated conditions. However, under the low light growth conditions, UV-B treatment reduced chlorophyll content, while no effect was observed under the light-saturated treatment.

MODIFICATION OF GENE EXPRESSION

Under environmental stress, cells can rapidly respond in a specific manner by selectively increasing or decreasing the expression of specific genes (Sachs and Ho 1986). Genes whose expression is increased during times of stress presumably are critical to the accommodation of the organism to adverse conditions (Matters and Scandalios 1986). The variety of proteins and mRNAs which are synthesized during stress indicates no exclusive simple group of "stress proteins" per se, although several genes, such as those encoding the heat shock proteins, are commonly expressed under more than one stress condition (Matters and Scandalios 1986). The function of many of the stress proteins is unknown, but the highly reproducible and specific nature of stress proteins indicates a specific role in plant metabolic systems. Other compounds, such as stress ethylene, are produced under a variety of stresses and may mediate plant processes (Gunderson and Taylor 1988).

There is substantial evidence suggesting that increased oxidative stress produced by environmental factors can alter superoxide dismutase (SOD) and catalase (CAT) gene expression in many organisms (Scandalios 1987). Oxidative stress induces specific isozymes of SOD, and enhances the expression of both SOD and CAT genes. A variety of environmental factors can create conditions of intracellular oxidative stress (redox-active compounds like paraquat, elevated O_2, hyperbaric O_2, air pollutants, and increased temperature) which result in these metabolic readjustments.

In animals, metal-binding proteins (metallothionein) have been identified, isolated, and implicated in the detoxification of and protection from metal toxicity as well as the storage and transport of metals (Cherian and Goyer 1978). Plants exposed to heavy metals such as Cd or Zn produce "metallothionein-like" proteins which bind heavy metals (Blakeley et al. 1986; Grünhage et al. 1985; Rauser 1984; Wagner and Trotter 1982). These proteins, rich in cysteine, are found in both the roots and leaves. Heavy metal complexing peptides are also formed (Grill et al. 1985; Robinson and Jackson 1986). The observations that the metal-binding proteins are metal inducible has led to the conclusion that these compounds function as a protective system against metal toxicity (Robinson and

Jackson 1986). The contention is supported by the observation that metal-resistant plants produce higher than normal levels of the metal-binding proteins. However, both metal-sensitive and metal-resistant plants produce the protein. In some genotypes, metal-binding proteins appear to be important avoidance/tolerance mechanisms (Rauser and Curvetto 1980; Robinson and Jackson 1986).

Plant protective mechanisms to UV light can influence the amount of UV radiation reaching the target site, which is an example of feedback control increasing stress avoidance. For example, UV induces the formation of flavonoids, anthocyanins, and phenols in plant tissue (Caldwell et al. 1983; Beggs et al. 1986; Robberecht and Caldwell 1986). The flavonoids and related phenols are wave-length selective filters for UV which allow the penetration of visible radiation but remove much of the UV radiation. Flavonoid synthesis is apparently mediated via UV-induced pyrimidine dimers (Beggs et al. 1985). UV treatment causes the coordinated induction of enzymes in the flavonoid biosynthetic path, including, phenylalanine ammonialyase, 4-coumarate: CoA ligase, chalcone synthase and UDP-apiose synthase (Kuhn et al. 1984). Using cDNA probes, it was found that the UV treatment increased the levels of the mRNA encoding these enzymes and that this increased the transcription rates of the genes (Kuhn et al. 1984; Chappell and Hahlbrock 1984). These compounds accumulate in the epidermis of plants exposed to UV and can significantly affect the amount of UV radiation reaching the mesophyll tissue. A UV-sensitive species, *Rumex patientia* L., has a higher epidermal transmission than *R. obtusifolius* L. (Robberecht and Caldwell 1986).

Ozone exposure also increases the activity of phenylalanine ammonialyase and induces the formation of phenols and anthocyanins in leaf tissue (e.g., Tingey et al. 1976; Koukol and Dugger 1967; Howell and Kremer 1973). The similar effect of O_3 and UV light on these metabolites leads to the suggestion that the O_3-induced increase in UV-B absorbing metabolites could provide a partial protection from UV-B. However, it is unlikely that UV-induced increase in phenols and anthocyanins will alter O_3 sensitivity as there is not evidence that they are involved in either O_3 tolerance or avoidance mechanisms.

Foliar received NO_2 can induce the formation of nitrate reductase in plant foliage (Zeevaart 1974; Wellburn 1982; Wingsle et al. 1987; Rowland 1986; Norby et al. 1989). Nitrogen dioxide reacts with extracellular H_2O and results in the formation of NO_3^- and NO_2^-. In addition, Lindberg et al. (1986) has shown that dry deposition of NO_3^- (primarily as HNO_3 vapor) is an extremely important source of anthropogenic N input into forested ecosystems. Nitrate reductase enzymes reduce NO_3^- to NH_4^+ which is subsequently metabolized into amino acids and proteins (Zeevaart 1976; Rogers et al. 1979; Rowland et al. 1987). This inducible enzyme system is an example of stressor-induced changes in gene expression. The consequence of the altered gene expression is enzyme induction with the subsequent metabolism of the stressor and its sequestering amino acids and protein.

Various inducible systems appear to be important in pollutant stress response (i.e., "stress proteins" or "stress metabolites"). At this time, no single factor (initiator) seems to be responsible for inducing "stress proteins" or "stress metabolites." Air pollutants stimulate the formation of stress ethylene, as do other abiotic and biotic stresses (e.g., Abeles 1973; Bressan et al. 1979; Tingey 1980). This hormone may function as a signal that triggers changes in gene expression during stress. An increase in abnormal or denatured proteins occurs under heat shock, which may serve as the trigger (Matters and Scandalios 1986). The altered proteins are preferentially degraded by cellular proteases, causing the build-up of the heat shock activating factor and the induction of heat shock gene expression.

Interactions of Changes in Atmospheric Quality

Although it is important to understand the basic mechanisms influencing plant response, it is equally important to consider how the basic response mechanism will be influenced by changes in other environmental factors such as altered rainfall (i.e., drought) or elevated CO_2 concentrations, as the various stress factors seldom occur individually. There have been extensive studies on the effects of individual factors on the mechanism of plant response, but there has been little concern for the interactive effects of several environmental stress factors.

Johnson (1989) developed a conceptual model for evaluating factors involved in forest decline in the northeastern United States (Fig. 2). This model, developed from concepts described earlier by Manion (1981), is based on the available energy for avoidance, repair, and compensation processes. As such it is conceptually similar to the proposed model for differential sensitivity (Fig. 1). At either the cellular or the individual level, a major factor in determining if repair/compensation occurs and a new stable metabolic state is reached, is the available energy (carbon). In Johnson's model, predisposing factors reduce the amount of carbon substrate available for defense (avoidance) and repair/compensation, but not to the extent that a response threshold level is reached. An inciting stress (either atmospheric pollutant or abiotic/biotic stress) shifts the response below a threshold level, resulting in the appearance of stress symptoms. If the inciting stress is sufficient to reduce available carbon for repair to a level below the critical level, the decline phase may directly result. If the stress is episodic, the plant may recover. However, if a secondary episodic stress (the same or different) occurs, it could delay the time to recovery or sufficiently reduce the available energy to shift the response below the critical level, resulting in decline. Each phase is a continuous process which involves metabolic readjustments, the magnitude of which are determined by the availability of carbon substrates. The decline stage in the Johnson model is analogous to the destabilized state in Fig. 1.

Two examples of physiologically interacting/offsetting events which may occur as a result of changes in ambient air quality will be discussed using the concepts from the conceptual models of differential sensitivity (Figs. 1, 2).

DROUGHT STRESS—POLLUTANT INTERACTIONS

Drought stress induces a broad range of physiological and biochemical changes (e.g., Hsiao 1973; Hanson and Hitz 1982; Kramer 1983). For example, drought stress alters hormonal levels, influences metabolic processes and enzyme activities, reduces carbon fixation, alters carbon allocation, reduces leaf conductance and H_2O potential, and may increase leaf temperature. These metabolic and physiological changes ultimately influence plant growth and response to other stresses. Typically drought stress favors root growth at the expense of producing/maintaining leaf tissue or defensive compounds (Kramer 1983). This change in allocation can cause an additional metabolic readjustment or further decline and an increase in the susceptibility to pest attack or abiotic stresses (Schoeneweiss 1978). In contrast, drought-stressed plants are less susceptible to the herbicide paraquat than well-watered ones (Burke et al. 1985). In field grown *Glycine max* (L.) Merr., UV-B exposure reduced photosynthesis, transpiration, plant growth, and number of pods (Murali and Teramura 1986), whereas, in drought-stressed plants, UV-B had no significant effects on these parameters. The drought-stressed plants had a higher specific leaf weight and methanolic extracts displayed a higher UV absorbance (300 nm).

Increased atmospheric CO_2 can alter plant response to H_2O deficits. The development of internal drought stress was delayed during a drying cycle in *Liquidambar styraciflua* L. grown at high CO_2 levels (Tolley and Strain 1984, 1985). Seedlings maintained higher photosynthetic rates during the drying cycle and exhibited increased H_2O use efficiency (WUE). The response may have a genetic basis since other species such as *Pinus taeda* L. did not respond differentially to H_2O deficits when grown at high CO_2. Regardless of the mechanisms, increased WUE and drought tolerance may enable *Liquidambar styraciflua* L. to extend its range under elevated levels of CO_2.

Drought stress reduces the impacts of air pollutants on plants through both avoidance and tolerance mechanisms. Field and controlled environment studies have established that drought-stressed plants display less foliar injury, and O_3 impairs growth/yield to a lesser extent than in well-watered plants (e.g., Dean and Davis 1967; Markowski and Grzesiak 1974; Moser et al. 1988; Olszyk and Tibbitts 1981). The drought-induced reduction in stomatal conductance (an avoidance mechanism) reduces the uptake of gaseous pollutants such as O_3 (Rich and Turner 1972; Tingey and Hogsett 1985). Drought stress also increases plant resistance to oxidant stress through modification of gene expression. The levels of cellular antioxidant systems and glutathione reductase increase under water stress (Burke et al. 1985; Gamble and Burke 1984).

Queiroz (1988) suggested that drought stress perturbed the metabolic balance so that the energy and/or enzyme capacity required to repair or compensate for the additional stress from an air pollutant was no longer available. In a study of the interactive effects of drought stress and SO_2, a low concentration of SO_2 (80 ppb up to 6 weeks) had no effect on the protein levels in well-watered *Picea abies* L. Karst. (5-month- and 5-year-old plants) (Pierre and Queiroz 1988; Queiroz 1988). Drought stress (H_2O withheld for 1 or 2 weeks) reduced foliar proteins by 17% to 28%. The impact of drought stress was intensified by SO_2 which reduced the protein level by 38% to 54%. The activities of isocitrate, glucose-6-phosphate, and glutamate dehydrogenases showed the same response patterns as the protein levels. Foliar injury developed only on the plants that were drought stressed and exposed to SO_2. When the water-stressed plants were re-watered, the protein and enzyme activities increased to normal levels within a few days. However, in young *Picea abies* L. Karst. plants exposed to SO_2 and water stress, only a portion of the plants recovered (11% of the plants died). In another study, SO_2 had no effect on photosynthesis after 5 weeks in well-watered *Picea abies* L. Karst. (Cornic 1987). When H_2O was withheld, reductions in photosynthesis were the greatest in SO_2-treated plants. Studies with field-grown *Glycine max* L. Merr. found that drought and O_3 stresses reduced yields by 4% and 5%, respectively, while the combined stresses reduced yields by 25% (Heggestad et al. 1985).

Interactions among various pollutants and natural stresses such as drought are complex. The above examples suggest a strong genetic basis for plant response. The examples clearly show that not all interactions will be detrimental. In fact, increasing the CO_2 concentration may ameliorate some of the detrimental effects of gaseous pollutants and in some species may increase drought tolerance.

CO_2—POLLUTANT INTERACTIONS

Increasing atmospheric concentrations of CO_2 reduces stomatal conductance (e.g., Acock and Allen 1985), decreasing the uptake of other pollutants. Stomatal frequency decreased when the partial pressure and mole fraction of CO_2 increased (Woodward and Bazzaz 1988). However, stomatal frequency was unchanged when the partial pressure of CO_2 was held constant and mole fraction was increased, suggesting stomatal frequency was directly controlled by the CO_2 partial pressure. An examination of herbarium species collected from the 18th, 19th, and 20th centuries supports the possibility that stomatal frequency has decreased as the ambient CO_2 concentrations have increased (Woodward 1987).

Despite effects of increased CO_2 on stomatal conductance and frequency, the overall effect of increased CO_2 on growth is stimulatory, particularly for C-3 plants. *Quercus alba* L. seedlings grown under elevated CO_2 concentrations (690 ppm) showed greater growth and H_2O use efficiency than those grown under current ambient levels (Norby et al. 1986). The photosynthetic rates of plants grown at lower CO_2 concentrations (225 to 260 ppm compared with 340 ppm) were lower for a given stomatal frequency, suggesting that biochemical factors affecting CO_2 fixation were altered at the lower concentrations (Woodward and Bazzaz 1988). Fetcher et al. (1988) found that elevated CO_2 increased growth of *Pinus taeda* L. and *Liguidambar styraciflua* L. even though carbon exchange rate (CER) of *Liquidambar styraciflua* L. was reduced. Further examination of the response revealed that the reduction in CER in *Liquidambar styraciflua* L. resulted from both stomatal limitations to CO_2 uptake and from a reduced biochemical capacity for CO_2 fixation, presumably from reduced ribulose bisphosphate carboxylase (rubisco) activity. The apparent conflict between an increase in net productivity and decreased CER in *Liquidambar styraciflua* L. may have resulted from a change in the CER:PPFD relationship or from increased leaf area under elevated CO_2.

Because carbon fixation and growth is enhanced under elevated CO_2, additional substrates may be available for maintenance and repair processes, enabling the plant to tolerate higher levels of other pollutant stresses (Figs. 1, 2). For example, the inhibitory effects of $SO_2 + NO_2$ on the photosynthetic rate of *Medicago sativa* L. "Ranger" was 50% less at high CO_2 than at low CO_2 (Hou et al. 1977). The protection resulted from both avoidance of $SO_2 + NO_2$ uptake via stomatal closure at high CO_2 and increased inactivation of the toxic SO_3^{-2}. In *Glycine max* (L.) Merr., photosynthesis declined with SO_2 fumigation; the magnitude of the reduction was greater at low than high CO_2 levels (Carlson 1983). Further study showed that at equivalent SO_2 uptake rates, the reductions in photosynthesis were less when the plants were grown at high CO_2. The results were similar to those of Carlson and Bazzaz (1982), which showed that high CO_2 protected plants via two mechanisms: stomatal closure increasing resistance to SO_2 uptake and altered processes at the biochemical level. It is possible that increased carbohydrate was available for maintenance and repair processes or detoxification reactions. Koziol and Jordan (1978) found a correlation between photosynthesis and respiration rates, leaf carbohydrate levels, and pollutant damage in plants grown at ambient CO_2 and suggested that the availability of carbon substrates for maintenance and repair may be related to pollutant sensitivity.

Carlson and Bazzaz (1982) studied both C-3 and C-4 plants to determine the effects of elevated CO_2 on SO_2 toxicity. Growth of the C-3 species was stimulated more than the growth of C-4 species under elevated CO_2. At low CO_2, C-3 plants were more sensitive to SO_2 than C-4 plants, while at high CO_2, C-4 plants were more sensitive to SO_2. Differences in stomatal response between C-3 and C-4 species were suggested as factors controlling the differential response.

The overall effect of elevated CO_2 shifts the threshold concentration necessary to elicit a plant response to higher levels of SO_2 and/or NO_2. If elevated CO_2 decreases plant susceptibility via detoxification of toxic SO_3^{-2}, the activation of specific enzyme systems may be involved. This would be an example of homeostatic readjustment through modification of gene expression (Fig. 1). Elevated CO_2 can provide the increased carbon substrate necessary for this metabolically expensive process, decreasing the susceptibility of a plant to an inciting stress (Fig. 2).

Conclusion

The conceptual model (Fig. 1) of plant response to pollutant stress is generic for various stresses regardless of their origin or site of action and has many features that are subject to genetic control. Plants have various mechanisms which enable them to (1) avoid (either through exclusion or detoxification systems) or (2) tolerate stress through metabolic readjustments. Metabolic readjustments involve utilization of reserve capacity with minimal energy cost or modification of gene expression at relatively greater energy cost.

Traits affecting sensitivity to pollutants may be induced by other natural stresses. For example, water stress may induce the formation of oxidant scavengers which are also important in detoxifying O_3 and its derivatives. Consequently, it is important to distinguish between response to the pollutant and actual evolution of resistance in response to selection by pollutants.

The few studies which have examined the interactions of multiple stresses simultaneously have shown that plant responses are not additive. This may result in part from the complex nature of genotype/environment interactions and the role of environment on genotypic expression. Differential plant sensitivity is influenced by genetically determined traits mediated through various mechanisms.

Differential plant sensitivity is a result of a range of physiological processes which are subject to genetic control. The relative importance of tolerance and avoidance mechanisms in controlling sensitivity varies within and among species. The magnitude of the response is determined by the amount of the pollutant reaching the target site and the ability of the plant to reestablish homeostatic equilibrium. In general, avoidance may be an effective accommodation to occasional acute episodes while tolerance is more likely with chronic pollutant exposure. However, mechanisms of response are dynamic and are influenced by a multitude of factors; tolerance and avoidance mechanisms likely do not act independently. Future research efforts should be directed at characterizing interactive roles of avoidance and tolerance processes and the interaction of various atmospheric stresses simultaneously rather than identifying responses to individual stresses.

References

Abeles FB (1973) Ethylene in plant biology. Academic Press, New York

Acock B, Allen LH Jr (1985) Crop responses to elevated carbon dioxide concentrations. In: Strain BR, Cure JD (eds) Direct effects of increasing carbon dioxide on vegetation. US Department of Energy, Office of Energy Research, Washington, DC, pp 53–97

Alscher RG, Amthor JS (1988) The physiology of free-radical scavenging: maintenance and repair processes. In: Schulte-Hostede S, Darrally NM, Blank LW, Wellburn AR (eds) Air pollution and plant metabolism. Elsevier Applied Science, New York

Alscher R, Bower JL, Zipfel W (1987) The basis for different sensitivities of photosynthesis to SO_2 in two cultivars of pea. Journal of Experimental Botany 38:99–108

Amthor JS (1988) Growth and maintenance respiration in leaves of bean (*Phaseolus vulgaris* L.) exposed to ozone in open-top chambers in the field. New Phytologist 100:319–325

Amthor JS, Cumming JR (1988) Low levels of ozone increase bean leaf maintenance respiration. Canadian Journal of Botany 66:724–726

Ariens EJ, Simonis AM, Offermerier J (1976) Introduction to general toxicology. Academic Press, New York

Ayazloo M, Garsed SG, Bell JNB (1982) Studies on the tolerance to sulfur dioxide of grass populations in polluted areas, II: Morphological and physiological investigations. New Phytologist 90:109–126

Barnes JD, Davison AW (1988) The Influence of ozone on the winter hardiness of Norway spruce [*Picea abies* (L.)]. New Phytologist 108:159–166

Barnes JD, Davison AW, Booth TA (1988) Ozone accelerates structural degradation of epicuticular wax on Norway spruce needles. New Phytologist 110:309–318

Barnes RL (1972) Effects of chronic exposure to ozone on photosynthesis and respiration of pines. Environmental Pollution 3:133–138

Beggs CJ, Schneider-Ziebert U, Wellman E (1986) UV-B radiation and adaptive mechanisms in plants. In: Worrest RC, Caldwell MM (eds) Stratospheric ozone reduction, solar radiation and plant life. Springer-Verlag, Berlin, pp 235–250

Beggs CJ, Stolzer-Jehle A, Wellmann E (1985) Isoflavonoid formation as an indicator of UV stress in bean (*Phaseolus vulgaris* L.) leaves. The significance of photorepair in assessing potential damage by increased solar UV-B radiation. Plant Physiology 79:630–634

Bennett JH, Lee EH, Heggestad HE (1984) Biochemical aspects of plant tolerance to ozone and oxyradicals: superoxide dismutase. In: Koziol MJ, Whatley FR (eds) Gaseous air pollutants and plant metabolism. Butterworths, London, pp 413–424

Black CR, Black VJ (1979a) The effects of low concentrations of sulphur dioxide on stomatal conductance and epidermal cell survival in field bean (*Vicia faba* L.). Journal of Experimental Botany 31:667–677

Black CR, Black VJ (1979b) Light and scanning electron microscopy of SO$_2$-induced injury to leaf surfaces of field bean (*Vicia faba* L.). Plant, Cell and Environment 2:329–333

Black VJ, Unsworth MH (1980) Stomatal responses to sulfur dioxide and vapour pressure deficit. Journal of Experimental Botany 31:667–677

Blakeley SD, Robaglia C, Brzezinski R, Thirion J-P (1986) Induction of low molecular weight cadmium-binding compound in soybean roots. Journal of Experimental Botany 37:956–964

Bressan RA, LeCureux L, Wilson LG, Filner P (1979) Emission of ethylene and ethane by leaf tissue exposed to injurious concentrations of sulfur dioxide or bisulfite ion. Plant Physiology 63:924–930

Bressan RA, LeCureux L, Wilson LG, Filner P, Baker LR (1981) Inheritance of resistance to sulfur dioxide in cucumber. HortScience 16:332–333

Burke JJ, Gamble PE, Hatfield JL, Quisenberry JE (1985) Plant morphological and biochemical responses to field water deficits. I. Responses of glutathione reductase activity and paraquat sensitivity. Plant Physiology 79:415–419

Butler LK, Tibbitts TW, Bliss FA (1979) Inheritance of resistance to ozone in *Phaseolus vulgaris* L. Journal of the American Society of Horticultural Science 104:211–213

Caldwell MM, Robberecht R, Flint SD (1983) Internal filters: prospects for UV-acclimation in higher plants. Physiologia Plantarum 5:445–450

Carlson RW (1983) The effect of SO$_2$ on photosynthesis and leaf resistance at varying concentrations of CO$_2$. Environmental Pollution (Series A) 30:309–321

Carlson RW, Bazzaz FA (1982) Photosynthetic and growth response to fumigation with SO$_2$ at elevated CO$_2$ for C$_3$ and C$_4$ plants. Oecologia 54:50–54

Castillo FJ, Greppin H (1986) Balance between anionic and cationic extracellular peroxidase activities in *Sedum album* leaves after ozone exposure. Analysis by high-performance liquid chromatography. Physiologia Plantarum 68:201–208

Chameides WL (1989) The chemistry of ozone deposition to plant leaves: role of ascorbic acid. Environmental Science and Technology 23:595–600

Chanway CP, Runeckles VC (1984a) The role of superoxide dismutase in the susceptibility of bean leaves to ozone injury. Canadian Journal of Botany 62:236–240

Chanway CP, Runeckles VC (1984b) Effect of ethylene diurea (EDU) on ozone tolerance and superoxide dismutase activity in bush bean. Environmental Pollution (Series A) 35:49–56

Chappell J, Hahlbrock K (1984) Transcription of plant defense genes in response to UV-light or fungal elicitor. Nature 311:76–78

Cherian MG, Goyer RA (1978) Metallothioneins and their role in the metabolism and toxicity of metals. Life Sciences 23:1–10

Cornic G (1987) Interaction between a sublethal pollution by SO_2 and water stress. The effect on photosynthetic capacity. Physiologia Plantarum 71:115–119

Dean CR, Davis DR (1967) Ozone and soil moisture in relation to the occurrence of weather fleck on Florida cigar-wrapper tobacco in 1966. Plant Disease Reporter 51:72–75

DeCormis L (1968) Dégagement d'hydrogène sulfuré par des plantes soumises à une atmosphère contenant de l'anhydride sulfureux. Compte rendu de l'Academie de Sciences (serie D) 266:683–685

Dodrill SA (1976) Fate of sulfur-dioxide absorbed by foliage of coleus and bean in relation to visible injury. Ph.D. thesis, University of West Virginia, USA

Dragoescu N, Hill RR Jr Pell EJ (1988) An autotetraploid model for genetic analysis of ozone tolerance in potato, *Solanum tuberosum* L. Genome 29:85–90

Engle RL, Gabelman WH (1966) Inheritance and mechanism for resistance to ozone damage in onion, *Allium cepa* L. Proceedings of the American Society for Horticultural Science 89:423–430

Fetcher N, Jaeger CH, Strain BR, Sionit N (1988) Long-term elevation of atmospheric CO_2 concentration and the carbon exchange rates of saplings on *Pinus taeda* L. and *Liquidambar styraciflua* L. Tree Physiology 4:255–262

Fridovich I (1978) The biology of oxygen radicals. Science 201:875–880

Gamble PE, Burke JJ (1984) Effect of water stress on the chloroplast antioxidant system. I. Alterations in glutathione reductase activity. Plant Physiology 76:615–621

Grill E, Winnacker E-L, Zenk MH (1985) Phytochelatins: the principal heavy-metal complexing peptides of higher plants. Science 230:674–676

Grünhage L, Weigel H-J, Ilge D, Jäger H-J (1985) Isolation and partial characterization of a cadmium-binding protein from *Pisum sativum*. Plant Physiology 119:327–334

Gunderson CA, Taylor GE Jr (1988) Kinetics of inhibition of foliar gas exchange by exogenous ethylene: an ultrasensitive response. New Phytologist 110:517–524

Hällgren JE, Fredriksson SA (1982) Emission of hydrogen sulfide from sulfur dioxide fumigated pine trees. Plant Physiology 70:456–459

Hanson AD, Hitz WD (1982) Metabolic responses of mesophytes to plant water deficits. Annual Review of Plant Physiology 33:163–203

Hanson GP, Addis DH, Thorne L (1976) Inheritance of photochemical air pollution tolerance in petunia. Canadian Journal of Genetics and Cytology 6:75–83

Hanson GP, Thorne L, Jativa CD (1971) Ozone tolerance of petunia leaves as related to their ascorbic acid concentrations. In: Englund HM, Beery WT (eds) Proceedings of the Second International Clean Air Congress, Academic Press, New York, pp 261–266

Heath RL (1980) Initial events in injury to plants by air pollutants. Annual Review of Plant Physiology 31:395–431

Heath RL, Chimiklis P, Frederick P (1974) The role of potassium and lipids in ozone injury to plant membranes. In: Dugger M (ed) ACS Symposium Series 3, American Chemical Society, Washington, DC, pp 58–75

Heggestad HE, Gish TJ, Lee EH, Bennett JH, Douglass LW (1985) Interaction of soil moisture stress and ambient ozone on growth and yields of soybeans. Phytopathology 75:472–477

Hou L-Y, Hill AC, Soleimani A (1977) Influence of CO_2 on the effects of SO_2 and NO_2 on alfalfa. Environmental Pollution 12:7–16

Houston DB, Stairs GR (1973) Genetic control of sulfur dioxide and ozone tolerance in eastern white pine. Forest Science 19:267–271

Howe TK, Woltz SS (1981) Resistance of tomato cultivars to sulfur dioxide and accumulation of foliar sulfite related to sulfur dioxide susceptibility. HortScience 16:413

Howell RK, Kremer DF (1973) The chemistry and physiology of pigment in leaves injured by air pollution. Journal of Environmental Quality 2:434–438

Hsiao TC (1973) Plant responses to water stress. Annual Review of Plant Physiology 24:519–570

Johnson AH (1989) Decline of red spruce in the northern Applachians: determining if air pollution is an important factor. In: Biologic Markers of Air Pollution Stress and Damage in Forests, National Academy Press, Washington, DC, pp 91–104

Karnosky DF (1977) Evidence for genetic control of response to sulfur dioxide and ozone in *Populus tremuloides*. Canadian Journal of Forest Research 7: 437–440

Karnosky DF, Berrang PC, Scholz F, Bennett JP (1989) Variation in and natural selection for air pollution tolerances in trees. In: Scholz F, Gregorius H-R, Rudin D (eds) Genetic effects of air pollutants in forest tree populations. Springer-Verlag, Berlin, pp 29–37

Kerstiens G, Lendzian KJ (1989) Interactions between ozone and plant cuticles. I. Ozone deposition and permeability. New Phytologist 112:13–19

Kondo N, Akiyama Y, Fujiwara M, Sugahara K (1980) Sulfite-oxidizing activities in plants. In: Studies on the effects of air pollutants on plants and mechanisms of phytotoxicity. Research Report No. 11, National Institute of Environmental Studies, Iburaki, Japan, pp 137–150

Koukol J, Dugger WM Jr (1967) Anthocyanin formation as a response to ozone and smog treatment in *Rumex crispus* L. Plant Physiology 42:1023–1024

Koziol MJ, Jordan CF (1978) Changes in carbohydrate levels in red kindey bean (*Phaseolus vulgaris* L.) exposed to sulphur dioxide. Journal of Experimental Botany 29:1037–1043

Kramer PJ (1983) Water relations of plants. Academic Press, New York

Kuhn DN, Chappel J, Boudet A, Hahlbrock K (1984) Induction of phenylahanine ammonia-lyase and 4-coumarate: CoA ligase mRNAs in cultured plant cells by UV light or fungal elicitor. Proceedings of the National Academy of Science USA 81:1102–1106

Larcher W (1975) Physiological plant ecology. Springer-Verlag, Berlin

Lee EH, Bennett JH (1982) Superoxide dismutase. A possible protective enzyme against ozone injury in snap beans (*Phaseolus vulgaris* L.). Plant Physiology 69:1444–1449

Lee EH, Jersey A, Gifford C, Bennett J (1984) Differential ozone tolerance of soybean and snapbeans: analysis of ascorbic acid in O_3-susceptible and O_3-resistant cultivars by high-performance liquid chromatography. Environmental and Experimental Botany 24:331–341

Leshem YY (1988) Plant senescence processes and free radicals. Free Radical Biology and Medicine 5:39–49

Levitt J (1972) Responses of plants to environmental stress. Academic Press, New York

Lindberg SE, Lovett GM, Richter DD, Johnson DW (1986) Atmospheric deposition and canopy interactions of major ions in a forest. Science 231:143–145

Manion PD (1981) Tree disease concepts. Prentice-Hall, Englewood Cliffs, NJ

Markowski A, Grzesiak S (1974) Influence of sulphur dioxide and ozone on vegetation of bean and barley plants under different soil moisture conditions. Bulletin L'Académie Polonaise Sciences Série sciences biologiques 22:875–887

Matters GL, Scandalios JG (1986) Changes in plant gene expression during stress. Developmental Genetics 7:167–175

McKersie BD, Beversdorf WD, Hucl P (1982) The relationship between ozone insensitivity, lipid soluble antioxidants, and superoxide dismutase in *Phaseolus vulgaris*. Canadian Journal of Botany 60:2686–2691

McLaughlin SB, Taylor GE (1981) Relative humidity: important modifier of pollutant uptake by plants. Science 211:167–168

Miller JE, Xerikos P (1979) Residence time of sulfite in SO_2 "sensitive" and "tolerant" soybean cultivars. Environmental Pollution 18:259–264

Mirecki RM, Teramura AH (1984) Effects of ultraviolet-B irradiance on soybean. V. The dependence of plant sensitivity on the photosynthetic photon flux density during and after leaf expansion. Plant Physiology 74:475–480

Moser TJ, Tingey DT, Rodecap KD, Rossi DJ, Clark CS (1988) Drought stress applied during the reproductive phase reduced ozone-induced effects in bush bean. In: Heck WW, Taylor OC, Tingey DT (eds) Assessment of crop loss from air pollutants. Elsevier Applied Science, London, pp 345–364

Mudd JB (1982) Effects of oxidants on metabolic function. In: Unsworth MH, Ormrod DP (eds) Effects of gaseous air pollution in agriculture and horticulture. Butterworths, London, pp 189–203

Murali NS, Teramura AH (1986) Effectiveness of UV-B radiation on the growth and physiology of field-grown soybean modified by water stress. Photochemistry and Photobiology 44:215–219

Noble PS (1974) Biophysical plant physiology. WH Freeman, San Francisco

Norby RJ, O'Neill EG, Luxmoore RJ (1986) Effects of atmospheric CO_2 enrichment on the growth and mineral nutrition of *Quercus alba* seedlings in nutrient-poor soil. Plant Physiology 82:83–89

Norby RJ, Weerasuriya Y, Hanson PJ (1989) Induction of nitrate reductase activity in red spruce needles by NO_2 and HNO_3 vapor. Canadian Journal of Forest Research 19:889–896

Olszyk DM, Tibbitts TW (1981) Stomatal response and leaf injury of *Pisum sativum* with SO_2 and O_3 exposures. II. Influence of moisture stress and time of exposure. Plant Physiology 67:545–549

Peiser G, Yang SF (1985) Biochemical and physiological effects of SO_2 on nonphotosynthetic processes in plants. In: Winner WE, Mooney HA, Goldstein RA (eds) Sulfur dioxide and vegetation, physiology, ecology, and policy issues Stanford University Press, Stanford, CA pp 148–161

Pell EJ (1974) The impact of ozone on the bioenergetics of plant systems. In: Dugger M (ed) Air pollution effects on plant growth. ACS Symposium Series 3, American Chemical Society, Washington, DC, pp 106–114

Pierre M, Queiroz O (1988) Air pollution by SO_2 amplifies the effects of water stress on enzymes and total soluble proteins of spruce needles. Physiologia Plantarum 73:412–417

Pitelka LF (1988) Evolutionary responses of plants to anthropogenic pollutants. Trends in Ecology and Evolution 3:233–236

Queiroz O (1988) Air pollution, gene expression and post-translational enzyme modifications. In: Schulte-Hostede S, Darrall NM, Blank LW, Wellburn AR (eds) Air pollution and plant metabolism. Elsevier Applied Science, London. pp 238–254

Raddi P, Rinallo C (1989) Variation in needle wax degradation in two silver fir provenances differentiated by tolerance to air pollution. In: Scholz F, Gregorius H-R, Rudin D (eds) Genetic effects of air pollutants in forest tree populations. Springer-Verlag, Berlin

Rauser WE (1984) Isolation and partial purification of cadmium-binding protein from roots of the grass *Agrostis gigantea*. Plant Physiology 74:1025–1029

Rauser WE, Curvetto NR (1980) Metallothionein occurs in roots of *Agrostis* tolerant to excess copper. Nature 287:563–564

Reich PB (1987) Quantifying plant response to ozone: a unifying theory. Tree Physiology 3:63–91

Rennenberg H (1982) Glutathione metabolism and possible biological roles in higher plants. Phytochemistry 21:2771–2781

Rennenberg H (1984) The fate of excess sulfur in higher plants. Annual Review of Plant Physiology 35:121–153

Rich S, Turner NC (1972) Importance of moisture on stomatal behavior of plants subjected to ozone. Journal of the Air Pollution Control Association 22:718–721

Robberecht R, Caldwell MM (1986) Leaf UV optical properties of *Rumex patientia* L. and *Rumex obtusifolius* L. in regard to a protective mechanism against solar UV-B radiation injury. In: Worrest RC, Caldwell MM (eds) Stratospheric ozone reduction, solar radiation and plant life. Springer-Verlag, Berlin, pp 251–259

Robinson NJ, Jackson PJ (1986) "Metallothionein-like" metal complexes in angiosperms; their structure and function. Physiologia plantarum 67:499–506

Rogers HH, Campbell JC, Volk RJ (1979) Nitrogen-15 dioxide uptake and incorporation by *Phaseolus vulgaris* (L.). Science 206:333–335

Roose ML, Bradshaw AD, Roberts TM (1982) Evolution of resistance to gaseous air pollutants. In: Unsworth MH, Ormrod DP (eds) Effects of gaseous air pollution in agriculture and horticulture. Butterworth Scientific, London, pp 379–409

Rowland AJ (1986) Nitrogen uptake, assimilation and transport in barley in the presence of atmospheric nitrogen dioxide. Plant and Soil 91:353–356

Rowland AJ, Drew MC, Wellburn AR (1987) Foliar entry and incorporation of atmospheric nitrogen dioxide into barley plants of different nitrogen status. New Phytologist 107:357–371

Sachs MM, Ho T-HD (1986) Alteration of gene expression during environmental stress in plants. Annual Review of Plant Physiology 37:363–376

Salin ML (1988) Toxic oxygen species and protective systems of the chloroplast. Physiologia Plantarum 72:681–689

Sauter JJ, Voss JV (1986) SEM-observations on the structural degradation of epistomatal waxes in *Picea abies* (L.) Karst. and its possible role in the "Fichtensterben." European Journal of Forest Pathology 16:408–423

Scandalios JG (1987) The antioxidant enzyme genes *Cat* and *Sod* of maize: regulation, functional significance, and molecular biology. Current Topics in Biological and Medical Research 14:19–44

Schoeneweiss DF (1978) Water stress as a predisposing factor in plant disease. In: Kozlowski TT (ed) Water deficits and plant growth, Vol 5. Academic Press. New York, 5:61–99

Scholz F, Gregorius H-R, Rudin D (eds) (1989) Genetic effects of air pollutants in forest tree populations. Springer-Verlag, Berlin

Silvius JE, Baer CH, Dodrill S, Patrick H (1976) Photoreduction of sulfur dioxide by spinach leaves and isolated spinach chloroplasts. Plant Physiology 57:799–801

Skärby L, Troeng E, Boström C-A (1987) Ozone uptake and effects on transpiration, net photosynthesis, and dark respiration in Scots pine. Forest Science 33:801–808

Steinmüller D, Tevini M (1986) UV-B-induced effects upon cuticular waxes of cucumber, bean and barley leaves. In: Worrest RC, Caldwell MM (eds) Stratospheric ozone reduction, solar radiation and plant life. Springer-Verlag, Berlin, pp 261–269

Sutton R, Ting IP (1977a) Evidence of repair of ozone induced membrane injury; alteration in sugar uptake. Atmospheric Environment 11:273–275

Sutton R, Ting IP (1977b) Evidence for the repair of ozone induced membrane injury. American Journal of Botany 64:404–411

Tanaka K, Suda Y, Kondo N, Sugahara K (1985) O_3 tolerance and the ascorbate-dependent H_2O_2 decomposing system in chloroplasts. Plant Cell Physiology 26:1425–1431

Taylor GE Jr (1978) Genetic analysis of ecotypic differentiation within an annual plant species, *Geranium carolinianum* L. in response to sulfur dioxide. Botanical Gazette 139:362–368

Taylor GE Jr, Tingey DT (1983) Sulfur dioxide flux into leaves of *Geranium carolinianum* L.: evidence for a nonstomatal or residual resistance. Plant Physiology 72:237–444

Taylor GE Jr, Tingey DT, Gunderson CA (1986) Photosynthesis, carbon allocation, and growth of sulfur dioxide ecotypes of *Geranium carolinianum* L. Oecologia 68:350–357

Thomas MD (1961) Effects of air pollution on plants. In: World Health Organization (ed) Air pollution. Columbia University Press, New York, pp 233–278

Thorne L, Hanson GP (1972) Species differences in rates of vegetal ozone absorption. Environmental Pollution 3:303–312

Thorne L, Hanson GP (1976) Relationship between genetically controlled ozone sensitivity and gas exchange rate in *Petunia hybrida* Vilm. Journal of the American Society for Horticultural Science 101:6–63

Tingey DT (1980) Stress ethylene production—a measure of plant response to stress. HortScience 15:630–633

Tingey DT, Hogsett WE (1985) Water stress reduces ozone injury via a stomatal mechanism. Plant Physiology 77:944–947

Tingey DT, Olszyk DM (1985) Intraspecific variability in metabolic responses to SO_2. In: Winner WE, Mooney HA, Goldstein RA (eds) Sulfur dioxide and vegetation. Stanford University Press, Stanford, CA, pp 178–205

Tingey DT, Taylor GE (1982) Variation in plant response to ozone: a conceptual model of physiological events. In: Unsworth MH, Ormrod DP (eds) Effects of gaseous air pollution in agriculture and horticulture. Butterworth Scientific, London, pp 113–138

Tingey DT, Fites RC, Wickliff C (1976) Differential foliar sensitivity of soybean cultivars to ozone associated with differential enzyme activities. Physiologia Plantarum 37:69–72

Tingey DT, Thutt GL, Gumpertz ML, Hogsett WE (1982) Plant water status influences ozone sensitivity of bean plants. Agriculture and Environment 7:243–254

Tolley LC, Strain BR (1984) Effects of CO_2 enrichment and water stress on growth of *Liquidambar styraciflua* and *Pinus taeda* seedlings. Canadian Journal of Botany 62:2135–2139

Tolley LC, Strain BR (1985) Effects of CO_2 enrichment and water stress on gas exchange of *Liquidambar styraciflua* and *Pinus taeda* seedlings grown under different irradiance levels. Oecologia 65:166–172

Wagner GJ, Trotter MM (1982) Inducible cadmium binding complexes of cabbage and tobacco. Plant Physiology 69:804–809

Warner CW, Caldwell MM (1983) Influence of photon flux density in the 400–700 nm waveband on inhibition of photosynthesis by UV-B (280–320 nm) irradiation in soybean leaves: separation of indirect and immediate effects. Photochemistry and Photobiology 38:341–346

Wellburn AR (1982) Effects of SO_2 and NO_2 on metabolic function. In: Unsworth MH, Ormrod DP (eds) Effects of gaseous air pollution in agriculture and horticulture. Butterworths, London, pp 169–187

Wellburn AR, Majernik O, Wellburn F (1972) Effects of SO_2 and NO_2 polluted air upon the ultrastructure of chloroplasts. Environmental Pollution 3:37–49

Wingsle G, Nasholm T, Lundmark T, Ericsson A (1987) Induction of nitrate reductase in needles of Scots pine by NO_x and NO_3^-. Physiologia Plantarum 70:399–403

Winner WE, Gillespie C, Shen W-S, Mooney HA (1988) Stomatal responses to SO_2 and O_3. In: Schulte-Hostede S, Darrall NM, Blank LW, Wellburn AR (eds) Air pollution and plant metabolism. Elsevier Applied Science, London, pp 255–271

Woodward FI (1987) Stomatal numbers are sensitive to increases in CO_2, from pre-industrial levels. Nature 327:617–618

Woodward FI, Bazzaz FA (1988) The responses of stomatal density to CO_2 partial pressure. Journal of Experimental Botany 39:1771–1781

Zeevaart AJ (1974) Induction of nitrate reductase by NO_2. Acta Botanica Nederlands 23:345–346

Zeevaart AJ (1976) Some effects of fumigating plants for short periods with NO_2. Environmental Pollution 11:97–100

Commentary to Chapter 8

Possible Genetic Effects of Continually Increasing Atmospheric CO_2

Boyd R. Strain

Introduction

This book is devoted to the analysis of the effects of atmospheric toxicity on plants. Specifically, the book considers the response of individual plants to toxic air pollutants and the consequences of differential intraspecific survival and reproduction on the genetic structure of plant populations. The goal is to consider the long-term effects of air pollution particularly as they may influence the ecological genetics of plant species in both intensively managed and natural ecosystems (Pitelka 1988).

Tingey and Andersen (Chapter 8, this volume) reviewed the physiological basis of differential sensitivity of individual plants to changes in atmospheric quality. Their review focused on toxic atmospheric stresses and emphasized the classic stress syndrome of declining plant physiological activity and growth.

In addition, however, Tingey and Andersen also explained interactions between toxic pollutants and other nontoxic global atmospheric changes (e.g., CO_2 increase) that may interact to ameliorate some of the toxic stresses (Coyne and Bingham 1977; Dahlman et al. 1986). This effect has been shown in alfalfa for SO_2 and NO_2 (Hou et al. 1977), in soybean for SO_2 (Carlson and Bazzaz 1985), in sugar beet under particulate aerosol deposits (Lister 1975) and theorized for O_3 effects on ponderosa pine (Green and Wright 1977). Two primary mechanisms have been proposed to explain the CO_2-toxic gas interaction. The first is the reduction of stomatal conductance with increasing atmospheric CO_2. Stomatal closure would impede diffusion of toxic gases into leaves. The second mechanism is the beneficial effect of more vigorous leaf metabolism associated with increased photosynthate production from CO_2 fertilization. Damage caused by a toxic pollutant would be partially repaired by increased leaf carbon and energy. The possibility of interaction among components of global atmospheric change as they affect plants must be understood and incorporated into projections of future biological response.

Objective of this Commentary

This commentary extends the discussion on differential plant sensitivity to air quality changes other than the toxic air pollutants which induce the classic stress syndrome of declining physiology and growth. Global atmospheric changes that differentially increase plant growth and vigor among genotypes of an ecosystem will induce ecological stresses associated with the differential competitive potentials and survival of individuals. If these changes proceed in a unidirectional manner for long periods of time, the genetic structure of populations and communities will likely change.

Review of Current Knowledge

Carbon dioxide is an essential compound for plant growth and ecosystem function (McLaughlin and Norby, Chapter 4, this volume). Increases of concentration of CO_2 in the atmosphere have been shown to induce differential growth and physiological changes within and among plant species (Zangerl and Bazzaz 1984; Marks and Strain 1989).

At the current atmospheric concentration of 350 ppm at Mauna Loa, Hawaii, CO_2 is limiting to many plants. Plants with the C-3 photosynthetic pathway increase 40% to 80% in biomass when ambient CO_2 is doubled and other requirements are optimal (Cure and Acock 1986). Plants with the C-4 photosynthetic pathway respond positively to CO_2 enrichment but generally less than C-3 species. If CO_2 continues to increase in the global atmosphere, we can expect differential plant response and eventual genetic changes in populations (Strain 1987).

Prior to the beginning of the industrial revolution (ca. 1870), the mean global CO_2 concentration was approximately 270 ppm. It has increased 23%, reaching 350 ppm in 1989. At the current annual increase of approximately 1.2% per year, doubling of the 1870 concentration to 540 ppm will occur by 2050 ± 20 years and will continue to track the CO_2 injection from the burning of fossil fuels (Edmonds and Reilly 1985). Consequently, it is appropriate to examine the possible genetic effects of a century or more of ever-increasing atmospheric CO_2 concentration and global atmospheric change on plants (Strain 1985, 1987)

Table 1 summarizes the physiological and growth responses of plants to atmospheric CO_2 increase and considers the possible genetic consequences. The primary and secondary physiological responses to CO_2 increase will definitely affect individual plants. The tertiary ecological responses will also impact individuals but, more importantly for this volume, the ecological changes working on differential sensitivity through time can be expected to induce changes in the genetic structure of plant populations.

TABLE 1. Physiological, genetic and ecological effects of atmospheric CO_2 enhancement.[a]

Physiological Responses Known to Be Genetically Variable

Primary physiological responses
    Photosynthesis
    Photorespiration
    Stomatal aperture
Secondary physiological responses
    Photosynthate concentration and composition
    Plant water status: transpiration, tissue water potential, water use efficiency, leaf
        temperature
    Plant nutrient status: carbon/nitrogen, carbon/phosphorus, other elemental ratios
    Tolerance to gaseous atmospheric pollutants
Tertiary whole-plant responses
    Growth rate: weight, height, leaf area, node formation, stem diameter, leaf senescence
    Growth form: height, branch number, leaf number and area, root/shoot weight, stem
        diameter, leaf weight/area
    Reproduction: flowering and fruiting time, size, and number, seed maturation rate and
        viability, number of seeds per plant

Ecology, Genetics, and Evolution

Primary organism interactions leading to differential survival
    Plant-plant: interference, competition, symbiosis
    Plant-animal: herbivory, pollination, shelter
    Plant-microbe: disease, decomposition, symbiosis
Secondary organism effects
    Genetic differentiation of populations, biodiversity changes in communities
Tertiary ecosystem responses
    Evolutionary feedback effects (integration of all biotic and abiotic interactions occurring
        through time)

Modified from Strain (1987).
[a] If responses occur in the primary processes, secondary and tertiary level responses are unavoidable.

Unidirectional environmental change (e.g., a century of continually increasing CO_2 and warming) can be expected to differentially affect survival and reproduction among the plant species of ecosystems (Zangerl and Bazzaz 1984). Differential survival within plant species can also be expected under persistent environmental change. Thirty-nine published reports of intraspecific variation in direct response to CO_2 increase were listed in Strain and Cure (1986). A few examples are cited below.

Intraspecific variability in response to CO_2 concentration has been reported in net carbon exchange rates (Potvin and Strain 1985), thermal stability of phospho-enol-pyruvate carboxylase (Simon et al. 1984), stomatal conductance of H_2O (Paez et al. 1984), stomatal conductance of SO_2 (Majernik and Mansfield 1972), total plant and economic crop yield (Peet and Willits 1982), phenology of flowering (Garbutt and Bazzaz 1984), seed weight and viability (Wulff and Alexander 1985), and other aspects of plant growth and reproduction.

Throughout the Pleistocene era (approximately 1 million years before the present), the mean global CO_2 concentration remained below 300 ppm (Fig. 1). The mean CO_2 concentration from the preindustrial period to 5000 years BP averaged 270 ppm. Thus, evolutionary selection of the last million years has selected plants that are most competitive under the low CO_2 supply. Prior to the Pleistocene, from 10^{-6} to 10^{-8} years BP, CO_2 concentration was higher, from 1000 to 5000 ppm. It is likely, therefore, that some residual genetic potential for growth response and increased competition potential under higher CO_2 levels is available.

If the "greenhouse effect" occurs, irradiance, air and soil temperatures, storm patterns and intensity, and atmospheric absolute humidity are expected to change on a global scale (MacCracken and Luther 1985). A century or more of gradual warming and drying of the global climate will undoubtedly initiate genetic change in plant populations. If stratospheric O_3 continues to decline, allowing increased levels of UVB to reach the ground, differential plant sensitivity to UVB indicates that additional genetic change may occur in plants (Teramura and Murali 1986).

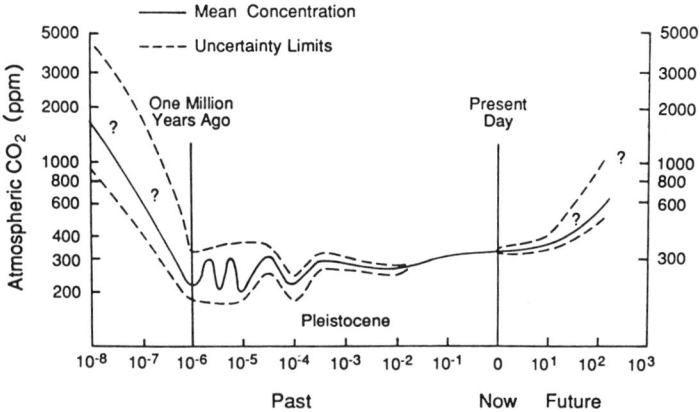

FIGURE 1. Overview of the history of atmospheric CO_2 from 10^8 years ago to 10^2 years into the future. *Solid line* is the estimated mean CO_2 concentration. *Dashed lines* are limits of uncertainty. The ice age cycling is only representative of the range of concentration, not the specific number of glaciations. Redrawn from Gammon et al. (1985).

Conclusions

With plenty of evidence for intra- and interspecific variation in individual plant response to CO_2 enrichment, CO_2-induced competitive changes, and the probability that atmospheric CO_2 will continue to increase for a century or more, it is hypothesized that the direct effects of CO_2 increase, with or without climatic change, will induce ecological changes that will bring about genetic changes. If plant populations change in biochemistry, phenology, and reproductive potential, coevolved microbes and animals should also begin to respond genetically. Thus, it appears that massive genetic change could occur in response to anthropogenic environmental change at the global level. The actual change to be experienced is not predictable without detailed information on the nature and the rate of global change.

Proposed Research Agenda

The research agenda provided below is directed to the acquisition of information required to predict physiological and genetic response of plants to future global environments. The agenda is updated from the extensive review and research recommendations of Strain and Cure (1985).

Crop Plants

The ten crop species (wheat, barley, rice, corn, sorghum, soybean, cotton, white potato, sweet potato, and alfalfa) reviewed by Cure and Acock (1986) should be examined for varietal differences in sensitivity to CO_2 enhancement under predicted future environments. Although all crop species cannot be examined, these are recommended because of their extensive data base and global economic importance. Understanding the genetic basis controlling the mechanisms of the interactions between toxic and nontoxic environmental changes (e.g., warmer air and soil temperatures, atmospheric H_2O vapor pressure, soil water, nutrient availability and salinity, toxic atmospheric pollutants, solar irradiance) is critical to future agriculture. Equally important is the need to identify physiological and biochemical characteristics that result in better growth and yield response, which can pinpoint specific characteristics to be incorporated into genetically superior stock (e.g., varieties of tillering grains have been found that increase tiller number and, therefore, head and seed number under CO_2 enrichment). Water-use and nutrient-use efficiency have both been shown to be differentially sensitive to CO_2 increase and climate change but neither issue has been addressed from the viewpoint of air pollution scientists or population geneticists.

Experiments must be conducted so that plants are continuously exposed (from germination to seed maturation) in CO_2-enriched atmospheres. Measurements made in short-term fumigations or at a given stage of a plant's development cycle are not necessarily extrapolatable to other times or stages. Appropriate selection of seed, hybridization, genetic engineering, and clonal propagation should all be used to obtain genotypes predicted to display superior yield under anticipated future environments.

Forest Species

Realizing that trees planted into plantations today will not be harvested until the mid-21st century, it is imperative to conduct an intensive review of the syndrome of physiological, biochemical, and growth dynamics most likely to preadapt individuals for the future environment. The United States Forest Service tree farms should make this determination for their species and region and begin selection and genetic stock development now. Midcourse corrections will, of course, be required but an adequate starting point has already been established (Shands and Hoffman 1987; Layton and Lucier, Commentary to Chapter 11, this volume). Of course, commercial timber companies should also conduct this research but very short planning cycles have prevented private companies from taking action.

Natural Ecosystems

Representative ecosystems should be studied for sufficient periods of time to determine the long-term, quasiequilibrium to continually changing environment. Some ecosystem-level information has been obtained from Arctic tundra and coastal salt marsh, but most studies have been autecological in controlled environments. Grasslands, forests, steppe, and chaparral remain untouched from an ecological or genetic point of view. It is possible that native annuals will remain relatively well adapted to their habitat if environmental change is not excessively rapid or extreme in amount. Community composition will change because of differential species and ecotype responses to the environmental changes. Perennial species will respond more slowly because of the resistance to change offered by the persistence of the established individuals (Wray and Strain 1987). This is speculation from isolated controlled environment studies, however, and each of the above issues must be experimentally examined.

Water-use efficiency may be more important in natural ecosystems than in agricultural systems. Plants with the largest net response to the combined effects of water saving and increased growth may outcompete, and presumably displace, less responsive genetic types. Nutrient-use efficiencies may become critical in natural ecosystems if the most vigorously growing genotypes tie up available nutrients in the standing biomass. On the other hand, if tissue becomes nutrient poor, feeding efficiencies of herbivores may decline. This has been shown to increase leaf-feeding rates in insects (Lincoln et al. 1984). Studies should be conducted on genetically based aspects of tissue quality and the effect of survival and reproduction of variants on the genetic structure of populations within representative eco-systems. Carbon and nutrient sequestering in the organic matter of litter and soil should be characterized as to genetic source to allow analysis of interaction between plant genotypes and microbial associates. The genetics of mycorrhizae, plant pathogens, and decomposers may be sensitive to genetic changes in the vegetation and experiments are needed in this critical area.

References

Carlson RW, Bazzaz FA (1985) Plant response to SO_2 and CO_2. In: Winner WE, Mooney HA, Goldstein RA (eds). Stanford University Press, Stanford, CA

Coyne PI, Bingham CE (1977) Carbon dioxide correlations with oxidant air pollutions in the San Bernardino Mountains of California. Air Pollution Control Association Journal 27:782–784

Cure J, Acock, B (1986) Crop responses to carbon dioxide doubling: A literature survey. Agricultural and Forest Meteorology 38:127–145

Dahlman RC, Reynolds JF, Strain BR (1986) Modeling the responses of eco-systems to fossil fuel emissions: carbon dioxide and pollutant interactions. Air Pollution Control Association Proceedings. Paper 86–9.5 15 pp

Edmonds JA, Reilly JM (1985) Future global energy and carbon dioxide emissions. In: Trabalka JR (ed) Atmospheric carbon dioxide and the global carbon cycle. DOE/ER-0239, U.S. Department of Energy. Available from NTIS, U.S. Department of Commerce, Springfield, VA

Gammon RH, Sundquist ET, Fraser PJ (1985) History of carbon dioxide in the atmosphere. In: Trabalka JR (ed) Atmospheric carbon dioxide and the global carbon cycle. DOE/ER-0239, U.S. Department of Energy. Available from NTIS, U.S. Department of Commerce, Springfield, VA

Garbutt K, Bazzaz FA (1984) The effects of elevated CO_2 on plants. III. Flower, fruit and seed production and abortion. New Phytologist 98:433–446

Green K, Wright RA (1977) Field response of photosynthesis to CO_2 enhancement in ponderosa pine. Ecology 58:687–692

Hou L-Y, Hill AC, Soleimani A (1977) Influence of CO_2 on the effects of SO_2 and NO_2 on alfalfa. Environmental Pollution 12:7–16

Lincoln DE, Sionit N, Strain BR (1984) Growth and feeding responses of *Pseudoplusia includens* (Lepidoptera: Noctuidae) to host plants growth in controlled carbon dioxide atmospheres. Environment Entomology 13:1527–1530

Lister RA (1975) The effects of increasing atmospheric carbon dioxide and particles on plant photosynthesis. Ph.D. thesis, University of North Carolina.

MacCracken MC, Luther FM (1985) Detecting the climatic effects of increasing carbon dioxide. DOE/ER-2035, U.S. Department of Energy. Available from NTIS, U.S. Department of Commerce, Springfield, VA

Majernik O, Mansfield TA (1972) Stomatal responses to raised atmospheric CO_2 concentrations during exposure of plants to SO_2 pollution. Environmental Pollution 3:1–7

Marks S, Strain BR (1989) Effects of drought and CO_2 enrichment on competition between two old-field perennials. New Phytologist 111:181–186

Paez A, Hellmers H, Strain BR (1984) Carbon dioxide enrichment and water stress interaction on growth of two tomato cultivars. Journal of the Society for Agricultural Science 102:687–693

Peet M, Willits DH (1982) The effect of density and postplanting fertilization on response of lettuce to CO_2 enrichment. HortScience 27:948–949

Pitelka LF (1988) Evolutionary responses of plants to anthropogenic pollutants. Trends in Ecology and Evolultion 3:233–236

Potvin C, Strain BR (1985) Effects of CO_2 enrichment an temperature on growth in two C_4 weeds, *Echinochloa crusgalli* and *Eleusine indica*. Canadian Journal of Botany 63:483–487

Shands WE, Hoffman JS (1987) The greenhouse effect, climate change and U.S. forests. The Conservation Foundation 1250 Twenty-fourth Street, N.W., Washington, DC

Simon J-P, Potvin C, Strain BR (1984) Effects of temperature and CO_2 enrichment on carbon translocation of plants of the C_4 grass species *Echinochloa crusgualli* (L.) Beauv. from cool and warm environments. Plant Physiology 75:1054–1057

Strain BR (1985) Physiological and ecological controls on carbon sequestering in terrestrial ecosystems. Biogeochemistry 1:219–232

Strain BR (1987) Direct effects of increasing atmospheric CO_2 on plants and ecosystems. Trends in Ecology and Evolution 2:18–21

Strain BR, Cure JD (eds) (1985) Direct effects of increasing carbon dioxide on vegetation. DOE/ER-0238, U.S. Department of Energy. Available from NTIS, U.S. Department of Commerce, Springfield, VA

Strain BR, Cure JD (eds) (1986) Direct effects of atmospheric CO_2 enrichment on plants and ecosystems: bibliography with abstracts. ORNL/CDIC-13. Available from NTIS, U.S. Department Commerce, Springfield, VA

Teramura AH, Murali NS (1986) Intraspecific differences in growth and yield of soybean exposed to ultraviolet-B radiation under greenhouse and field conditions. Environmental and Experimental Botany 26:89–95

Wray SM, Strain BR (1987) Interaction of age and competition under CO_2 enrichment. Ecology 68:1116–1120

Wulff RD, Alexander HM (1985) Intraspecific variation in the response to CO_2 enrichment in seeds and seedlings of *Plantago lanceolata*. Oecologia 66:458–460

Zangerl AR, Bazzaz FA (1984) The response of plants to elevated CO_2. II. Competitive interactions among annual plants under varying light and nutrients. Oecologia 62:412–417

9
Molecular Characterization of Plant Responses to Stress

CHRIS A. CULLIS

Introduction

The advances in DNA technology over the past few decades have significantly altered the range of investigations with all organisms. Genetic engineering is clearly still in its developmental stages, but a number of the new technologies can be incorporated into existing strategies for identifying important and desirable new material. In the case of higher plants, both the traditional breeding and the biotechnological approaches are directed towards the same goals. These are:

1. To increase the knowledge about the way in which genes determine the observed phenotype
2. To use the knowledge to identify and manipulate plants to meet particular needs.

Much of the plant material of interest is essentially uncharacterized and the range of variation unknown. The use of DNA markers can start to unravel the relationships between individuals and populations and so identify individuals or families of special interest. In addition to the characterization of material, the transformation techniques offer the possibility of introducing new genes for specific traits, in a much shorter timescale than any conventional breeding techniques. However, a corollary is the need to identify and isolate the genes which are needed. With particular reference to this book, what genes are important in determining the response to anthropogenic changes in the atmosphere?

Much molecular characterization will be needed to identify the response of current populations to stresses in terms of gene expression and resistance. Genetic responses to environmental stresses can also occur and the best described example is the induction of heritable changes in flax. However, even in this case, the effect of alterations in atmospheric chemistry have not been documented.

The Biotechnological Approach to Characterization of Plant Responses to Stress

This approach uses sophisticated molecular techniques to identify, produce, and subsequently analyze new gene combinations. These new gene combinations can be produced by classical crossing programs or by genetic engineering. The basis for genetic engineering has been established for model systems (Wilke-Douglas et al. 1986; Perani et al. 1986) and for a number of crop species.

There are three parts to a successful genetic engineering strategy. First, is the ability to isolate genes of interest from various organisms. The major successes in this area have been in the isolation of useful genes from organisms with simple genomes, such as viruses (Abel et al. 1986) and bacteria (Vaek et al. 1987). The isolated genes are then tailored to allow their appropriate function in the plant. Second, the newly tailored gene or gene complex has to be transferred back into the plant of interest. Once again, model systems have been used very successfully for the characterization of the steps involved. Third, transgenic plants have to be characterized as to the phenotype caused by the insertion of the new information. The major obstacle in the use of biotechnology for the improvement, or tailoring, of plants for specific purposes is the lack of information about the genes involved in the control of many traits. In conventional manipulations of plant genomes by breeding, the knowledge that a gene is present and scorable is sufficient for its incorporation into the appropriate material, although the process could be very prolonged. For the biotechnological methods to work, the gene of interest has to be identified at the biochemical level and then isolated. Unfortunately, for many important traits, this information is not available. Thus, although the new technology is very powerful, the use to which it can be put is limited by the quantity and quality of the biological knowledge that has already been accumulated.

Location and Identification of Genes

The location of a gene can be done in a number of ways. Clearly the most desirable would be to directly identify the actual piece of DNA which encodes the gene. However, this may not be possible, especially as a first step. The strategies for approaching the gene of interest include the identification of the biochemical mechanism for the trait and then moving to the gene itself; the use of a gene from another species, which is thought to be analagous to the desired trait, and using this to identify the gene, or by isolating a fragment of DNA linked to the gene and walking along the chromosome until the gene is reached.

Many important heritable characters, including air pollution (Roose, Chapter 5, this volume), are governed by the joint action of a number of genes. The theory and techniques of quantitative genetics have provided much useful information about many quantitative traits, but if the determinants could be resolved into single components then the relative importance of each could be discerned. The new molecular techniques offer the potential to undertake this type of genetic analysis and the mapping of quantitative trait loci has been carried out in the tomato (Patterson et al. 1988).

Restriction Fragment Length Polymorphisms

The digestion of the genomic DNA with a restriction enzyme and subsequent electrophoresis results in the DNA being separated into a reproducible set of fragments. The comparison of these fragments between genetically distinct individuals can result in the differences in some of these fragments. The term restriction fragment length polymorphism (RFLP) has been coined to describe this variation. The potential usefulness of RFLP markers in basic plant genetics and plant improvement programs has been reviewed by Tanksley (1983), Beckmann and Soller (1983), and Tanksley et al. (1989).

Basically, mapping with RFLPs involves the application of molecular biological techniques to the concepts of transmission genetics. These RFLPs are virtually limitless and are unaffected by any interactions which may occur between the genotype and the environment. Thus, they can be used to construct a genetic map and so mark the position of any gene which can be recognized phenotypically. However, the association between an RFLP and a trait does not give any information about the underlying biochemical basis for the trait since it is not necessarily the actual gene. The average relationship between a recombination distance and the distance in nucleotides along the DNA strand is that 1 centimorgan is approximately equal to 10^6 base pairs, although it may be less in a small genome-sized plant like *Arabidopsis thaliana* than in conifers in which the genome size is very much larger.

Two types of polymorphic markers can be generated. One type, which gives a small number of bands, is frequently generated from cDNAs or low copy number sequences. These probes are particularly useful in genetic mapping as the small number of bands are easily placed into allelic groups. This type of result is shown in Fig. 1 (a and b), although this example is not a low copy number probe. The ribosomal RNA genes occur in clusters and usually segregate as a single unit, so they can be used as markers. However, a large number of single-band probes need to be isolated to get a "fingerprint" of any lineage. However, once such a low copy number marker has been identified and placed near a gene of interest, it can be used as a starting point for the isolation of that gene.

The second type of RFLP marker is the so-called hypervariable sequence markers. A number of simple sequence repeats have been shown to hybridize to the DNA from a wide range of organisms. The two best-studied of these are the tandem repeat from gene III from the *E. coli*

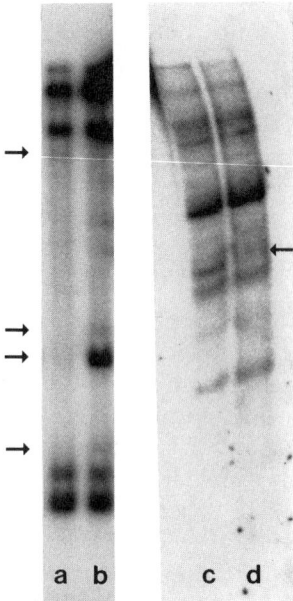

FIGURE 1 (a–d). Hybridization profiles demonstrating RFLPs involving single band differences and fingerprint-type patterns. DNA from two individuals of *Pinus radiata* digested with the restriction enzyme HindIII (a, b) or DraI (c, d), separated on a 1% agarose gel, transferred to a nylon membrane and hybridized with a ribosomal RNA probe (a, b) or the tandem repeat sequence from M13 (c, d). The bands differing between the two individuals are indicated by *arrows*.

bacteriophage M13 and the core sequence from the human minisatellite family (Jeffreys et al. 1985). Both of these probes have been shown to hybridize with plant DNAs (Dallas 1988; Rogstad et al. 1988; Zimmerman et al. 1989). The pattern of bands highlighted by these probes is complicated, and although differences exist between individuals it is difficult to ascribed relationships between the bands (a pattern is shown in Fig. 1, c and d). Thus, using these probes the information is useful in the typing of individuals rather than as a marker of a specific chromosome fragment. Two particular areas into which the fingerprinting technique is being recruited are in the identification of the parentage of progeny and determination of the degree of relatedness between individuals.

The combination of these two types of markers can help to identify individuals which show marked variations of a trait. Atmospheric changes can cause the selective loss of parts of a natural stand. How do those individuals that survive differ from those that do not? The use of these markers can give information about the genetic structure and the number of genes (or perhaps in this case the number of distinct chromosomal regions) involved in the ability to tolerate particular conditions. Since these DNA markers can be screened in nondestructive way on seedlings, they can be used as "early markers" for a trait of interest. However, only polymorphic sequences which show a high degree of correlation with the trait of interest can be used as early markers. As the number of RFLPs increases and the information concerning the molecular architecture of the genome of tree species increases, the combined data from physiological and molecular methods will make an important contribution to the identification of genotypes which are tolerant to particular stresses.

Creation of New Genotypes

The use of DNA markers will only identify individuals within a population which have the appropriate genetic constitution but will not allow the production of new, novel genotypes. The processes of transformation and regeneration now promise the ability to tailor genotypes for particular purposes. Specific gene transfer has been achieved in model systems, in some crop species and some trees (Rogers and Klee 1987; Sederoff et al. 1986). The range of plants which can be transformed by the many protocols available, including *Agrobacterium tumefaciens*, *A. rhizogenes*, and the introduction of DNA ballistic projectiles is ever-increasing. The most likely limiting steps in the ability to produce useful novel combinations by genetic engineering are the lack of genes controlling important pathways and the limitations on the size of the DNA which can be transferred. Thus, the ability to transfer only single genes or relatively small pieces of DNA precludes the ability to affect polygenic characters. This is compounded by our lack of understanding of the biochemical basis for any of the mechanisms which are involved in stress tolerance. However, as described above, the use of molecular markers to identify the number and location of important genes may eventually lead to their molecular isolation and characterization.

One important consideration in the transfer of genes by genetic engineering is the potential to greatly reduce the genetic diversity. Thus, the transfer, selection, and propagation of a line with a novel gene which performs better in the desired circumstances is the major goal. However, once one such line is achieved, the pressure to place the same gene in other backgrounds will be reduced and that initial line is likely to become widespread. The introduced gene may also work much better in some genetic backgrounds than in others, so even if the will is there to get a diversity of genotypes it may not be practical. Thus, the narrowing of the genetic base must be an ever-present consideration in the development of novel combinations, especially in the case of long-lived plants.

Identifying Stress-Related Genes

The use of DNA markers to identify the number of genes and their chromosomal location and then to subsequently isolate them is a long-term set of goals. A number of stress-related genes have been isolated and characterized in higher plants. Among the most intensively studied are the heat shock genes (Nagao et al. 1986: see Tingey and Andersen Chapter 8, this volume). The organization of these genes within the genome is known for a number of plant species and the control of their function in response to a range inducing stimuli determined. Stress-related genes can be coinduced by number of environmental perturbations, and the responses of these sets of genes to atmospheric pollutants could be determined. This would define the reaction of plants to specific atmospheric changes and allow comparisons between tolerant and sensitive genotypes. The overall effect of the atmospheric changes could be due to the sensitization of the individuals to other environmental variables, so the occurrence, or lack of an effect on a single gene may lead to false conclusions.

One of the possible routes for the effects of atmospheric changes may be on the extent of free radicals within the plant cell, especially through the production of excess active oxygen species (Tingey and Andersen, Chapter 8, this volume). In this case the activity of enzymes which act to eliminate or minimize these effects may be important for the response to atmospheric pollutants. Two enzymes, superoxide dismutase (Bowler et al. 1989) and catalase (Chandlee et al. 1983), are involved in the protection of cells against the damaging effects of active oxygen species. These two enzyme systems act in concert to reduce the formation of hydroxyl radicals, which are thought to be in large part responsible for oxygen toxicity in vivo (Halliwell 1984). Superoxide dismutases are a class of metalloproteins which convert the active oxygen species to oxygen and hydrogen peroxide. Catalase then functions to destroy the peroxide before it can react with active oxygen to form the destructive hydroxyl radical. Superoxide dismutase mutants in yeast (Bilinski et al. 1985) and *E. coli* (Carlioz and Touati 1986) are hypersensitive to oxygen. Catalase acts in the protection of plants from excess peroxide during photorespiration (Halliwell 1976). Its absence has been associated with injury from chilling (Wise and Naylor 1987) and dehydration (Dhindsa and Matowe 1981).

Representative cDNA clones have been produced for both superoxide dismutase and catalase genes. A manganese superoxide dismutase cDNA from *Nicotiana plumbaginifolia* has been isolated and characterized (Bowler et al. 1989). Full length cDNA clones for three catalase genes from maize have been isolated (Bethards et al. 1987; Redinbaugh et al. 1988). As the characterization of these gene families proceeds, they may have use as heterologous probes, both for activity and genetic organization in other species. The use of either heterologous probe or the isolation and subsequent use of the homologous sequences from tree species to characterize populations which show differential sensitivity to atmospheric pollutants should indicate whether or not these systems are involved in the differential sensitivity. The demonstration of the importance of these oxygen detoxification mechanisms can be followed in, at least, two ways. First, populations can be screened to determine the extent of genetic variability in the activity for these enzyme systems. This must be done both for the basal activity as well as the degree by which the activity can be increased under air pollution stress. Second the introduction of a foreign gene with higher activity, or more copies of the homologous gene with a different promoter could then pave the way for developing more tolerant individuals and families. However, this has to be viewed as a long-term strategy, as it has yet to be demonstrated that either, or both, of these enzymes are involved in the responses to air pollution.

The molecular techniques described above can be used to characterize the genetic make-up of populations and differentiate between individuals which can be distinguished by their sensitivity to pollutants at three different levels. First, the enzymic activity of detoxifying functions can be determined in individuals within and between populations. Second, the control of gene expression at the RNA level and the inducibility of these important activities in response to atmospheric pollution stress can be compared. Finally, if tolerance is essentially a quantitative trait, the use of DNA markers, as RFLPs and fingerprints, may identify especially tolerant genotypic combinations, without the necessity to identify the individual components contributing to the resistance. However, the foregoing discussion is based on the assumption that the genetic constitution is unaffected by the environment itself. There are examples where this is clearly not the case, and some of these are considered below.

Environmentally Induced Variation in Plant DNA and Associated Phenotypic Consequences

The interpretation of the genetic behavior of higher plants in response to air pollution must include considerations of the life cycle (Tonsor and Kalisz, Chapter 10, this volume). The higher plant is, in reality, an assemblage of competing units, the meristems, each of which is capable of differentiating gametes and so contributing to the next generation. The interaction between these competing meristems determines the final form of the plant, with one of the most important controls being apical dominance. However, if the dominant meristem is damaged, alternative meristems can be released from their repression of growth and subsequently contribute to the next generation. Thus, in higher plants there is no clear separation of the germ line and the soma. This lack of separation means that genetic variation arising from a mutation in any somatic cell, has the potential to be transmitted to the next generation. This form of life strategy and its consequences for variation in higher plants has been reviewed recently (Walbot and Cullis 1985; Walbot 1986).

When organisms, whether single-celled or multicellular, are subjected to certain forms of strong selection pressure such as air pollution, variants emerge bearing appropriate changes in phenotype. In many cases the cause of the altered phenotype can be shown to be a change in the DNA sequence. The initial event in this process is the generation of genetic variation by mutation. How many of these variants arose as a direct and specific response to the selection pressure (not occurring in its absence) and how many are "spontaneous" (occurring even in the absence of the selection)? Although mutations are frequently considered in terms of base pair substitutions, they may include any changes in the structure of the genome. A significant proportion of the mutational repertoire of cells is mediated by transposable elements, while in higher organisms rearrangements, which move genes between euchromatic and heterochromatic regions, can dramatically affect levels of expression. Recent results considering the origin of mutations in *E. coli* (Cairns et al. 1988; Hall 1988) have demonstrated that the range and type of mutation arising can be influenced by the environment. It is not necessary to invoke anticipatory evolution to explain this situation, if one considers the possibility that regulatory feedback loops could modulate the probability of mutations at specific loci when a cell is under stress (Cullis 1987; Hall 1988). It is generally accepted that gene expression can be regulated by environmental conditions and now the possibility that mutation is subject to regulation by environmental factors must be examined.

The plant genome, or at a least part of it, is in a state of flux, and a range of phenomena have been proposed to account for the rapid changes. These include the activity of transposable elements and frequent amplification and deletion events. The frequency at which these restructuring events occur can be influenced by a number of factors including the disruption of chromosomes (Dellaporta and Chomet 1985), formation of particular F_1 hybrids (Gerstel and Burns 1966), external environment in which growth takes place (Durrant 1962), and passage of cells through rounds of tissue culture and regeneration (Scowcroft and Larkin 1985). A characterization of examples where rapid genomic changes occur will help elucidate the molecular mechanisms causing these genomic alterations. Three such instances of conditions under which rapid genomic changes occur are considered here.

The Plant Genome

The DNA content per diploid nucleus varies greatly among eukaryotic organisms (Bennett and Smith 1976). Within the angiosperms there is a nearly thousandfold range of variation although the organismal complexity does not appear to be correlated with genome size. This range in genome size leads to questions concerning the function and origin of the "additional" DNA and whether it exerts an effect independent of its sequence or is it merely "selfish." The vast majority of the genome in most higher plants cannot be responsible for direct coding sequences. The bulk of the non-coding DNA is made up of repetitive sequences which can be arranged either in tandem arrays of a simple, or complex, repeating unit or as dispersed repeating sequences (Flavell 1980). The separation into clustered tandem arrays and dispersed repeats is not always clear-cut. The 5S ribosomal RNA genes in flax are present as tandem arrays, but these are dispersed through the genome on a number of chromosomes (Cullis and Creissen 1987; Schneeberger et al. 1989).

How and why has this wide range of variation in DNA content arisen? Although a large genome size and heterogeneity seem to be indicatiors of evolutionary flexibility and progressivity, the decline in total DNA content apparently accompanies evolutionary specialization and adaptation to certain ecological niches (Nagl et al. 1983). Genome evolution proceeds by sequence amplification, divergence, dispersal, and loss. Thus, the state of the genome, which is characterized at a single time point in a continuing process, is a result of the relative frequencies with which these events have occurred over evolutionary time.

Higher eukaryotes have very complex nuclear genomes. An under-standing of the origins of this complexity is beginning to emerge with the recognition that the DNA of higher organisms is subject to a variety of sequence rearrangements including amplification, deletion, and trans-location events both within and between chromosomes (Flavell 1980). The net result of these processes is the interspersion of the repeated and single copy sequences into a characteristic pattern. This turns out to be a short interspersion pattern (in which most of the single copy sequences are about 1000 base-pairs long) in most higher plants (Sorenson 1984). Those plants with small genomes (<0.5 pg single copy DNA/2C nucleus) such as flax (Cullis 1981) and *Arabidopsis* (Leutwiler et al. 1984) have a pattern which can be classified as long period (much of the single copy DNA > 10 kilobases long). However, in most higher plants the families of sequences which have the highest redundancy are organized in long tandem arrays located at a small number of sites, while the intermediately repetitive sequences are more widely dispersed around the genome. The different proportions of a genome which are organized as either tandem arrays or interspersed sequences may be a consequence of the relative frequency of amplification events versus transposition events during the evolution of that particular genome. In all the cases where it has been investigated, the blocks of heterochromatin which can be observed cytologically are com-posed of tandemly repeating sequences of varying complexity (Nagl et al. 1983), but the converse is not necessarily true as not all the tandemly repeating sequence families form recognizable heterochromatic structures. In general, heterochromatin appears not to be transcribed. An exception is the rRNA genes in *Drosophila*, which are transcribed, but still act as a block of heterochromatin in respect to their effect on the activity of neighboring genes (Hilliker and Appels 1982). Thus, a clear distinction cannot be made between heterochromatin and tandemly arrayed families in their affect on linked loci.

A higher order of organization of the genome is the structure of the chromosomes. It is now suggested that the chromosomes of *Triticum* spp have a fixed arrangement in the nucleus resulting from their relative sizes (Bennett 1983). Changes in the heterochromatin content, inversions, and other rearrangements can alter this physical distribution of the chro-mosomes. How changes in repetitive DNA sequence type, amount, or arrangement mediate changes in chromosomal behavior and perhaps gene expression is a current topic of investigation. One extreme example of the effect of heterochromatin on the activities of neighboring genes is that of inactivation of genes involved in X-autosome translocations. In *Drosophila* general effects of heterochromatin have been demonstrated by long-range influences which modify the expression of genes in other chromosomal regions (Hilliker and Appels 1982). In rye, there is a wide range of values for the amount of telomeric heterochromatin without any gross effects on the plant phenotype. In some Triticales, however, the rye telomeric heterochromatin does have an obvious effect in its role in causing aberrant endosperm development (Gustafson et al. 1983).

The existence of changes occurring over a short time-scale has been observed in a number of systems which are described below. The characterization of these rapidly varying systems facilitates experimental investigation into the precise mechanisms by which these genomic alterations can occur.

Environmentally Induced Heritable Changes in Flax

One of the most extensively studied systems in which rapid DNA changes can be generated in response to the environment is that in the flax variety "Stormont Cirrus" (Durrant 1962; Cullis 1977, 1983). Changes can occur in the genome when plants are exposed to various environmental stimuli with the generation of stable, genetically different lines. These lines have been termed "genotrophs" (Durrant 1962), and they differ from each other and the original line from which they were derived in a number of characters. These include plant weight, height (Durrant 1962), total nuclear DNA content as determined by Feulgen staining (Evans et al. 1966) and the isozyme band patterns for peroxidase and acid phosphatase (Fieldes and Tyson 1973). The alteration of the peroxidase isozyme band pattern was controlled at a single locus, with a dominant and a recessive allele. This gene showed the expected dominance in the F_1 and and a 3:1 segregation of the dominant: recessive pattern in the F_2 in crosses between genotrophs.

NUCLEAR DNA VARIATION

The nuclear DNA changes associated with the environmentally induced heritable changes have been extensively characterized. The majority of the variation occurred in the highly repeated fraction of the genome (Cullis 1983). Representative members from all the highly repeated sequence families have been cloned and characterized (Cullis and Cleary 1986a). This set of cloned probes has been used to detail the DNA differences between the genomes of the genotrophs. All but one of the highly repetitive sequence families varied to some extent, and the variation in any one family was independent of the variation of any other family. The quantitative variation of repeated sequences was detailed, although there was also evidence for qualitative variation (Cullis 1983).

The genes coding for the 5S ribosomal RNA (5S DNA) are one of the variable, highly repetitive sequence families. Flax 5S DNA is arranged in tandem arrays with a repeating unit of 350 base pairs. There can be more than 100,000 copies of this sequence in the genome, which amounts to 3% of the total nuclear DNA. The variation was more than twofold, from 1.17×10^5 copies in Pl (the original line from which the genotrophs were generated) to 4.96×10^4 in LH (Goldsbrough et al. 1981). In situ hybridization has shown that the 5S RNA genes are dispersed through the genome at many chromosomal sites (Cullis and Creissen 1987; Schneeberger et al. 1989). In addition the tandem arrays in which they are organized are of varying length, flanked by non-5S DNA. This latter conclusion is based on the characterization of a dozen 5S DNA containing recombinants isolated from a flax library (Schneberger et al. 1989). The characterization of these clones has demonstrated the variability which exists for this family in flax and has identified subsets which appear to be differentially labile. Thus, within a repetitive sequence family it appears that there are subsets which can be differentially modulated. However, the mechanism by which these subsets are recognized within the genome is not known.

One of the families of 5S RNA genes has been characterized further. This family has a 700 base pair repeating unit which only has a single 5S RNA coding region within the repeat. There are about 10^3 copies of this family in the flax genome (Schneeberger and Cullis 1991). When the organization of this family was determined in a range of genotrophs a number of polymorphisms were observed. The number and position of the new bands varied among the genotrophs. However, one consistent pattern was observed for four, independently derived small genotrophs showing that there was some common determinant in each of these sets of changes. In addition, the pattern remained stable within the genotrophs through subsequent generations, and even through a cycle of tissue culture and regeneration. Preliminary analysis of the segregation of these polymorphic bands indicated that most of them were linked (less than 10 map units spanning the whole set). Further studies in natural populations of flax and linseed indicated that a number of cultivars showed most of the polymorphic bands could be observed in a range of cultivars (Schneeberger and Cullis 1991). Thus, it appears that during the induction of heritable changes, the variation generated is part of that which occurs naturally.

The data support the contention of stable and labile fractions within the flax genome. All the rapid variations appear to take place in the labile subset, and their frequency appears to be affected by the environment. The results obtained from flax, and in particular Pl, which has been through a passage of tissue culture and regeneration, add support to this conclusion. Changes in the nuclear DNA were observed and found to be localized in the same labile subset of genome as that identified for the environmentally induced changes. Some of the rearrangements observed in the generation of the genotrophs are also found in natural populations. Yet to be resolved is whether or not theses variants seen in the cultivars were derived by "induction" events, such as those seen with Stormont Cirrus, or by some other mechanism. It will take the characterization of both naturally occurring variants and the induced changes at the sequence level to determine whether or not a single mechanism is responsible for all these changes.

Induction of Environmentally Caused Changes

The genotrophs are stable lines resulting from environmentally induced heritable changes. Therefore, a comparison between the phenotypic and DNA characteristics measures the end result of the induction process, but neither how nor when the changes actually occur during growth under inducing conditions. The environmental perturbations, themselves, affect the phenotypic characters making them unsuitable for following the induction process. However, the nuclear DNA, or some component thereof, is an appropriate marker for following changes occurring during growth under inducing conditions since it is unaffected by normal developmental processes.

The induction has been followed using both the total nuclear DNA and the number of ribosomal RNA genes (Evans et al. 1966; Cullis and Charlton 1981). These experiments demonstrated that the changes occured during the vegetative growth of these plants under inducing conditions. By the time the plants flower, the changes which will be observed in the next generation are already apparent in the apical cells. Thus, these changes do not appear to be generated during meiosis, although events at this stage of the life cycle may be important in the stabilization and subsequent heritability of the induced changes (Durrant and Jones 1971).

The observations that all the seed from inbred individuals grown under one set of inducing conditions are similar to each other but different from the parental line has to be explained in terms of the composition of the meristem and the life cycle. The developmental patterns of plants grown under inducing conditions suggest that there may be some form of selection in the generation of the variation. When the inducing conditions are set for low growth with subsequent reductions in the rDNA, the plants grow suboptimally although all individuals survive and contribute to the next generation. However, at some stage under these conditions, the growth rate of some part of the plant can increase dramatically. This portion of the plant grows as if it was under more favorable conditions and generally contributes most to the next generation (Cullis 1986, 1987). It is proposed that there has been a change in the composition of the cells in either the apical and/or adventitious meristems (the meristem which is the basis for the increased growth which can be any of those present), such that the cells are now more fitted to grow under the prevailing environmental conditions. Since all the progeny of a single individual in which heritable changes have been induced are similar, the cells in the meristem must contain the same changes.

Stability of the Genome in Inbreds and Hybrids

Extensive quantitative variation in DNA content exists among species, but within a species large differences are rare although there are exceptions (Cavallini et al. 1989). What controls the variation in those cases where there is limited variation within a species? Is it due to a lack of variation in the repetitive sequence families or some mechanism which constrains the absolute amount of DNA while allowing the individual sequence families to float freely? A comparison of ten inbred maize lines for nine repetitive sequences families was carried out (Rivin et al. 1986). These families represented about 50% of the most highly repetitive sequence families of the maize genome, but only a very small proportion of the midrepetitive sequence families. Variation was found for most of the sequence families tested, with the ribosomal RNA genes being the most variable. In contrast to the variation observed between inbred lines, the individual members of an inbred line were not detectably different.

If each of the inbred lines has a stable set of repeated sequence families, then how are the differences between lines generated and how quickly can these changes occur? One mechanism would be the destabilization of sequence families on outcrossing the inbred lines. The altered sets could then be reset to new, different values on the repeated inbreeding required to establish new inbred lines. Evidence for such a mechanism has been found in maize, *Microseris* spp. and flax. The range of genomic restructuring events observed in intra- and interspecific crosses covers the activation of transposable elements (McClintock 1950), alterations in the 2C nuclear DNA content (Price et al. 1981), and variation in the number of rRNA genes (Cullis 1979). Thus, it would appear that intra- and interspecific crossing between inbred lines is a factor which can precipitate genomic variation. It is not yet clear whether or not this form of genomic instability is restricted to a particular subset of the genome. If it transpires that there is a labile set of sequences involved in the genomic variability, this hybrid effect may be generated by a mechanism analagous to that responsible for the environmentally induced changes in flax.

Somaclonal Variation

Many of the mechanisms by which the genome is reorganized have been observed in cells in tissue culture or in plants regenerated from cultures; this source of genetic variability is termed "somaclonal variation." A detailed understanding of the mechanisms by which this variation can occur may allow the manipulation of the system to generate desired phenotypes (e.g., in response to air pollution).

The genomic changes that have been observed to occur in tissue culture include aneuploidy, chromosomal rearrangements such as translocations, inversions, deletions, gene amplification and deamplification, activation of transposable elements, point mutations, cytoplasmic genome rearrangements, and changes in ploidy level (Larkin and Scowcroft 1983; Orton 1983, 1984; Evans et al. 1984). There is also accumulating evidence that specific DNA sequences can be altered in copy number (Zheng et al. 1987).

Changes in the copy number of specific genes can also be selected in tissue culture. Cells with amplified number of genes have been selected using resistance to the herbicides phosphinothricin (Donn et al. 1984) and glyphosate (Shah et al. 1986, Ye et al. 1987). These results show that gene amplifications can, indeed, be selected in tissue culture. However, it has not been determined if such amplifications are restricted to tissue culture systems.

Although specific instances of gene amplification have been described, most of the DNA alterations are found in the highly repetitive sequences. In studies with oats, Johnson et al. (1987) showed that the heterochromatic regions of root tip chromosomes were late replicating. If chromatid separation occurred during anaphase, before replication was complete, then these heterochromatic regions were the sites of chromosome breakage. In maize cultures, meiotic analysis showed that no chromosomal abnormalities were present in plants regenerated after 3 to 4 months in culture, while about half those regenerated after 8 to 9 months were cytologically abnormal (Lee and Phillips 1987). The breakpoints of these abnormalities were primarily on chromosome arms containing large blocks of heterochromatin such as knobs. Studies on regenerated wheat-rye hybrids showed a number of chromosomal abnormalities (Lapitan et al. 1984). Twelve of the 13 breakpoints involved in the deletions and interchanges were located in heterochromatin.

From the results such as those described above, Jonson et al. (1987) and Lee and Phillips (1987) proposed that one of the mechanisms by which chromosome breaks may be induced in culture is by the late replication of heterochromatin. Thus, if the replication of heterochromatin extended into anaphase, chromosome breaks could be induced. Such late replication of heterochromatin has been demonstrated in whole plants but not in tissue cultures.

One of the consequences of a chromosome breakage in maize crosses is the activation of transposable elements (McClintock 1950), a possible cause of somaclonal variants. Apparent activation of transposable elements has been found in studies with maize (Pesche et al. 1987) and alfalfa. Lee and Phillips (1987) proposed a model in maize whereby the incomplete replication of knob heterochromatin leads to the formation of chromosome bridges and, subsequently, chromosome breakage. This breakage cycle would activate transposable elements resulting in chromosome rearrangements, duplications, and deletions. It is still not clear whether the activation of transposable elements is the major cause of somaclonal variation in systems other than maize or even in maize itself.

Qualitative changes in the genome, unassociated with the action of transposable elements, can give rise to restriction fragment length polymorphisms. Evidence has been obtained in soybean cell cultures that these polymorphisms can arise in the process of cell culture (Roth et al. 1989). One of the remarkable observations is the types of polymorphic bands that arose. In all the cases, the new band already appeared to exist in natural populations. Thus, as in the case of the polymorphisms in the flax genotrophs, at least some of the changes are those which have already occurred. Here again, a subset of sites has shown evidence of greater lability under conditions which can be regarded as stressful. A second intriguing observation in this system was that for all the polymorphic loci, there appeared to be two alternative forms. In the few cases where the alternative "alleles" have been isolated, the polymorphism was caused by a small insertion event. Thus, to recreate the types of events and sizes of fragments, a very specific rearrangement must be taking place at these positions in the genome. Although their contribution to the overall variation is presently unknown, they clearly have the potential to be the cause of specific, repeatable changes within the genome.

Conclusions

The plant genome can be considered to be in a state of flux with many changes occurring during both the mitotic and meiotic cycles. Some of the variation is related to normal developmental processes, such as the endoreduplication in endosperm or cotyledon cells and appears to affect the whole the genome equally. Clearly this type of variation does not change the genotype of the individual concerned. However, modulations of the genome associated with the environmental induction of heritable changes in flax and with somaclonal variation do alter the genotype of the plants. The rapid changes in the genome involve a variety of processes including transpositions, amplifications, deletions, and translocations.

The genomic alterations which have been described take place under conditions which may be considered to be stressful to the individual and consequently applicable to air pollution stress. McClintock (1984) has proposed that the genomic response to stress, with the production of new genotypes, is related to an ability to generate diversity under stress conditions. The relationship between the stress-related DNA alterations and the new phenotype is not clear. Most of the DNA changes have been found in the highly repetitive fraction of the genome, much of which is not transcribed (the exception being the genes coding for the ribosomal RNAs). Under these circumstances, the altered sequences cannot mediate their effect through products encoded by their sequence but must generate diversity in some other way. One possible mechanism would be through some form of position-effect variegation, where change in the DNA at a specific site alters the expression of adjacent sequences. Examples of such effects are in the pleiotropic effects of the distal X heterochromatin in *Drosophila* (Hilliker and Appels 1982) and the effects of telomeric heterochromatin on seed development in triticale (Bennett and Gustafson 1982).

The accumulating evidence that bacteria can also undergo a stimulation of the frequency of advantageous mutations, and the restriction fragment length polymorphism changes in soybean callus, widens the range of possible mechanisms. Since *E. coli* does not have blocks of heterochromatin and the mutation was shown to be in the gene itself, all the environmentally —induced genetic changes need not be expressed in an indirect manner. A much larger number of examples will have to be analyzed to gain an understanding of the extent of these types of changes and of the range of stimuli which can induce an effective adaptive response. In addition, will all types of species respond in the same way and to the same extent? If the required result is an increase in the variability, then the evidence with soybean (i.e., only two alternative states) means that any alteration in a hetrozygote will result in the extent of diversity being reduced. Thus, this mechanism may only be applicable to homogeneous, inbreeding plants where the opportunities for generating phenotypic diversity are severely limited and will be unavailable for the heterozygous, outbreeding species.

In conclusion, it is clear that the plant genome is variable, and the rate at which variation is generated can be increased by certain stressful perturbations; the role for changes in atmospheric chemistry is not known. The alterations generated in response to perturbations are limited to a subset of the genome. However, how this variation is generated, its direct phenotypic consequences, and the way in which those phenotypic alterations are mediated by the changes in DNA are currently not understood. All of the information on genomic instabilities has been derived from annuals, and the situation in long-lived plants is unknown. In addition to the generation of changes, the establishment of altered genotypes within a stand of long-lived plants, would be a second bottleneck. Moreover, none of the atmospheric pollutants have been shown to be inducers of variability, even in those organisms where such a mechanism is known to exist. The use of molecular techniques to sample the range of genotypic variation available should allow the identification of useful genotypes for specific applications. The coupling of this approach with the characterization of detoxifying systems, whether natural of introduced via genetic engineering, may prove successful at ameliorating some of the deleterious effects of pollution.

References

Abel P, Nelson RS, De B, Hoffman N, Rogers SG, Fraley RT, Beachy RN (1986) Delay of disease development in transgenic plants that express the tobacco masaic virus coat protein gene. Science 232:738–743

Beckman JS, Soller M (1983) Restriction fragment length polymorphisms in genetic improvement: methodologies, mapping and costs. Theoretical and Applied Genetics 67:35–43

Bennett MD (1983) The spatial distribution of chromosomes. In: Brandham P, Bennett MD (eds) Proceedings of the 2nd Kew Chromosome Conference. G. Allen and Unwin, London. pp 71–79

Bennett MD, Gustafson JP (1982) The effect of telomeric heterochromatin from *Secale cereale* on triticale (X Triticosecale Wittmack). II. The presence or absence of blocks of heterochromatin in isogenic backgrounds. Canadian Journal of Genetics and Cytology 24:93–100

Bennett MD, Smith JB (1976) Nuclear DNA amounts in angiosperms Phil. Transactions of the Royal Society (London) B 274:227–240

Bethards LA, Skadsen RW, Scandalious JG (1987) Isolation and characterization of a cDNA clone for *Cat2* gene in maize and its homology with other catalases. Proceedings of the National Academy Science, USA 84:6830–6834

Bilinski T, Krawiec Z, Liczmanski A, Litwinska J (1985) Is hydroxyl radical generated by the fenton reaction in vivo? Biochemical and Biophysical Research Communications 130:533–539

Bowler C, Alliotte T, De Loose M, Van Montagu M, Inze D (1989) The induction of manganese superoxide dismutase in response to stress in *Nicotiana plumbaginifolia*. EMBO Journal 8:31–38

Cairns J, Overbaugh J, Miller S (1988) The origin of mutants. Nature 335: 142–145

Carlioz A, Touati D (1986) Isolation of superoxide dismutase mutants in *Escherichia coli*: is superoxide dismutase necessary for aerobic life? EMBO Journal 5:623–630

Cavallini A, Zolfino C, Natali L, Cionini G, Cionini PG (1989) Nuclear DNA changes within *Helianthus annus* L: origin and control mechanism. Theoretical and Applied Genetics 77:12–16

Chandlee JM, Tsaftaris AS, Scandalious JG (1983) Purification and partial characterization of three genetically defined catalases of maize. Plant Science Letters 29:117–131

Cullis CA (1977) Molecular aspects of the environmental induction of heritable changes in flax. Heredity 38:129–154

Cullis CA (1979) Quantitative variation of ribosomal RNA genes in flax genotrophs. Heredity 42:237–246

Cullis CA (1981) DNA sequence organization in the flax genome. Biochimica et Biophysica Acta 652:1–15

Cullis CA (1983) Environmentally induced DNA changes in plants. CRC Critical Reviews in Plant Science 1:117–129

Cullis CA (1986) Phenotypic consequences of environmentally induced changes plant DNA. Trends in Genetics 2:307–309

Cullis CA (1987) The generation of somatic and heritable variation in response to stress. American Naturalist 130:S62–S73

Cullis CA, Charlton LM (1981) The induction of ribosomal DNA changes in flax. Plant Science Letters 20:213–217

Cullis CA, Cleary W (1986a) Rapidly varying DNA sequences in flax. Canadian Journal of Genetics and Cytology 28:252–259

Cullis CA, Cleary W (1986b) DNA variation in flax tissue culture. Canadian Journal of Genetics and Cytology 28:247–251

Cullis CA, Creissen GP (1987) Genomic variation in plants. Annals of Botany 60 [Supplement 4]:103–113

Dallas JF (1988) Detection of DNA "fingerprints" of cultivated rice by hybridization with a human minisatellite DNA probe. Proceedings of the *National Academy Science USA* 85:6831–6835

Dellaporta SL, Chomet PS (1985) The activation of maize controlling elements. Plant Gene Research 2:169–216

Dhindsa RS, Matowe W (1981) Drought tolerance in 2 mosses: Correlated with enzymatic defense against lipid peroxidation. Journal of Experimental Botany 32:79–92

Donn G, Tischer E, Smith JA, Goodman HM (1984) Herbicide resistant alfalfa cells: an example of gene amplification in plants. Journal of Molecular and Applied Genetics 2:621–635

Durrant A (1962) The environmental induction of heritable changes in *Linum*. Heredity 17:27–61

Durrant A, Jones TWA (1971) Reversion of the induced changes in the amount of DNA in *Linum*. Heredity 27:431–439

Evans GM, Durrant A, Rees H (1966) Associated nuclear changes in the induction of flax genotrophs. Nature 212:697–699

Evans DA, Sharp WR, Medina-Filho HP (1984) Somaclonal and gametoclonal variation. American Journal of Botany 71:759–774

Fieldes MA, Tyson H (1973) Activity and relative mobility of peroxidase and esterase isozymes of flax (*Linum usitatissimum*) genotrophs. I. Developing main stems. Canadian Journal of Genetics and Cytology 15:731–744

Flavell RB (1980) The molecular characterization and organization of plant chromosomal DNA. Annual Review of Plant Physiology 31:596–596

Gerstel DU, Burns JA (1966) Chromosomes of unusual length in hybrids between two species of *Nicotiana*. Chromosomes Today 1:41–56

Goldsbrough PB, Ellis THN, Cullis CA (1981) Organization of the 5S RNA genes in flax. Nucleic Acids Research 9:5895–5904

Gustafson JP, Lukaszewski AJ, Bennett MD (1983) Somatic deletion and redistribution of telomeric heterochromatin in the genus *Secale* and in *Triticale*. Chromosoma (Berlin) 88:293–298

Hall BG (1988) Adaptive evolution that requires multiple spontaneous mutations. I. Mutations involving an insertion sequence. Genetics 120:887–897

Halliwell B (1984) Oxygen-derived species and herbicide action. Physiologia Plantarum 15:21–24

Hilliker AJ, Appels R (1982) Pleitropic effects associated with the deletion of heterochromatin surrounding rDNA on the X chromosome of *Drosophila*. Chromosoma (Berlin) 86:469–490

Jeffreys AJ, Wilson V, Thein SL (1985) Hypervariable "minisatellite" regions in human DNA. Nature 314:67–73

Johnson SS, Phillips RL, Rines HW (1987) Possible role of heterochromatin in chromosome breakage induced by tissue culture in oats (*Avena sativa* L). Genome 29:439–446

Lapitan NLV, Sears RG, Gill BS (1984) Translocations and other karyotypic structural changes in wheat × rye hybrids regenerated from tissue culture. Theoretical and Applied Genetics 68:547–554

Larkin PJ, Scowcroft W (1983) Somaclonal variation and crop improvement. In: Kosuge T, Meredith CP, Hollaender A (eds) Genetic engineering of plants. Plenum, New York. pp 289–314

Lee M, Phillips RL (1987) Genomic rearrangements in maize induced by tissue culture. Genome 29:122–128

Leutwiler LS, Hough-Evans BR Myerowitz EM (1984) The DNA of *Arabidopsis thaliana*. Molecular and General Genetics 194:15–23

McClintock B (1950) The origin and behaviour of mutable loci in maize. Proceedings of the National Academy of Science USA 36:344–355

McClintock B (1984) The significance of responses of the genome to challenge. Science 26:792–801

Nagao RT, Kimpel JA, Vierling E, Key JL (1986) The heat shock response: A comparative analysis. Oxford Surveys in Plant Molecular and Cell Biology 3:384–438

Nagl W, Jeanjour M, Kling H, Kuhner S, Michels I, Muller T, Stein B (1983) Genome and chromatin organization in higher plants. Biologische Zentralblatt 102:129–148

Orton TJ (1983) Experimental approaches to the study of somaclonal variation. Plant Molecular Biology Reporter 1:67–76

Orton TJ (1984) Somaclonal variation: theoretical and practical considerations. In: Gustafson JP (ed). Genetic manipulation and plant improvement. Plenum Press, New York, p 427

Patterson AH, Lander ES, Hewitt JD, Peterson S, Lincoln SE, Tanksley SD (1988) Resolution of quantitative traits into Mendelian factors by using a complete RFLP linkage map. Nature 335:721–726

Perani L, Radke S, Wilke-Douglas M, Bossert M (1986) Gene transfer methods for crop improvement: introduction of foreign DNA into plants. Physiologica Plantarum 68:566–570

Pesche VM, Phillips RL, Gegenbach BG (1987) Discovery of transposable element activity among progeny of tissue culture-derived maize plants. Science 238:804–807

Price HJ, Chambers KL, Bachmann K, Riggs J (1983) Inheritance of nuclear 2C DNA content in intraspecific and interspecific hybrids of *Microseris* (Asteraceae). American Journal of Botany 70:1133–1138

Redinbaugh HD, Wadsworth GJ, Scandalious JG (1988) Characterization of catalase transcripts and their differential expression in maize. Biochimica et Biophysica Acta 951:104–116

Rivin CJ, Cullis CA, Walbot V (1986) Evaluating quantitative variation in the genome of *Zea mays*. Genetics 113:1009–1019

Rogers SG, Klee H (1987) Pathways to plant genetic engineering employing

Agrobacterium. In: Hohn T, Schell J (eds) Plant DNA infectious agents. Springer-Verlag, New York, pp 179–203

Rogstad SH, Patton JC, Schaal BA (1988) M13 repeat probe detects DNA minisatellite-like sequences in gymnosperms and angiosperms. Proceedings of the National Academy of Science USA 85:9176–9178

Roth EJ, Frazier BL, Apuya NR, Lark KG (1989) Genetic variation in an inbred plant: variation in tissue cultures of soybean (*Glycine max* (L) Merrill). Genetics 121:359–368

Schneeberger RS, Creissen GP, Cullis CA (1990) Chromosomal and molecular analysis of 5S RNA gene organization in the flax *Linum usitstissimum*. Gene 83:75–84

Schneeberger RS, Cullis CA (1991) Specific DNA alterations associated with the environmental induction of heritable changes in flax. Genetics (in press)

Scowcroft WR, Larkin PJ (1985) Somaclonal variation, cell selection and genotype improvement. In: Moo-Young M (ed) Comprehensive biotechnology. Pergamon Press, London pp 153–168

Sederoff R, Stomp A-M, Chilton WS, Moore LW (1986) Gene transfer into lobloooy pine by *Agrobacterium tumefaciens*. Bio/Technology 4:647–649

Shah DM, Horsch RB, Klee HJ, Kishore GM, Winter JA, Tumer NE, Hironaka CM, Sanders PR, Gasser CS, Aykent S, Siegel NR, Rogers SG, Fraley RT (1986) Engineering herbicide tolerance in transgenic plants. Science 233:478–481

Sorenson, IC (1984) The structure and expression of nuclear genes in higher plants, Advances in Genetics 22:109–166

Tanksley SD (1983) Molecular markers in plant breeding. Plant Molecular Biology Reporter 1:3–8

Tanksley SD, Young ND, Paterson AH, Bonierbale MW (1989) RFLP mapping in plant breeding: new tools for an old science. Bio/Technology 7:257–264

Vaek M, Reynaerts A, Hofte H, Jansens S, De Beucheleer M, Zabeau M, Van Montagu M, Leemans J (1987) Transgenic plants protected from insect attack. Nature 328:33–37

Walbot V (1986) On the life strategies of plants and animals. Trends in Genetics 1:165–169

Walbot V, Cullis CA (1985) Rapid genomic change in higher plants Annual Review of Plant Physiology 36:367–396

Wilke-Douglas M, Perani L, Radke S, Bossert M (1986) The application of recombinant DNA technology toward crop improvement. Physiology Plantarum 68:560–565

Wise RR, Naylor AW (1987) Chilling enhanced photooxidation: The peroxidative destruction of lipids during chilling injury to photosynthesis and ultrastructure. Plant Physiology 83:272–277

Ye J, Hauptman RM, Smith AG, Widholm JM (1987) Selection of a *Nicotiana plumbaginifolia* universal hybridizer and its use in intergenic somatic hybrid formation. Molecular and General Genetics 208:474–480

Zheng KL, Castiglione S, Biasini MG, Biroli A, Morandi C, Sala F (1987) Nuclear DNA amplification in cultured cells of *Oryza sativa* L. Theoretical and Applied Genetics 74:65–70

Zimmerman PA, Lang-Unnasch N, Cullis CA (1989) Polymorphic regions in plant genomes detected by an M13 probe. Genome 32:824–828

Commentary to Chapter 9(A)

Common Mechanisms of Intracellular Stress Induction by Atmospheric Pollutants and the Role of Genes and Mutations in Damage Alleviation

Duncan R. Talbot, Ronald T. Nagao, and Eva J. Pell

Introduction

Important considerations for the practical assessment of direction, rate and efficacy of evolution which could occur as a result of exposure to anthropogenic atmospheric pollutants include the following:

1. A mutation conferring stress resistance will affect a gene encoding an activity already present in a cell, altering its properties to render the cell more optimally suited for shifted environmental conditions. Some mutations, when studied further, have been shown to have costs because the mutations provide a new function at the expense of the original activity. Thus, single gene mutations identified in species of agronomic interest may not necessarily reflect evolutionary directions taken by natural populations subjected to pullution stress. More detailed research on stress resistance mechanisms in plants is needed in order that specific costs, benefits and evolutionary directions can be evaluated.
2. The atmospheric pollutants O_3, SO_2, NO_x, and heavy metals all participate in intracellular generation of chemical oxidants. Therefore, if sources of new evolutionary directions in plants exposed to these pollutants are to be examined, oxidative stress defensive systems already in place would be logical candidates for continued assessment of available useful genetic variation in natural populations.
3. Understanding the mechanisms of stress resistance is extremely important if the research goal is more than the identification of superior phenotypes. If the mechanism of stress resistance is understood, then it is feasible to design experiments not only for testable predictions and modeling of potential plant evolution in response to stress, but also for the beneficial exploitation of that mechanism. Mechanisms involving structural modifications of complex pathways would be difficult to exploit with current technology. However, the identification of natural

mechanisms which involve simple traits (for example, an enzyme to modify/detoxify a pollutant) encoded by a single gene or gene family has the potential for beneficial manipulation using molecular biology techniques.

Commentary

Study of systems where single genes (or single classes of genes) determine stress resistance allows their mechanisms of action to be dissected at the molecular level. Additionally, detailed determinations of specific benefits and costs to plants containing mutations in these genes are feasible once mechanisms are known.

Plant mutations conferring resistance to the triazine herbicides (e.g., atrazine) have been a subject of intense study concerning mechanisms of action of the photosynthetic apparatus, as well as assessment of the agronomic benefits offered by exploiting these mutations. Atrazine is a photosynthetic inhibitor which has been found to bind the 32,000 Dalton molecular weight protein encoded by the *psbA* gene (which is found in the chloroplast genome; consequently there are approximately 3×10^3 copies of this gene in a higher plant leaf mesophyll cell). This protein is a thylakoid membrane-localized component of the PSII reaction center. An important class of mutations in this protein, found in plant populations growing in fields repeatedly exposed to triazine herbicides (Bettini et al. 1987), involves single DNA base changes at the 264th codon of the gene, which change glycine to serine or threonine, resulting in dramatically increased resistance to atrazine as well as other herbicides (Hirshberg and McIntosh 1983; Sigematsu et al. 1989). In laboratory experiments chloroplasts containing mutant psbA proteins have been found to be functional, as judged by performance of the Hill reaction, in the presence of 600-fold more atrazine than wild-type chloroplasts (Sigematsu et al. 1989). These *psbA* mutations are maintained in natural populations of weed species, and efforts are under way to incorporate this type of mutation into crop plant improvement programs by breeding or by genetic engineering. Integration of *psbA* mutant genes into the nuclear genomes of test plants has resulted in partial atrazine resistance under laboratory conditions (Cheung et al. 1988).

These mutations to atrazine resistance involve alterations of an existing activity. Costs to the host plants reflect the nature of the atrazine binding sites on the psbA protein molecules. These sites naturally function, in an as yet incompletely characterized way, in binding of quinone electron acceptors to the PSII complex as measured by the criterion of electron transfer efficiency among components of the photosynthetic reaction center (Sigematsu et al. 1989). Thus, binding site alterations which decrease affinity for atrazine, which blocks electron transfer within PSII, also decrease binding affinities for natural quinones (which are required for efficient electron transfer). The result has been decreased photosynthetic efficiency. Under sets of environmental conditions where photosynthesis rates could be limited (e.g., oxidative stress or drought; Tingey and Andersen, Chapter 8, this volume), *psbA* mutations might prove to be detrimental.

Further work, using molecular and biotechnological approaches, has shown that plants are somewhat flexible concerning the types of mutations able to confer atrazine tolerance. For example, additional *psbA* mutations have been examined which do not extensively alter the electron-accepting quinone sites (Brusslan and Haselkorn 1989). Another class of triazine herbicide-resistant mutants, which inactivates these compounds before they reach their psbA protein targets, has been constructed using plant gene-transfer technology (Anonymous 1986). These mutants overproduce the detoxifying enzyme glutathione-S-transferase (GST; see Edwards and Owen 1986; Stephenson et al. 1983), resulting in accelerated metabolic degradation of the herbicide in a manner which leaves photosynthesis unimpaired. Thus, either mutations in structural genes which alter interactions between stress-producing agents and intracellular targets, or regulatory mutations which avoid stress by altering deterrence systems, are capable of conferring resistance.

Although molecular biology/biotechnology research can elucidate genetic mechanisms of stress resistance, as shown by this example involving herbicide resistance, a multidisciplinary approach also including genetics, biochemistry, physiology, and ecology is really needed to evaluate the benefits and costs of mutations conferring stress resistance. This type of research is now being done in agriculture to ensure continued availability of prospects for crop plant improvement. The importance of this multi-disciplinary approach to evolutionary biology is that it will provide further opportunities for defining characteristics of mutants or genetic variants which would potentially predominate in plant populations subjected to selection pressures imposed by atmospheric pollutants.

Evolutionary targets: each component of a cellular stress response system is encoded by a gene.

As stated above, the tools and perspectives of molecular biology and biotechnology can strongly impact pollution research, especially where single-gene systems are involved. However, the majority of the genetic evidence available indicates that stress resistance, for example to atmospheric pollutants such as O_3, is a function of a large number of genes which are collectively expressed as quantitative traits (see, for example, Dragoescu et al. 1987; Pell 1987; reviewed by Roose Chapter 5, this volume). How can a practical assessment be made of the modes of action of these genes and of their evolutionary rates and directions? In the following section, two possible ways of approaching this question are discussed. First, it is pointed out that research aimed at understanding oxidative stress responses to atmospheric pollution will be critical: pollutants such as O_3, SO_2, and heavy metals participate in common metabolic pathways which generate toxic oxygen species. Second, it is suggested that genes responsible for signalling functions, along with genes involved in maintenance of polypeptide structures, are important candidates for understanding mechanisms and evolution of plant responses to stress.

Deterrent systems are already in place (and have evolved long ago) to combat oxidative stress conditions which a plant encounters on a daily basis, for example: (a) closure of stomates under high light and/or high transpiration conditions which results in depletion of $NADP^+$ levels, subsequent reduction of molecular oxygen, and generation of reactive chemical species intracellularly; (b) production of hydrogen peroxide through metabolism of glycollate under photorespiratory conditions, followed by generation of reactive chemical species; (c) high rates of mitochondrial or symbiotic bacterial aerobic respiration, resulting in production of toxic oxygen species via reduction of molecular oxygen.

Defense systems include inducible enzymatic deterrents (see Fig. 1) such as superoxide dismutase, catalase, glutathione reductase, and ascorbate peroxidase, as well as chemical scavengers including GSH, carotenoids, and ascorbate. Additionally, production of toxic oxygen species as by-products of plant metabolism may initiate "redox cycling" of electrophilic quinones and phenolic compounds (Sies 1985), which are found at high levels in plant cells, resulting in further generation of toxic and radical chemical species. Enzymatic deterrent systems for these chemical species include glutathione-S-transferase and cytochrome P-450, which have been extensively studied in mammalian systems but less so in plants (see, for example, Mozer et al. 1983).

These scavenger and enzymatic deterrent systems act in concert with metabolic regulatory mechanisms and physiological adjustments (for example, stomatal closure) to optimize cellular functions under stress conditions (see Tingey and Andersen, Chapter 8, this volume, for an organization of these systems into a conceptual model for plant stress responses). Every component of each system is encoded by a gene which therefore may be termed a "stress resistance gene." It is thus these gene systems which will be subjected to selection in plants exposed to atmospheric pollutants.

Genetic and molecular biological data indicate that many stress resistance genes are present in plant systems. The functions of most of them are currently unknown. One way to describe their roles in stress responses is to construct three categories of potential functions: (a) structural genes; (b) genes involved in signalling; and (c) "maintenance" genes responsible for correct folding, processing, and delivery of protein gene products to their appropriate intracellular compartments. The last category is especially important to plant cells because of the necessity to import many proteins essential for metabolism into chloroplasts as well as mitochondria (see Della Cioppa et al. 1987).

Structural Genes

Included are a great many genes shown in plants and other systems to encode enzymes having stress deterrent activities. Some involved in oxidative stress deterrence are listed in Fig. 1. Their roles are more thoroughly discussed by Cullis (Chapter 9, this volume) and by Tingey and Andersen (Chapter 8, this volume). Structural genes encoding cellular components which act as targets for stress agents include *psbA*, discussed above. Additionally, structural gene regions involving domains with biochemical regulatory properties on metabolic enzymes, or domains specifying potentially alterable protein folding conformations (Huner 1985), are possible targets for inactivation or for favorable mutation events.

Signalling Genes

Regulatory regions of structural genes or structural genes which encode stress signal transduction components, such as separate regulatory molecules (Lawton et al. 1989), are involved in intracellular sensing of stress signals. These systems can serve to induce defenses against naturally occurring stress agents; they would be expected to evolve further where anthropogenic atmospheric pollutants are acting as selective agents.

Signal locus systems have not been extensively studied because of the difficulties encountered in assaying specifically for their functions. They act to modulate gene activities or to initiate regulatory cascades outside of nuclei by binding to DNA, RNA, proteins, or small molecules. In eukaryotic systems, most studies of regulatory genes have focused on direct physiological or biochemical characterizations, along with analyses of DNA or protein sequences involved in regulatory factor binding interactions (Czarnecka et al. 1988; Sorger and Pelham 1988). In bacterial systems, where detailed genetic analyses of stress responses have been performed, important aspects of oxidative stress signalling mechanisms have been elucidated, including:

1. Genes involved in heat shock and oxidative stress responses are organized into batteries which are termed "regulons" (Christman et al. 1985; Morgan et al. 1986; Teo et al. 1986; Van Bogelen et al. 1987). Regulons are themselves controlled by genetic regulatory loci which activate or repress appropriate genes to achieve a given response. These regulatory loci encode gene products which induce individual genes singly or in groups from multiple regulons; cells lacking regulatory locus gene products are hypersensitive to stress-inducing agents such as hydrogen

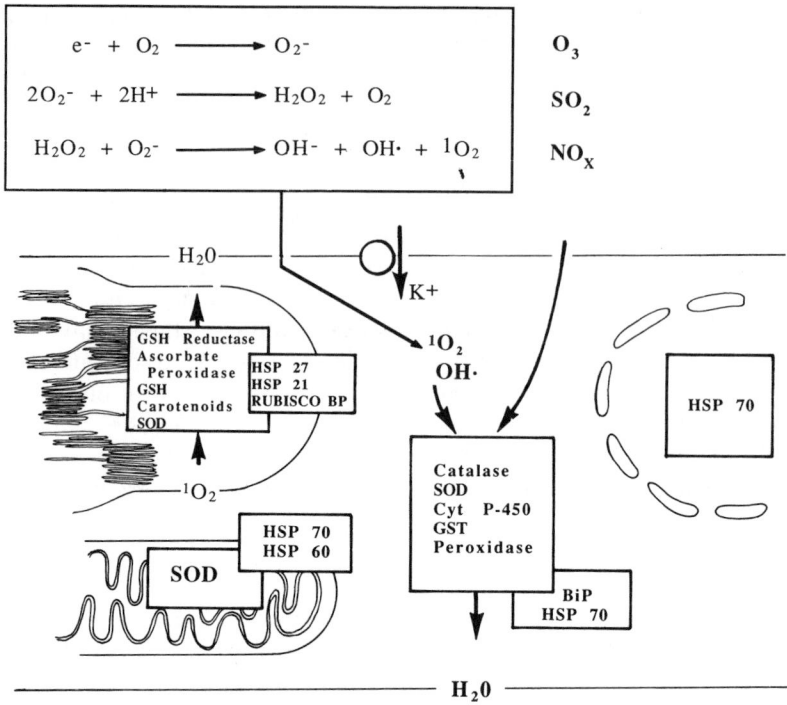

FIGURE 1. Cellular localizations of stress deterrent activities concerned with responses to chemical oxidants. Metabolically derived oxidants produce the damaging species singlet oxygen and hydroxyl radical via Haber-Weiss chemistry outlined in the *upper left* portion (Sies 1985). Dissolved atmospheric pollutants (*upper right*) will also enter the damaging oxidant-producing pathways. Stress systems in one or more compartments will be activated, depending on the initial site of perception. Initial sensing may take place at the cell surface, as in the case of a cadmium-inactivated potassium channel in *E. coli* (R. Vinopal and D. Talbot, unpublished data). Defensive systems include, in *chloroplasts*: glutathione (GSH) reductase, ascorbate peroxidase, and GSH (Halliwell 1984); SOD (Scioli and Zilinskas 1988); carotenoids (Mayfield et al. 1986). HSPs and ribulose 1,5-bisphosphate carboxylase binding proteins are included to indicate their possible roles in assembly and maintenance of chloroplast protein structures (Vierling et al. 1986, 1988; Ellis 1987). In *mitochondria*: SOD (White and Scandalios 1987; Bowler et al. 1989), HSPs 70 and 60 (Deshaies et al. 1989; Pelham 1986). In the *cytoplasm*: catalase (Havir and McHale 1987); SOD (Cannon et al. 1987); peroxidase, cytochrome (Cyt) p-450, and glutathione-S-transferase (Sies 1985). In the *nucleus*: HSP 70 (Lin et al. 1984).

peroxide. It is notable, in terms of the actions of atmospheric pollutants described above, that the control of these systems may depend upon the redox status of the cells (Christman et al. 1989). Results of molecular biological and biochemical experiments in plants and mammals suggest that they also have overlapping genetic systems controlling oxidative stress, heat shock, and other stress responses (Czarnecka et al. 1984). The underlying lesion responsible for triggering these responses may be partial denaturation of a critical polypeptide by heat or by oxidative conditions (Ananthan et al. 1986; Edington et al. 1989; Hightower et al. 1980; Pelham 1986).

2. There are extensive backup systems in oxidative stress response gene batteries. Where one deterrent system, such as glutathione metabolism, is inactivated by a mutation, another system will be able to compensate (Fuchs et al. 1983).

3. Antioxidant systems are balanced in terms of their responses to stress. In experiments where superoxide dismutase (SOD) overproducers were constructed by recombinant DNA methods, mutant strains were hypersensitive to external oxidative stress-generating agents (Scott et al. 1987). This effect is likely due to increased levels of hydrogen peroxide produced in these cells as a product of SOD activity. Thus, enhancements in activity of one antioxidant system, such as SOD, cannot necessarily be beneficial unless responses of other systems, such as catalase or peroxidase, are adjusted as well.

Evidence from prokaryotic and eukaryotic systems suggests that protein factors and DNA regulatory regions, all scorable as stress genes, are involved in sensing stress conditions and triggering appropriate responses. The complexities involved in registering conditions in multicompartmented cells and controlling responses of multigene families indicate that a considerable number of signalling genes must be present in plant cells.

"Protein Maintenance" Genes

Recent research has demonstrated that eukaryotic cells expend considerable energy assembling proteins and transporting them to the correct intracellular compartments. In plant cells, which, depending on their developmental state, can synthesize up to 5×10^4 different proteins, all but about 150 are made in the cytosol and then shuttled to final destinations. Examples of compartments targeted for importation of polypeptides include chloroplasts (Chua and Schmidt 1978), mitochondria (Lin et al. 1984; Unger et al. 1990), golgi-derived vacuoles (Beachy et al. 1985), cell wall spaces (Averyhart-Fullard et al. 1988), and nuclei (Lin et al. 1984).

In addition to mechanisms required for transport of proteins across membranes, cells possess a number of genes which mediate maintenance of correct three-dimensional polypeptide configurations, aid in assembly of multisubunit protein complexes, and facilitate precise protein translocation into correct compartments. Many of these "maintenance" proteins were initially defined as stress proteins; they include several heat shock proteins (HSPs; Chirico et al. 1989; Pelham 1986), HSP-related proteins and "chaperonins" (e.g., Rubisco binding protein and yeast HSP 60; Deshaies et al. 1989; Ellis 1987; Hemmingsen et al. 1988). As a group, these proteins bind denatured (or incompletely folded) proteins and incompletely assembled protein complexes. These activities, which are required by cells on a routine basis, understandably might therefore be needed at elevated levels under conditions (altered cellular redox balance, high temperature) where proteins with altered configurations (i.e., nonfunctional proteins) might otherwise accumulate. Under these conditions "maintenance" proteins apparently have the important functions of repairing or, alternatively, facilitating turnover of denatured proteins.

Subcellular compartmentation: specific responses to stress may be determined by the location of initial stress signal sensing. A number of stress deterrent systems are preferentially located in particular compartments.

In order to assess the range of selective pressures on cells resulting from exposure to atmospheric pollutants, it is useful to visualize the locations of stress targets, defenses, and potential signalling mechanisms within cells. Atmospheric pollutants O_3, SO_2, NO_x, and heavy metals produce parallel effects on cells via oxidative stress in that they initiate chemical reactions which form toxic, reactive oxygen species that damage cellular components such as membranes, proteins and DNA (Halliwell 1987; Pell 1987; Sies 1985).

A diagram of a cell, depicting nucleus, cytoplasm, chloroplast, and mitochondrion, and the stress defense systems localized in each of these compartments, is presented in Fig. 1. The figure incorporates underlying parallels observed in responses of plant, fungal, mammalian, and bacterial systems to oxidants generated either as byproducts of metabolism or resulting from dissolution of external pollutants, such as O_3 (Pell 1987), in cellular fluids. For reference, a modified Haber-Weiss reaction scheme (Sies 1985) is included in Fig. 1 to show chemical pathways involved in producing damaging oxidative species (hydroxyl radical, $OH\cdot$; singlet oxygen, $^1O_2$) from intermediates such as superoxide (O_2^-) and hydrogen peroxide. Heavy metals such as cadmium can also initiate these reactions through chemical steps which allow donations of electrons to reactive oxygen species (Halliwell 1987).

Given the number of possible defense systems to oxidative stress found in multicompartmented cells, such as plant cells, responses will probably ensue depending on the location of the initial sensing of stress conditions. This sensing may involve signalling by a damaged or partially unfolded cellular component or by leakage of compartmentalized ions such as Ca^{2+} (Bellomo et al. 1987; Christman et al. 1989; Edington et al. 1989; Goff and Goldberg 1985). Compartmentation of stress defense systems may be explained on the basis of their proximities to other systems (i.e., genetic; metabolic) which are essential for cellular functions. Compartmentalized antioxidant systems are found in chloroplasts and mitochondria, whose electron transport systems can, under some conditions, utilize molecular oxygen as an alternate electron acceptor. This results in formation of active chemical species such as superoxide radical (Salin 1987). In chloroplasts, glutathione (GSH) is present in high concentrations; its roles include inactivation of singlet oxygen (which forms as a byproduct of photosynthesis) and chemical reduction of ascorbate, which functions in hydrogen peroxide detoxification (Halliwell 1984). Glutathione also may function in regulation of CO_2 fixation enzymes, activities of which are controlled by their protein thiol redox status (Ziegler 1985).

The additional levels of complexity imposed by cellular compartmentation are not necessarily insurmountable hurdles to determining mechanisms involved in stress responses or to assessing modes of their potential evolution. A significant amount of work remains to be done on the development of compartmentalized stress deterrent systems in cells of particular tissues and regulation of genes encoding compartmented products under conditions of stress. This research is needed to show the time frame in the life cycle of a plant in which a particular stress deterrence system is required. It is equally important to improve our understanding of intracellular stress signal transduction. From an evolutionary standpoint, a multidisciplinary approach, as mentioned above, will be needed to translate mechanistic principles derived at the cellular level into useful models for studying stress at the organismal and population levels.

References

Ananthan J, Goldberg AL, Voellmy R (1986) Abnormal proteins serve as eukaryotic stress signals and trigger the activation of heat shock genes. Science 232:522–524

Anonymous (1986) Engineered tobacco takes to the field. Bio/Technology 4:1047

Averyhart-Fullard V, Datta K, Marcus A (1988) A hydroxyproline-rich protein in the soybean cell wall. Proceedings of the National Academy of Science USA 85:1082–1085

Beachy RN, Chen Z-L, Horsch RB, Rogers SG, Hoffman NJ, Fraley RT (1985) Accumulation and assembly of soybean B-conglycinin in seeds of transformed petunia plants. EMBO Journal 4:3047–3053

Bellomo G, Thor H, Orrenius S (1987) Alterations in inositol phosphate production during oxidative stress in isolated hepatocytes. Journal of Biological Chemistry 262:1530–1534

Bettini P, McNally S, Sevignac M, Darmency H, Gasquez J, Dron M (1987) Atrazine resistance in *Chenopodium album*: low and high levels of resistance to the herbicide are related to the same chloroplast psbA mutation. Plant Physiology 84:1442–1446

Bochner BR, Lee PC, Wilson SW, Cutler CW, Ames BN (1984) AppppA and related adenylylated nucleotides are synthesized as a consequence of oxidation stress. Cell 37:225–232

Bowler C, Alliotte T, Van den Bulcke M, Bauw G, Vanderckhove J, Van Montagu M, Inze D (1989) A plant manganese superoxide dismutase is efficiently imported and correctly processed by yeast mitochondria. Proceedings of the National Academy of USA 86:3237–3241

Brusslan J, Haselkorn R (1989) Resistance to the photosystem II herbicide diuron is dominant to sensitivity in the cyanobacterium *Synechococcus* sp. PCC7942. EMBO Journal 8:1237–1245

Cannon RE, White JA, Scandalios JG (1987) Cloning of a cDNA for maize superoxide dismutase 2 (SOD2). Proceedings of the National Academy of Science USA 84:179–183

Cheung AY, Bogorad L, Van Montagu M, Schell J (1988) Relocating a gene for herbicide tolerance: a chloroplast gene is converted into a nuclear gene. Proceedings of the National Academy of Science USA 85:391–395

Chirico WJ, Waters MG, Blobel G (1989) 70K Heat shock related proteins stimulate protein translocation into microsomes. Nature 332:805–810

Christman MF, Morgan RW, Jacobsen FS, Ames BN (1985) Positive control of a regulon for defenses against oxidative stress and some heat shock proteins in *Salmonella typhimurium*. Cell 41:753–762

Christman MF, Storz G, Ames BN (1989) OxyR, a positive regulator of hydrogen peroxide-inducible genes in *Escherichia coli* and *Salmonella typhimurium*, is homologous to a family of bacterial regulatory proteins. Proceedings of the National Academy of Science USA 86:3484–3488

Chua N-H, Schmidt GW (1978) Post-translational transport into intact chloroplasts of a precursor to the small subunit of ribulose-1,5-bisphosphate carboxylase. Proceedings of the National Academy of Science USA 75:6110–6114

Czarnecka E, Edelman L, Schoffl F, Key JL (1984) Comparative analysis of physical stress responses in soybean seedlings using cloned heat shock cDNAs. Plant Molecular Biology 3:45–58

Czarnecka E, Nagao RT, Key JL, Gurley WB (1988) Characterization of Gmhsp26-A, a stress gene encoding a divergent heat shock protein of soybean: heavy metal-induced inhibition of intron processing. Molecular Cell. Biology 3:45–58

Della-Cioppa G, Kishore GM, Beachy RN, Fraley RT (1987) Protein trafficking in plant cells. Plant Physiology 84:965–968

Deshaies RJ, Koch BD, Werner-Washburne M, Craig EA, Schekman R (1989) A subfamily of stress proteins facilitates translocation of secretory and mitochondrial precursor polypeptides. Nature 332:800–805

Dragoescu N, Hill RR Jr, Pell EJ (1987) An autotetraploid model for genetic analysis of ozone tolerance in potato, *Solanum tuberosum* L. Genome 29:85–90

Edington BV, Whelan SA, Hightower LE (1989) Inhibition of heat shock (stress) protein induction by deuterium oxide and glycerol: additional support for the abnormal protein hypothesis of induction. Journal of Cell Physiology 139:219–228

Edwards R, Owen WJ (1986) Comparison of glutathione-S-transferases of *Zea mays* responsible for herbicide detoxification in plants and suspension-cultured cells. Planta 169:208–215

Ellis RJ (1987) Proteins as molecular chaperones. Nature 328:378–379

Fuchs JA, Haller B, Tuggle CK (1983) Mutants of *Escherichia coli* altered in glutathione metabolism. In: Larsson A (ed) Functions of glutathione:

biochemical, physiological, toxicological, and clinical aspects. Raven Press, NY, pp 385–393

Goff SA, Goldberg AL (1985) Production of abnormal proteins in *E. coli* stimulates transcription of *lon* and other heat shock genes. Cell 41:587–595

Halliwell B (1984) Chloroplast metabolism. Oxford University Press, Oxford

Halliwell B (1987) Oxidants and human disease: some new concepts. FASEB Journal 1:358–364

Havir EA, McHale NA (1987) Biochemical and developmental characterization of multiple forms of catalase in tobacco leaves. Plant Physiology 84:450–455

Hemmingsen SM, Woolford C, Vanderlies SM, Tilly K, Dennis DT, Georgopoulos CP, Hendrix RW, Ellis RJ (1988) Homologous plant and bacterial proteins chaperone oligomeric protein assembly. Nature 333:330–334

Hightower LE (1980) Cultured animal cells exposed to amino acid analogues or puromycin rapidly synthesize several polypeptides. Journal of Cell Physiology 102:407–427

Hirshberg J, McIntosh L (1983) Molecular basis of herbicide resistance in *Amaranthus hybridus*. Science 222:1346–1349

Huner NPA et al. (1985) Morphological, amatomical, and molecular consequences of growth and development at low temperature in *Secale cereale* cultivar Puma. American Journal of Botany 72:1290–1306

Lawton MA, Yamamoto RT, Hanks SK, Lamb CJ (1989) Molecular cloning of plant transcripts encoding protein kinase homologs. Proceedings of the National Academy of Science USA 86:3140–3144

Lin C-Y, Roberts JK, Key JL (1984) Acquisition of thermotolerance in soybean seedlings. Plant Physiology 74:152–160

Mayfield SP, Nelson T, Taylor WC (1986) The fate of chloroplast proteins during photooxidation in carotenoid-deficient maize leaves. Plant Physiology 82:760–764

Morgan RW, Christman MF, Jacobsen FS, Storz G, Ames BN (1986) Hydrogen peroxide-inducible proteins in *Salmonella typhimurium* overlap with heat shock and other stress proteins. Proceedings of the National Academy of Science USA 83:8059–8063

Mozer TJ, Tiemeyer DC, Jaworski EG (1983) Purification and characterization of corn glutathione-S-transferase. Biochemistry (USA) 22:1068–1072

Pelham HRB (1986) Speculations on the functions of the major heat shock and glucose-regulated proteins. Cell 46:959–961

Pell E (1987) Ozone toxicity—Is there more than one mechanism of action? In: Hutchinson TC, Meema KM (eds) Effects of atomspheric pollutants on forests, wetlands, and agricultural ecosystems. Springer-Verlag, Berlin, pp 229–240

Salin ML (1987) Toxic oxygen species and protective systems of the chloroplast. Physiology of Plants 72:681–689

Scioli JR, Zilinskas BA (1988) Cloning and characterization of a cDNA encoding the chloroplastic copper/zinc-superoxide dismutase from pea. Proceedings of the National Academy of Science USA 85:7661–7665

Scott MD, Meshnick SR, Eaton JW (1987) Superoxide dismutase-rich bacteria. Journal of Biological Chemistry 262:3640–3645

Sies H (ed) (1985) Oxidative stress. Academic Press, London

Sigematsu, Y, Sato F, Yamada Y (1989) The mechanism of herbicide resistance in tobacco cells with a new mutation in the Q_B protein. Plant Physiology 89:986–992

Stephenson GR, Ali A, Ashton FM (1983) In: Miyamoto JM, Kearney PC (eds), Pesticide chemistry, human welfare, and the environment, vol 3. Pergamon Press, NY, pp 219–224

Sorger PK, Pelham HRB (1988) Yeast heat shock factor is an essential DNA-binding protein that exhibits temperature-dependent phosphorylation. Cell 54:855–864

Teo I, Sedgewick B, Kilpatrick MW, McCarthy TV, Lindahl T (1986) The intracellular signal for induction of resistance to alkylating agents in *E. coli*. Cell 45:315–324

Unger EA, Hand JM, Cashmore AR, Vasconcelos AC (1990) Isolation of a cDNA encoding mitochondrial citrate synthase from *Arabidopsis thaliana*. Plant Molecular Biology 13:411–418

Van Bogelen RA, Kelley PM, Neidhardt FC (1987) Differential induction of heat shock, SOS, and oxidative stress regulons and accumulation of nucleotides in *Escherichia coli*. Journal of Bacteriology 169:26–32

Vierling E, Mishkind ML, Schmidt GW, Key JL (1986) Specific heat shock proteins are transported into chloroplasts. Proceedings of the National Academy of Science USA 83:361–365

Vierling E, Nagao RT, DeRocher AE, Harris LM (1988) A heat shock protein localized to chloroplasts is a member of a eukaryotic superfamily of heat shock proteins. EMBO Journal 7:575–581

White JA , Scandalios JG (1987) In vitro synthesis, importation, and processing of Mn-superoxide dismutase (SOD-3) into maize mitochondria. Biochimische Biophysische Acta 926:16–25

Ziegler DM (1985) Role of reversible oxidation-reduction of enzyme thiols-disulfides in metabolic regulation. Annual Review of Biochemistry 54:305–329

Commentary to Chapter 9 (B)

Genomic Stress, Genome Size, and Plant Adaptation

H. James Price

Introduction

The concept that the plant genome is fluid and dynamic in regard to the quantity of certain DNA sequences is presented by Cullis (Chapter 9, this volume). The generation, dispersion, and deletion of repetitive sequences apparently provided the quantitative variability that evolutionary processes have acted on to generate the several hundredfold range in genome size observed among angiosperms (Flavell 1980; 1982; Thompson et al. 1980; Price 1976, 1988a; Bennett 1985). The processes resulting in turnover of DNA sequences in the genome are not well understood. However, the activity of transposable elements, unequal crossing-over, and saltatory amplifications are generally considered as likely mechanisms.

At least some of the mechanisms involved are apparently sensitive to genetic and environmental parameters; genomic stress imposed by a stressful physical environment or by hybridization may trigger deletion and/or amplification events (McClintock 1984; Walbot and Cullis 1985; Cullis Chapter 9, this volume). The role of the environment in inducing heritable quantitative DNA changes has been well documented in flax (Cullis and Cleary 1986; Cullis, Chapter 9, this volume).

The objective of this commentary is to further develop the concept that quantitative variation in DNA content is important to plant evolution, including adaptation to stressed environments (including air pollution), and that this variation is more common and widespread than previously believed. A consideration of plant adaptation to polluted or stressed environments should discuss the roles of both variation in DNA amount and the genotype.

The following aspects of DNA content will be discussed: (1) the magnitude of the initial events generating quantitative differences; (2) variation within species; (3) inheritance; (4) biophysical or nucleotypic effects; and (5) model systems.

Magnitude of Amplification and Deletion Events

Amplification and deletion events have been directly detected in only a few species. However, the copious variation in DNA content observed within and among species (Price 1988a, 1988b) suggests that they are not rare. Available data indicate that the "mutational" events may involve a substantial amount of DNA. In flax, induced changes in DNA content are typically in the 7%–9% range (Evans et al. 1966; Evans 1968). Variation in DNA content exceeding 12% has been detected among plants grown from achenes of single heads of plants from a highly inbred line of sunflower, *Helianthus annuus*, L. (Cavallini et al. 1989). For the sunflower, this is approximately 1.2 picograms or nearly two and one-half times more DNA than the total 2C value of a low DNA content species such as *Arabidopsis thaliana*.

Variation in Genome Size Within Species

Intraspecific variability in diploid DNA amount was once considered to be very rare. However, over the past decade, substantial genome size variation within species has been well documented (see Bennett 1985; Price 1988a). Examples include *Microseris douglasii* and *M. bigelovii*, 25% (Price et al. 1980, 1981a, 1981b), *Zea mays*, 37% (Laurie and Bennett 1985; Rayburn et al. 1985), *Gibasis venustala*, 60% (Kenton 1983), *Poa annua*, 80% (Grime 1983), and *Collinsea verna*, 288% (Greenlee et al. 1984). The sunflower, *H. annuus*, possesses abundant variation in DNA amount, possibly up to 60% (Nagl and Capesius 1976; Cavallini et al. 1986). Michaelson (unpublished data), has been able to detect at least 28% variation in genome size among cultivated lines of sunflower using both scanning microspectrophotometry and flow cytometry.

Genomic Stress and the Inheritance of DNA Content

Genomic stress generated by either the physical, chemical, or the genetic environment may cause destabilization and rearrangement in the plant genome (McClintock 1984; Walbot and Cullis 1983). Such stresses have been proposed to generate variability and diversity within the plant or plant population as part of an evolutionary strategy for adaptation to a changing environment (Walbot and Cullis 1983).

In hybrids of *Nicotiana tabacum* with *N. otophora* and *N. plumbaginifolia*, the formation of one or more very large chromosomes (up to 20× normal size) occurs (Gerstel and Burns 1966; Moav et al. 1968). Since the DNA content in megachromosomes has been spectrophotometrically determined to be equal to, or greater than, that of equivalent areas of normal chromosomes in the same cell, the production of megachromosomes has been attributed to increases in DNA amount rather than to chromosome relaxation (Collins et al. 1970). The increased DNA amount occurs primarily in heterochromatic regions (Collins et al. 1970). The enlarged chromosomes in hybrids of *N. tabacum* with *N. plumbaginifolia* are passed on to progeny whereas the megachromosomes induced in the hybrids with *N. otophora* are not.

The inheritance of DNA content is important to an understanding of the evolution of genome size. Few studies have been made on the genetics of DNA content but results so far indicate that both stability and instability occur in crosses. Hutchinson et al. (1979) crossed *Lolium* species differing by 40% in DNA amount. The F_1s had the expected intermediate DNA amount; the F_2 and backcross progeny had normal distributions of DNA values that are compatible with Mendelian segregation of many dispersed differences.

Instability is apparent in flax and *Microseris* hybrids. Reciprocal crosses were obtained involving high and low DNA content plants (16% differences) of the flax variety Stormant Cirris with those of Liral Monarch, which had an intermediate amount. The F_1 DNA values were not the sum of the contribution of each parent as expected but were similar to the Liral Monarch parent (Durrant 1981). These results indicate that a quantitative change in DNA amount occurred in these hybrids.

Further evidence for genomic instability is apparent from crosses of annual *Microseris* differing in DNA amount (Price et al. 1983, 1985). The F_1 progeny of some crosses displayed unpredictable genomic readjustments, amplifications, and deletions. This is depicted most clearly in an interspecific cross between biotypes of *M. douglasii* and *M. bigelovii* differing in DNA amount by ca. 10% (Price et al. 1983). The F_1 progeny did not cluster around the expected parental midpoint but encompassed nearly the whole range between the parental means. Families of F_2 progeny had a mean DNA content corresponding to that of the particular F_1 from which they were derived, indicating that the F_1 plants were not of identical DNA content and therefore had undergone relatively stable alterations in DNA amount.

The observations of Price et al. (1983, 1985) are important not only in regard to development of the concept of a plastic genome, but also from a genetic analysis viewpoint. A trait with a strong nucleotypic basis may not display classical inheritance patterns due to instability and readjustment of DNA content in progeny following hybridization.

Biophysical or Nucleotypic Effects

Our current understanding of genome organization and function relegates only a relatively small percentage of the DNA sequences in eukaryotes to genes and coding sequences (Flavell 1980). The apparently huge excess of

DNA, the widespread variability in genome size among closely related taxa, and the lack of an overall positive correlation between genome size and genetic complexity and phylogenetic advancement have been called the DNA C-value paradox (Thomas 1971; Raff and Kaufman 1983).

It has been considered that the variation in DNA content is due to parasitic or selfish DNA sequences (Orgel and Crick 1980; Doolittle and Sapienza 1980) that are without significant phenotypic consequences or adaptive value. Although this hypothesis proposes a mechanism to generate variability, it dodges the issue of function. The concept that DNA has roles in plant development and adaptation independent of any coding function has been gaining strength (Price 1976, 1988b). It has been proposed that the mere bulk of the nuclear DNA exerts an influence on the phenotype so that under some circumstances selection favors accumulation of nuclear DNA, regardless of its nucleotide sequence, whereas, under different conditions the loss of sequences not necessary to survival may confer considerable adaptive advantage (Bennett 1972; Price 1976; Bachmann et al. 1985). The biophysical effects of the mere bulk of the DNA on the cell or organism have been called the "nucleotype" (Bennett 1972).

It is well-documented that several cellular parameters are influenced by the amount of DNA. Strong positive correlations exist between DNA content and nuclear volume (Baetcke et al. 1967), cell volume (Price et al. 1973; Cavalier-Smith 1978), mitotic cycle time (Van't Hof 1965; Evans and Rees 1971), and the duration of meiosis (Bennett 1971, 1977).

Cell size and the rate of cell division are apparently of adaptive significance. Bennett (1972) suggested that DNA content is causally correlated with plant developmental rate and that differentiation and development of annual species in a time-limited environment require a low DNA content. Price (1988b) proposed that DNA content determines fundamental cell volume upon which genetic and environmental factors interact.

Cell size, mitotic cycle time, and DNA content may be influenced by climatic selection. Grime and Mowforth (1982) suggested that selection may operate on genome size through a differential effect of temperature upon cell division and cell expansion, where at low temperature cell expansion is inhibited to a lesser degree than cell division. They contend that plants growing in cooler environments should have growth dominated by cell enlargement that would favor larger cells and high DNA content. Conversely, under warmer conditions where temperature is not adversely affecting the mitotic cycle time, growth should be facilitated by more rapidly dividing cells with less DNA (Grime and Mowforth 1982).

The theory of climatic selection has initial support from phenological data collected from the British flora (Grime and Mowforth 1982; Grime et al. 1985). In the Sheffield region, from early spring to midsummer, growth rates (based on time of obtaining 50% maximum shoot biomass) were significantly and negatively correlated with genome size (Grime and Mowforth 1982). In a species-rich limestone grassland community in Northern England, DNA content and rates of leaf elongation of grasses, sedges, forbs, and small shrubs were compared in relationship to temperature and chronology (Grime et al. 1985). During the cold condition of early spring the most rapid rates of leaf expansion occurred in species with relatively high DNA values. The differences disappeared under the warmer conditions of early summer.

Grime (1983) presented data on mean relative growth rates of families of seedling *Poa annua* grown from seed collected from established individuals in a single pasture and varying in DNA amount by over 80%. A significant negative correlation was detected between relative rate of dry matter production and DNA amount among plants grown at warm temperatures. No data were presented for plants grown at low temperatures.

The effects of cell size on plant physiology and adaptation have not received adequate research attention. Cell size may be an important parameter in the adaptation of plants to predictably stressed environments (Price 1988b). Reduced cell volume and decreased growth rates of leaves resulting in increased water use efficiency are major responses to water deficiency (Cutler et al. 1975; Nobel 1980). Iljin (1957) reported an inverse correlation between cell volume and the ability of a plant to survive drought.

An initial set of experiments to measure the relationship between cell size and the anatomical and physiological responses to experimentally induced drought stress has been conducted on two *Microseris* species differing 2.6-fold in DNA content and originating from habitats contrasting in water availability (Castro-Jiminez et al. 1989). The perennial *Microseris lacinata* has high DNA content, large cells, and is adapted to cooler more mesic habitats. The annual *Microseris bigelovii* has low DNA content, smaller cells, and is adapted to drier, ephemeral habitats (Chambers 1955; Price and Bachmann 1975). These species respond to low water availability by maintaining turgor with small cell volumes, elastic tissues, and osmotic adjustment (Castro-Jiminez et al. 1989). Enhanced tissue elasticity and small cell volume are inherent characteristics in *M. bigelovii* and drought induced responses in *M. laciniata*.

The ability to differentiate smaller cells in plants when subjected to water deficiencies is an apparent adaptation for coping with drought stress. However, this phenotypic plasticity may result in cells with less than optimal volume relative to DNA content and place the cell under developmental stress (Castro-Jiminez et al. 1989). If this hypothesis is correct, plants with smaller genomes within a population may be better adapted to drought stress because they already have smaller cells. Selection under drought conditions should favor low DNA content via its nucleotypic effect on cell size.

Model Systems

Microseris

In the Microseridinae (Asteraceae) of western North America there are correlations between ecological adaptation and DNA amount at the interspecific and intraspecific level. The more primitive, diploid species of *Microseris* occupy cooler and more mesic habitats than do the specialized annual species adapted to warmer, more xeric, and time-limited habitats. The evolution of these diploid annuals from their perennial ancestor(s) resulted in massive diminition of DNA sequences so that the annuals have only about 30% to 60% of the DNA amount of the perennials (Price and Bachmann 1975; Bachman and Price 1977; Bachmann et al. 1979).

DNA amount varies over 20% within the annuals *M. bigelovii* (Price et al. 1981a) and *M. douglasii* (Price et al. 1980; Price et al. 1981b). In *M. bigelovii*, the lower DNA values were detected in plants of geographically disjunct populations at the low-rainfall, southern extreme of the species range, and at the northern high-rainfall extreme on thin soil over a barren rock outcrop. The higher DNA values were from habitats at the center of the species range where intermediate amounts of precipitation occur. Price et al. (1981b) suggested that low DNA content resulted from selection in stressful and/or time limited habitats at the extremes of the ecological adaptation of the species.

In *M. douglasii*, from which over 24 populations have been sampled, high DNA values are restricted to plants growing in more mesic habitats, generally in well-developed soil (Price et al. 1981b). In addition, at some population sites, temporal changes in DNA content have been observed over a few generations that correlate with environmental changes, i.e., amount of precipitation and the length of time available to complete the life cycle (Price et al. 1986). The observations on the distribution of DNA amounts among species of *Microseris* and within *M. douglasii* and *M. bigelovii* imply an adaptive role of DNA amount, presumably through nucleotypic effects that influence cell size, mitotic cycle time, and the rate of development.

An alternative hypothesis is that the pattern in DNA content among and within *Microseris* populations is a direct response to environmental factors, although supporting empirical data are not available. High and low DNA content biotypes have been maintained as inbred lines for many generations under greenhouse and/or growth-chamber conditions. They have bred true for the DNA amounts that are present in the nuclei of plants grown directly from field collected achenes (Price et al. 1981b). Furthermore, when plants from six relatively high DNA content biotypes were grown under severely water and heat stressed growth-chamber conditions, their progeny retained DNA values similar to those of nonstressed controls (Price et al. 1986). These data, coupled with the previously discussed hybridization experiments suggest that the initial alterations of DNA content in *Microseris* annuals are not due to environmental stress but rather to genetic factors that can be induced by hybridization.

Zea mays

Zea mays L. is an obligate cultigen that apparently originated in Mexico or Central America and was taken northward by man to where the short, cool, growing season limited its maturation (Galinat and Gunnerson 1963). Maize varieties possess abundant variability in the number (0 to >20) and size of heterochromatic knobs (Longley 1938; Kato-Y 1976). Knob number has been negatively correlated with increasing latitude (Brown 1949) in North America and altitude in Mexico (Wellhausen et al. 1952; Bennett 1976; McClintock et al. 1981). Knob number and the amount of hetero-chromatin positively correlate with DNA content (Rayburn et al. 1985). Significant negative correlations exist between increasing latitude and DNA amount (Rayburn et al. 1985; Bennett 1985). Plants from Mexico have up to 37% more DNA than those from the northernmost regions of cultivation (Rayburn et al. 1985; Laurie and Bennett 1985). At least part of the variation in DNA amount is attributed to differing amounts of a 185 base pair satellite DNA sequence that resides in maize knob hetero-chromatin (Peacock et al. 1981).

The low DNA values at high latitudes apparently resulted from selection for rapid development, high yield and the largest plant size permitted by climatic constraints (Rayburn et al. 1985). This is compatible with the nucleotype concept where selection for more cells and rapid development could result from the shorter mitotic cycle time that correlates with reduced DNA amount.

Concluding Remarks

The current view of genome organization and evolution considers two major roles for DNA. The first is the genotype that has coding or sequence specific functions. It comprises only a relatively small percentage of the DNA base pairs. The second role of DNA is dictated by its mere bulk and is called the nucleotype. The sum of all the DNA sequences comprise the nucleotype and influences cellular parameters such as chromosome size, nuclear volume, cell volume, mitotic cycle time, and the duration of meiosis.

Among the factors that can induce amplification or deletion of DNA sequences in the nucleus is genomic stress due to the environment or to hybridization. These may be necessary to generate variability, both qualitative and quantitative. Quantitative differences in DNA amount are apparently adaptive via nucleotypic effects.

Studies of the adaptation of plants (particularly to stressed environments including polluted atmospheres) should involve genetic, DNA content, and nucleotypic analyses.

References

Bachmann K, Chambers KL, Price HJ (1979) Genome size and phenotypic evolution in *Microseris* (Asteraceae, Cichoriaeae). Plant System Evolution 2: 41–66

Bachmann K, Chambers KL, Price HJ (1985) Genome size and natural selection: observations and experiments in plants. In: Cavalier-Smith T (ed) The evolution of genome size. Wiley, New York, pp 267–276

Bachmann K, Price HJ (1977) Repetitive DNA in *Cichorieae* (Compositae). Chromosoma 61:267–275

Baetcke KP, Sparrow AH, Nauman CH, Schwemmer SS (1967) The relationship of DNA content to nuclear and chromosome volume and to radiosensitivity. Proceedings of the National Academy of Science 58:553–540

Bennett MD (1971) The duration of meiosis. Proceedings of the Royal Society of London (B) 178:277–299

Bennett MD (1972) Nuclear characters in plants. Brookhaven Symposium on Biology 25:344–366

Bennett MD (1976) DNA amount, latitude and crop plant distribution. In: Jones K, Brandham PE (eds) Current chromosome research. North-Holland, Amsterdam, pp 151–158

Bennett MD (1977) The time and duration of meiosis. Philosophical Transactions of the Royal Society London (B) 277:201–226

Bennett MD (1985) Intraspecific variation in DNA amount and the nucleotypic dimension in plant genetics. In: Freeling M (ed) Plant genetics. Alan R Liss, New York, pp 283–302

Brown WL (1949) Numbers and distribution of chromosome knobs in United States maize. Genetics 34:524–536

Castro-Jiminez Y, Newton RJ, Price HJ, Halliwell RS (1989) Drought stress responses of *Microseris* species differing in nuclear DNA content. American Journal of Botany 76:789–795

Cavalier-Smith T (1978) Nuclear volume control by nucleoskeletal DNA, selection for cell volume and cell growth rate, and the solution of the DNA C-value paradox. Journal of Cell Science 34:247–278

Cavallini A, Zolfino C, Cionini G, Cremonini R, Natali L, Sassoli O, Cionini PG (1986) Nuclear changes within *Helianthus annuus* L.: cytophotometric, karyological and biochemical analyses. Theoretical and Applied Genetics 73:20–26

Cavallini A, Zolfino C, Natali L, Cionini G, Cionini PG (1989) Nuclear DNA changes within *Helianthus annuus* L.: origin and control mechanism. Theoretical and Applied Genetics 77:12–16

Chambers KL (1955) A biosystematic study of the annual species of *Microseris*. Contributions of the Dudley Herbarium. 4:207–312

Collins GB, Legg PD, Anderson MP (1970) Cytophotometric determination of DNA content in *Nicotiana* megachromosomes. Canadian Journal of Genetics and Cytology 12:769–778

Cullis CA, Clearly W (1986) Rapidly varying DNA sequences in flax. Canadian Journal of Genetics and Cytology 28:252–259

Cutler JM, Rains DM, Loomis RS (1975) The importance of cell size in the water relations of plants. Physiologia Plantarum 40:255–260

Doolittle WF, Sapienza C (1980) Selfish genes, the phenotype paradigm and genome evolution. Nature 284:601–603

Durrant A (1981) Unstable genotypes. Philosophical Transactions of the Royal Society of London (B) 292:467–474

Evans GM (1968) Nuclear changes in flax. Heredity 23:25–38

Evans GM, Durrant A, Rees H (1966) Associated nuclear changes in the induction of flax genotrophs. Nature 212:697–699

Evans GM, Rees H (1971) Mitotic cycles in dicotyledons and monocotyledons. Nature 233:350–351

Flavell RB (1980) The molecular characterization and organization of plant chromosomal DNA sequences. Annual Review of Plant Physiology 31:569–596

Flavell R (1982) Sequence amplification, deletion and rearrangement. In: Dover GA, Flavell RB (eds) Genome evolution. Academic Press, New York, pp 301–323

Galinat WC, Gunnerson JH (1963) Spread of eight-rowed maize from the prehistoric Southwest. Botany Museum Leaflet, Harvard University 20:117–160

Gerstel DU, Burns JA (1966) Chromosomes of unusual length in hybrids between two species of *Nicotiana*. In: Darlington CD, Lewis KR (eds) Chromosomes today, vol 1. Plenum Press, New York, pp 41–56

Greenlee JK, Rai KS, FLoyd AD (1984) Intraspecific variation in nuclear DNA content in *Collinsea verna* Nutt (Scrophulariaceae). Heredity 52:235–242

Grime JP (1983) Prediction of weed and crop response to climate based upon measurements of nuclear DNA content. Aspects of Applied Biology 4:87–98

Grime JP, Mowforth MA (1982) Variation in genome size—an ecological interpretation. Nature 299:151–153

Grime JP, Shacklock JML, Band SR (1985) Nuclear DNA content, shoot phenology and species co-existence in a limestone grassland community. New Phytologist 100:435–445

Hutchinson J, Rees H, Seal AG (1979) An assay of the activity of supplementary DNA in *Lolium*. Heredity 43:411–421

Iljin WS (1957) Drought resistance in plants and physiological processes. Annual Review of Plant Physiology 8:257–274

Kato-Y TA (1976) Cytological studies of maize (*Zea mays* L.) and teosinte (*Zea mexicana*) (Schrader; Kuntze) in relation to their origin and evolution. Massachusetts Agricultural Experiment Station Bulletin 635:1–186

Kenton A (1983) Qualitative and quantitative chromosome change in the evolution of *Gibasis*. In: Brandham PE, Bennett MD (eds) Kew chromosome Conference II. George Allen and Unwin, London, pp 273–282

Laurie DA, Bennett MD (1985) Nuclear DNA content in the genera *Zea* and *Sorghum*. Intergeneric, interspecific and intraspecific variation. Heredity 55:307–313

Longley AE (1938) Chromosomes of maize from North America. Journal of Agricultural Research 56:177–195

McClintock B (1984) The significance of responses of the genome to challenge. Science 26:792–801

McClintock B, Kato-Y TA, Blumenschein A (1981) Chromosome constitution of races of maize. Colegio de Postgraduados, Chapingo, Mexico

Moav J, Moav R, Zohary D (1968) Spontaneous morphological alterations of chromosomes in *Nicotiana* hybrids. Genetics 59:57–63

Nagl W, Capesius I (1976) Molecular and cytological characteristics of nuclear DNA and chromatin for angiosperm systematics: basic data for *Helianthus annuus* (Asteraceae). Plant Systematics Evolution 126:221–237

Nobel PS (1980) Leaf anatomy and water use efficiency. In: Turner NC, Kramer PJ (eds), Adaptation of plants to water and high temperature stress. Wiley, New York, pp 43–55

Orgel LE, Crick FHC (1980) Selfish DNA: the ultimate parasite. Nature 284:604–607

Peacock WJ, Dennis ES, Rhoades MM, Pryor AJ (1981) Highly repeated DNA sequence limited to knob heterochromatin in maize. Proceedings of the National Academy of Science 78:4490–4494

Price HJ (1976) Evolution of DNA content in higher plants. Botanical Review 42:27–52

Price HJ (1988a) Nuclear DNA content variation within angiosperm species. Evolutionary Trends in Plants 2:53–60

Price HJ (1988b) DNA content variation among higher plants. Annuals of the Missouri Botanical Garden 75:1248–1257

Price HJ, Bachmann K (1975) DNA content and evolution in the Microseridinae. American Journal of Botany 62:262–267

Price HJ, Bachmann K, Chambers KL, Riggs J (1980) Detection of intraspecific variation in nuclear DNA content in *Microseris douglasii*. Botanical Gazette 141:195–198

Price HJ, Chambers KL, Bachmann K (1981a) Genome size variation in diploid *Microseris bigelovii* (Asteraceae). Botanical Gazette 142:156–159

Price HJ, Chambers KL, Bachmann K (1981b) Geographic and ecological distribution of genomic DNA content variation in *Microseris douglasii* (Asteraceae). Botanical Gazette 142:415–426

Price HJ, Chambers KL, Bachmann K, Riggs J (1983) Inheritance of nuclear 2C DNA content variation in intraspecific and interspecific hybrids of *Microseris* (Asteraceae) American Journal of Botany 70:1133–1138

Price HJ, Chambers KL, Bachmann K, Riggs J (1985) Inheritance of nuclear 2C DNA content in a cross between *Microseris douglasii* and *M. bigelovii* (Asteraceae). Biologisches Zentralblatt 104:269–276

Price HJ, Chambers KL, Bachmann K, Riggs J (1986) Patterns of mean nuclear DNA content in *Microseris douglasii* (Asteraceae) populations. Botanical Gazette 147:496–507

Price HJ, Sparrow AH, Nauman AF (1973) Correlations between nuclear volume, cell volume and DNA content in meristematic cells of herbaceous angiosperms. Experientia 29:1028–1029

Raff RA, Kaufman TC (1983) Embryos, genes, and evolution. Macmillan, New York

Rayburn AL, Price HJ, Smith JD, Gold JR (1985) C-band heterochromatin and DNA content in *Zea mays*. American Journal of Botany 72:1610–1617

Thomas CA (1971) The genetic organization of chromosome. Annual Review of Genetics 5:237–256

Thompson WF, Murray MG, Cuellar RE (1980) Contrasting patterns of DNA sequence organization in plants. In: Leaver CJ (ed) Genome organization and expression in plants. Plenum, New York, pp 1–15

Van't Hof J (1965) Relationships between mitotic cycle duration, S period duration and the average rate of DNA synthesis in the root meristem of several plants. Experimental Cell Research 39:48–58

Walbot V, Cullis CA (1983) The plasticity of the plant genome—is it a requirement for success? Plant Molecular Biology Reporter 1:3–11

Walbot V, Cullis CA (1985) Rapid genomic change in higher plants. Annual Review Plant Physiology 36:367–396

Wellhausen EJ, Roberts LM, Hernandez XE, Mangelsdorf PC (1952) Races of maize in Mexico: their origin, characteristics and distribution. Bussey Institute, Harvard University, Cambridge, MA

10
Population-Level Techniques for Measuring Microevolutionary Change in Response to Air Pollution

STEPHEN J. TONSOR and SUSAN KALISZ

Introduction

It has been clear for some time that airborne pollutants are a major cause of plant population decline in many parts of the world (Chapters 2 and 3, this volume; Krause, et al. 1986). A large body of work documents effects ranging from foliar damage to widespread mortality (reviewed by Treshow 1984; Troyanowsky 1985; Winner et al. 1985; Roose et al. 1982). In addition, more recent evidence indicates that not all individuals within a population are equally sensitive to anthropogenic atmospheric perturbations (e.g., Taylor 1978; Ernst et al. 1985; Pitelka 1988). The inference from such data is that populations demonstrated to have genetically based differential sensitivity will eventually recover from the deleterious effects of anthropogenic atmospheric changes through the evolution of resistant genotypes. The extent to which populations are adapting to the stress of air pollution remains to be established (Pitelka 1988).

Confident inference about the long-term effects of differential sensitivity requires considerably more than a knowledge of genetic differences in sensitivity. For example, a population exhibiting genotypically based variation in mortality in response to pollutants will surely change in its genotypic composition with long-term exposure to those pollutants. However, if demographic studies indicate that even the least sensitive genotypes are failing to replace themselves, it does not seem appropriate to predict that the population is "adapting" to the pollutants. It is important to distinguish between the general case of evolutionary change in response to atmospheric changes (i.e., genetically based changes in the mean phenotype) and the more restricted case of evolution which obviates the effects of atmospheric changes on the population.

In this chapter, methodologies are described for documenting potential microevolutionary changes in plant populations as a result of air pollution. A combination of methods is suggested, including basic population monitoring and demographic analyses, and more recent statistical methods for quantifying selection in the field.

The chapter starts with a discussion of methods available for measuring past evolution in response to atmospheric pollution. In many cases, it will not be possible to accurately assess whether evolution has already occurred, and instead it is necessary to ask about the potential for evolutionary change and its consequences. Artificial selection experiments are therefore discussed, since they can be used for estimation of the response to selection which might be experienced in nature. The limitations of this method are also discussed. Next, the three independent elements of a response to selection are covered: phenotypic variation in fitness associated with pollution, heritability of phenotype (Roose, Chapter 5, this volume), and phenotypic and genetic correlations among traits which may hinder the progress of adaptation. Finally, because the response means little if mean fitness remains below 1.0, the measurement of the population's demographic parameters in a microevolutionary context is discussed. Although it is impossible in the space of one chapter to cover all of the literature on field approaches to studying the microevolutionary process, key references are provided which should help interested researchers carefully address these questions.

Throughout the chapter, it is assumed that the traits which are of interest (survivorship, fecundity, yield, quality of the leaf or fruit or wood, disease susceptibility, pollution resistance, etc.) have a multigenic basis (e.g., Taylor 1968; Aycock 1972; Karnosky 1977; Devos et al. 1982; Engle and Gableman 1989), and that, at least at the outset, the only information one has about the trait is derived from direct field or laboratory measurement of extant phenotypes. Most of this chapter is an exploration of methods of inference about the genetic basis for phenotypic differences seen between populations, cohorts or groups of relatives. The methods of inference require knowledge of both the mean and the variance among individuals in the measures of the traits of interest. In this chapter, the term "fitness" is used to mean the survival and reproduction of individuals.

Measuring Past Evolutionary Change

The most fundamental problem one encounters in measuring the change which has already occurred in a population is that the past phenotypic distribution for that population no longer exists. There are various ways of recovering or estimating that distribution. One approach is to compare populations which differ in pollution history but are otherwise likely to have had a similar evolutionary history. The idea is that the phenotypic distribution of the trait of interest in the populations in unpolluted environments are a reasonable estimate of what the distribution was in the polluted population prior to the onset of pollution. This is the technique Taylor (1978) used in studying evolution of resistance to SO_2 in *Geranium carolinianum*. A potential problem with this method is that the differences could result from the chance comparison of two populations that are randomly differentiated with regard to the trait of interest. Without more information, we cannot assess the likelihood that by chance we could have just as easily sampled two populations that did not differ or that differed as much in the opposing direction. One solution to this problem is to bring to bear additional biological information. In some studies, the biology of the situation makes implausible an explanation which rests on the vagaries of single-sample comparisons. In the case of Taylor's (1978) study for example, it is quite reasonable to assume that the differences in susceptibility to SO_2 between the populations result from differential selection on SO_2 resistance. Suppose, however, that instead of measuring a trait directly related to the immediate effect of the pollutant on the plant, we measure pounds of fruits produced, or board feet accumulated, where we suspect that the pollutant has its effect of economic interest. In such cases, it is not so immediately obvious that the pollutant is the cause for the differences between the populations. It still may be possible to establish the link between the observed comparison and the effect of selection for pollution resistance. Detailed information about the mode(s) of biological action of the pollutant, and about how the individuals in the two populations differ in the magnitude(s) or mode(s) of action will be necessary to establish the causal link between the geographic differences in pollution levels and the observed genetic differences between the populations. Detailed knowledge of the mode(s) of action is also necessary for determining which traits should be measured. This is discussed in the section labelled "the measurement of natural selection."

A more direct method for establishing a statistically meaningful pattern of evolutionary change is to make the comparison over a number of populations, including both polluted and unpolluted sites. There are a number of statistical designs that might be appropriate. All of them are essentially correlational in their approach; by making repeated comparisons of level of pollutant exposure and mean and variance in phenotype within the population, one can establish whether a correlation exists between the phenotype and the supposed selective pressures exerted by the pollutant. While this does not establish a causal connection, it makes such a connection plausible. Plausibility depends critically on the ability to separate environmental and genetic effects. This separation cannot be made if individuals are measured in their population of origin under natural conditions. Only by manipulation of pollutant levels across sites or movement of genotypes between sites can genetic and environmental effects be separated.

One can use either an analysis of variance (ANOVA) design or a linear regression design. For those who do not ordinarily use these designs, some simple examples are presented. If one has two regions, one which is fairly uniformly polluted, and one which is not polluted, then an analysis of variance design is called for. In this case, one would measure each of a set of populations from each of the two regions. The goal is to detect any genetically based differences in resistance associated with the two regions. To remove environmental effects of the site of origin for the populations, the comparisons must be made in a common environment, where individuals have been transplanted from all sites. This can be accomplished in common gardens in the polluted regions, in reciprocal transplant gardens, or in an artificial environment. A one-way ANOVA comparing population means will establish if populations collected in the polluted region differ genetically on average from populations collected in the unpolluted regions. In the case of reciprocal transplants, a two-way factorial design with a region of origin mean square and "treatment" (garden location) mean square as well as an interaction mean square should be used.

In contrast, suppose one's interest is confined to a single region in which the extent of airborne pollutants differs from location to location. In this case, one should use an experimental design incorporating linear regression. Again, a common environment is needed, in which the phenotypes of individuals from all the sampled populations are measured after growth at a known level of pollutant. A significant slope for the regression of mean phenotype in the population on the level of pollutant at the population's site of origin establishes that among-population evolution has occurred. This pattern is consistent with pollution as the causal selective agent. Sokal and Rohlf (1981) provide detailed discussions and examples of the applications of analysis of variance and regression techniques.

When one is interested in global pollutants like CO_2, there may not be populations which remain in relatively unperturbed conditions. In such cases, there may be no way to assess the extent of evolutionary change which has already occurred.

In plants with a long-lived seed bank or other dormant organs, it may be possible to reconstruct the past genetic composition of the population. One experimental approach is to compare the mean phenotype of germinated seeds from the seed bank with the seeds produced by the current reproductive cohort. This could be done under either controlled conditions or in the natural habitat. Tonsor et al. (1990) have compared a seed bank's genetic composition to the composition of the adults. They found a significant difference between the electrophoretic genotypes represented in the seed bank seeds as compared with the extant population in *Plantago lanceolata*.

This method can provide evidence that the population has evolved, but does not provide a good estimate of the extent of evolutionary change. The seed bank is a mix of seeds from potentially many rounds of reproduction. The average phenotype of the individuals germinated from the seed bank represents the average of the phenotypes contributed from each round of reproduction, weighted by the relative size of their contributions (relative numbers of seeds) and by the survivorship of seeds in the soil times the time elapsed since their deposition. Thus, without detailed knowledge of the life expectancy of seeds in the soil, and of variation in mean fecundity from season to season, one cannot calculate a per-generation response to selection from the observed difference in mean phenotype between seed bank and "new" seeds.

In some highly stratified soils, such as peat soils, it may be possible to assign an approximate age to seeds based on the sediments in which they are found. It may also be possible to age some seeds by carbon dating of the seed coat (J. McGraw, personal communication). In such cases, an informative history of phenotypic variation for the population could be reconstructed. One must be aware of the (remote) possibility that the seed bank seeds have suffered mortality while in the soil which is nonrandom with regard to the phenotypic trait of interest, possibly through its correlation with other attributes of the seed.

In summary, in most circumstances, reconstruction of the past distribution of the population using the seed bank will be only approximate. It is best used to ask if there has been a change in the population, rather than to ask how large a change has occurred or how fast that change has taken place. Data on the extent of genetically based changes in phenotype since the onset of selection by the anthropogenically modified atmosphere will often be unavailable or sketchy.

Measuring Current Selection and Evolution

Even when the past change can be accurately measured, it is of interest to estimate the rate at which future change will occur, and to ask if the predicted response is sufficient to produce an eventual amelioration of the effects of the changes. Recent advances in the methodology for quantifying the evolutionary process in natural populations make this a less formidable task than was once thought. The methods of Lande and Arnold (1983) and Arnold and Wade (1984a, 1984b) are applicable to the measurement of evolution of continuously varying or quantitative traits. Mitchell-Olds and Shaw (1987) and Wade and Kalisz (1990) discuss some statistical refinements of the methods. The techniques are an extension of the classic methods of quantiative genetic breeders who employ artificial selection to improve plant and animal stock. Quantitative geneticists are interested in the *phenotype*, and how it evolves. The genetic basis of traits is of interest only insofar as it affects the rate and mode of evolution. Lande, Arnold, and Wade's methods show statistically the relationship between phenotype and fitness, and between genotype and phenotype, and do not demonstrate causal relationships between an assumed agent of selection and fitness, or between the measured trait and fitness (Wade and Kalisz 1990). The accuracy of any causal inferences that are drawn depends on knowledge from other sources, notably physiological, biochemical, and ecological knowledge. It also requires the measurement of large numbers of individuals. The level at which accuracy in measurement is crucial in this technique is not that of the individual, but the level of groups of individuals, notably groups of relatives and populations. Thus, it may be important, for example, to sacrifice a few millimoles of $CO_2\,m^{-2}\,s^{-1}$ accuracy in an individual measure of photosynthetic rate, for the sake of increasing the number of individuals to be measured.

The method that will be described relies on the partitioning of the evolutionary process into a set of independent relationships, each of which can then be measured. The terms "selection" and "response to selection," will be defined both verbally and mathematically. The use of these definitions is crucial to this method.

Selection and Response to Selection: Partitioning the Evolutionary Process

As discussed by RA Fisher (1930), evolution by selection, or the change in gene frequency due to selection, can be defined as:

$$R = h^2 s$$

Here, R is the response of any measurable (i.e., phenotypic) trait to selection. That is to say, it is the change in mean phenotype for the population from one generation to the next. The response is measured between generations. The ratio (h^2) of the additive genetic variance in a trait to the total phenotypic variance is (V_A/V_P). The variable s is the selection acting on that trait within a generation. It is measured by calculating the mean phenotype contributing to the next generation (the mean phenotype where each phenotypic value is weighted by product of that phenotype's survivorship and fecundity). Thus, the change in the distribution of a trait in the offspring is a function of selection on the trait in the parental generation and the extent to which the phenotypes breed true. Clearly, the rate of change in a phenotypic trait will be a function of both the strength and direction of selection, s, acting on it, and the magnitude of the additive genetic variance for that trait, h^2. An h^2 value greater than zero is required for a trait to respond to selection. Partitioning the phenotypic variation in a trait into genetic and environmental components is discussed by Roose (Chapter 5, this volume). To predict evolutionary change (i.e., response to selection), one can measure both the strength of selection on the trait and its heritability, or observe the evolutionary response to selection directly.

Direct Observation of Response to Selection

Observing a response to selection initially involves measuring the phenotypic distribution of the trait(s) of interest in a parental generation, then either measuring the strength of natural selection (see next section) or imposing selection of a known magnitude, and finally measuring the phenotypic distribution in the offspring (Falconer 1981). The change in the population mean of the phenotypic trait between the parental and offspring generations is the evolutionary response to selection. Endler (1986) lists a number of ways of calculating a response to selection and discusses their merits (note that Endler uses the term "natural selection" to refer to what is called "response to natural selection" in this chapter).

There are two general methods for estimating a response to selection. The method which has been used most frequently is to impose artificial selection of a known intensity and observe the response. Under some circumstances one may instead measure the response to natural selection under field conditions.

Observing a response to artificial selection has the advantage of being a direct measure of the evolutionary plasticity of a phenotypic trait. For plants in which large numbers of controlled crosses are impossible or extremely expensive (e.g., where each flower produces few seeds and must be emasculated to prevent selfing), the best obtainable heritability estimate will have associated with it a large standard error. Any estimates of response to selection based on the heritability estimate will therefore have very broad confidence intervals. For example, in an experimental design with 100 pollen parents, 10 seed parents mated to each, and 5 offspring measured from each cross, the 95% confidence interval on the heritability estimate is likely to be in the range of $\pm 10-20\%$; see Becker (1984) for methods of calculation of confidence intervals. It may be easier to get a statistically meaningful description of the response to selection than it is to estimate it based on heritability measures. Furthermore, for some methods of estimating the strength of selection, the predictions of response which result assume that the selection connected with the pollutant is the only source of selection acting on the trait. Falconer (1981) gives examples from an extensive body of work demonstrating that the response to artificial selection is often limited by opposing (but often unanticipated and causally unidentified sources of natural selection, even in a laboratory setting). Directly measuring the response to selection incorporates the effects of both the source of selection of interest to the researcher and unknown potentially counterbalancing sources of selection.

The major drawback to measuring response to selection in nature is that one must measure two generations to calculate a response. In trees and other long-lived perennials, this will sometimes be undesirable. However, not all traits need to be measured on the same age or stage-class in both of the generations. For example, if one is interested in the evolution of resistance to foliar damage by O_3 in white pine, the needles produced by a 3-year-old tree will probably be very similar in response to needles produced by a 30-year-old tree. When it is possible to measure the same trait near the two ends of the life history, the time involved in calculating a response can be greatly decreased. This approach should be used with caution, since many plant responses to pollutants are stage specific (see demography section, this chapter).

A second drawback to measuring response directly is that one must determine to what the change in phenotypic distribution is a response.

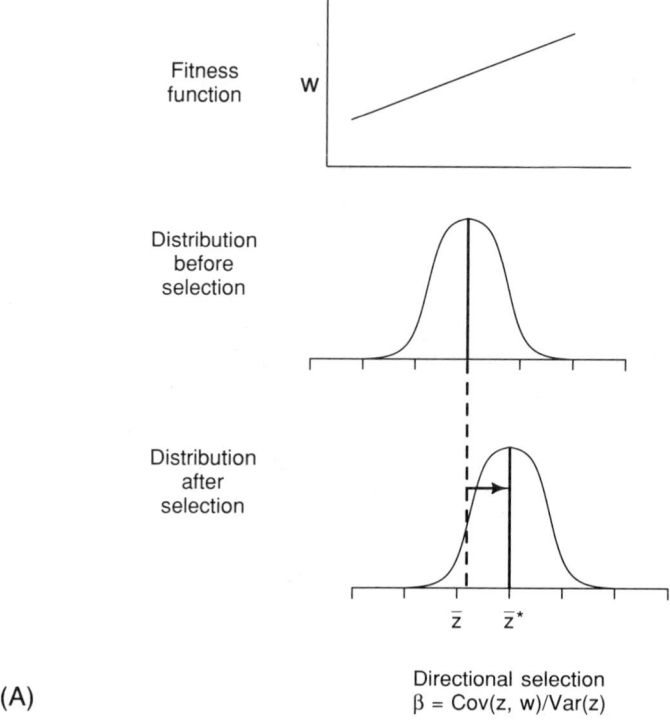

FIGURE 1. (A, B). Graphical depiction of linear and nonlinear selection. $\bar{z}$ indicates the mean of the population before selection, $\bar{z}^*$, the mean of the population after selection. (A) *Top*: Fitness function showing pure directional selection favoring large values of z. *Middle*: Phenotypic distribution of z in the population before selection. *Bottom*: Shift in the mean of the phenotypic distribution within a generation due to the action of pure directional selection without a change in the variance of the distribution.

Perhaps leaf damage in white pine is a correlated response to some foliar trait whose fitness depends on the successional stage of the forest, to give a hypothetical example. Although an evolutionary shift in the mean might be expected in a midsuccessional forest, it is not ozone that is causing the evolution. The solution to this problem is to impose the selection artificially (Wade and Kalisz 1990). One can then calculate the response to a precise level of selection, but it is unclear how applicable that response is to predictions of evolutionary change in the unmanipulated environment of the population. Ascribing a causal agent to a measured fitness differential for a trait can be a difficult problem. It requires observation and especially experimentation, the design of which is highly dependent on the ecology and biology of the study organism (Wade and Kalisz 1990). Nevertheless, by choosing an artificial selection method that mimics real conditions as

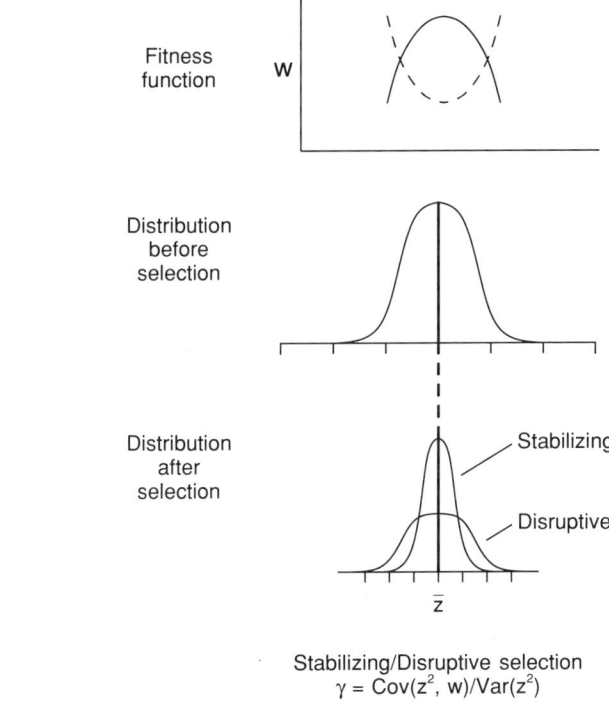

Stabilizing/Disruptive selection
$\gamma = \text{Cov}(z^2, w)/\text{Var}(z^2)$

(B)

FIGURE 1. (B) *Top*: Fitness function showing both pure stabilizing selection where the mean value has the highest fitness (*solid line*) and disruptive selection where the extremes of the distribution have the highest fitness (*dotted line*). *Middle*: Phenotypic distribution of z in the population before selection. *Bottom*: Change in the variance of the phenotypic distribution within a generation due to the action of pure stabilizing or disruptive selection without a change in the mean of the distribution. Stabilizing selection decreases the variance while disruptive selection increases the variance. Multiplication of the phenotypic distribution before selection by the fitness function describes the change in the population's phenotypic distribution after selection.

closely as possible, one maximizes the utility of the method. Even knowing if a trait is evolutionarily vagile, sluggish, or static can be of real value, so this method can be an important tool for predicting evolutionary change in response to pollution.

Natural Selection

Selection is an association between the value ("value" in the sense of "measure") of a phenotypic trait and the fitness of individuals possessing that value. For example, if foliar damage in the presence of SO_2 was measured for a population, one might expect that those with the lowest damage would be the most fit. Linear, or directional, selection occurs when individuals with either the highest or the lowest values for the trait have the highest fitness (Fig. 1A). The magnitude and sign of directional selection on a trait (z) is measured by the linear regression coefficient in a regression of relative fitness (w) on that trait (z). If selection favors high values of z, the slope of the regression is positive and the linear coefficient is positive. If selection favors low values of z, the slope is negative and the linear coefficient is negative. Directional selection changes the mean value of the phenotype within a generation (Fig. 1A, *bottom*). Quadratic, or nonlinear (Phillips and Arnold 1989) selection occurs when either individuals of an intermediate phenotype or individuals at *both* tails of the distribution show the highest fitness (Fig. 1B). For example, stomatal conductance must be under selection for intermediate values in most plants, since most species have stomata of finite size, and they must be open some of the time for CO_2 assimilation. The magnitude and sign of selection of this kind is measured by regressing relative fitness (w) on the square of the phenotypic value (z^2). If the sign of the quadratic regression coefficient is positive, selection is disruptive and the variance in the distribution is increased. If the sign is negative, selection is stabilizing and the variance is decreased. Pure stabilizing or disruptive selection do not cause a shift in the mean of the distribution (Fig. 1B, *bottom*); see Lande and Arnold (1983); Arnold and Wade (1984a, 1984b), and Phillips and Arnold (1989) for details of these methods.

The fitness of a phenotype is a result of the relationship between the individual and the environment, and is completely dependent on biotic and abiotic environmental conditions. The function that describes the relationship of fitness and the phenotypic distribution in a particular environment is called a fitness function. The fitness function applies only to the specific environment in which it was measured. The fitness function associated with a trait in any real population will rarely be purely linear or purely quadratic. Rather, it is likely to be a combination of both linear and quadratic selection components, e.g., Kalisz (1986).

The graphical representation of the fitness function is useful in that it depicts a population's performance in an environment. Imagine for the moment that one has available all possible phenotypes for a species. Suppose it is also possible to grow a number of replicates of each which is sufficient to describe their fitnesses accurately in a particular environment. One could then describe a fitness function since representative data on fitness and phenotype exist for all possible phenotypic values in that environment. Further suppose that this fitness function favored an intermediate phenotype (Fig. 2). Since it is unlikely that any single natural population will contain all possible phenotypes and those phenotypes present will vary in their proportional representations, the fitness function for any real population will differ from the fitness function imagined above.

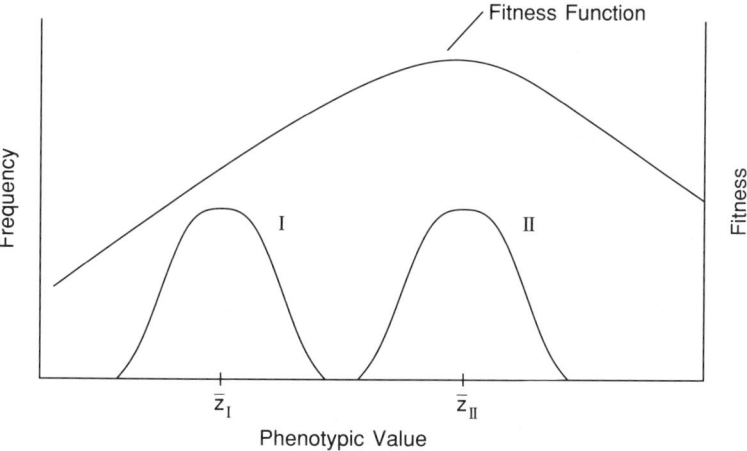

FIGURE 2. Relationship of fitness function to two populations' phenotypic distributions. The fitness function depicts the environment-specific association between phenotype and fitness. In this fitness function, individuals with an intermediate value of z have the highest fitness. The normal curves labeled I and II under the fitness function depict two (of many) possible phenotypic frequency distributions for a population. In population I, selection is directional, with individual having highest values of z having the highest fitnesses. In population II, selection is stabilizing, with individuals away from the mean having lower fitness than those close to the mean.

Clearly, the portion of the fitness function that a real population experiences will depend on the range of phenotypes present in the population. In Fig. 2, two populations are compared which differ in the position of their mean value of z and hence their position under the fitness function. Population I, whose phenotypes are mostly in the lower tail of the fitness function, experiences almost purely directional selection, favoring the highest values of the trait exhibited in the population. Population II, whose phenotypes are mostly in the middle section of the fitness function, experiences mostly stabilizing selection, with its intermediate phenotypes showing the highest fitness.

The realized fitness function that one measures depends not only on the location, i.e., the central moment, of the phenotypic distribution, but on the other moments as well (variance, skewness, kurtosis). As an extreme example, one could compare a population with a uniform distribution of phenotypes to a population with the same mean and range of phenotypic values but a normal distribution. Relative to each other they can have a very different slopes to their linear component and a shift in the relative magnitudes of linear and nonlinear selection. The relationship between fitness and phenotype has not changed between these two populations, only the relative representation of the phenotypes. Thus, a fitness function calculated for any particular population applies only to that population, even when the fitness associated with any particular phenotype is not expected to change from population to population.

In this chapter, univariate examples of multivariate regression analysis methods (Lande and Arnold 1983; Arnold and Wade 1984a, 1984b) are used to exemplify the quantification of natural selection in the field. Application of these methods for the measurement of phenotypic selection in natural plant population can be found in Kalisz (1986) and in Scheiner (1989). Many other methods for quantifying selection or estimating the form of the fitness function have been proposed. *The Statistics of Natural Selection on Animal Populations* by Manly (1985) and *Natural Selection in the Wild* by Endler (1986) are recommended as sources for the theoretical underpinnings of the study of natural selection as well as sources of practical methods. Endler's book starts with a discussion of the evolutionary process, in which he defines the components of the process differently; the reader will have to make a translation. The reader is also referred to Schluter (1988) for a nonparametric estimate of the fitness function, Crespi and Bookstein (1989) for the use of path analysis, and Phillips and Arnold (1989) for the use of canonical analysis.

These methods are new; both their full potential and limitations remain unknown since only a handful of researchers have applied them. The potential evolution of resistance to air pollutants is an ideal system for application of these methods. Extant knowledge of the source of selection, its time of onset and modes of action are key components in the description of evolutionary change. Both basic and applied scientists would benefit greatly from such studies.

Evolution of Correlated Traits

Thus far, selection has been discussed as though it always acted directly on the trait of interest, in the way that stomatal conductance might affect survivorship in a period of drought. However, the fitness associated with a trait also depends on how selection acts on traits which are themselves more or less directly tied to the trait of interest (e.g., Winner and Mooney 1980a, 1980b). For example, if one looked only at transpiration rate as estimated by stomatal conductance and fitness, one would undoubtedly find a fitness surface which was stabilizing. However, if carbon assimilation was measured as well, one might find that selection acting directly on conductance always favors the minimum transpiration rate, while selection on carbon gain always favors maximum carbon gain and high transpiration rates. It is because of the inexorable ties between transpiration and assimilation that when only one is measured, selection nearly always seems to be stabilizing. In comparing the two approaches, measurement of conductance only, versus measurement of both conductance and carbon gain, nothing has changed about the underlying relationship between the traits of the plant and how well the plant manages to survive and reproduce. But because in the first case only conductance is measured, its direct effects on fitness and the indirect correlated effects of the unmeasured carbon gain have been combined. Clearly, when traits are strongly correlated, but have inverse relationships to fitness, measuring only one of the traits can lead to serious errors in the mechanistic interpretation of the action of natural selection. A concrete example of these ideas can be found in Pell (1987), where multiple effects of ozone are discussed and the multiple, potentially correlated plant responses can be understood.

Information about the morphological and physiological correlated traits involved in air pollution resistance should be carefully considered, in addition to their correlations with economically important traits, in choosing traits to measure in the field. The information gained by examining selection and response to selection in a particular application will strongly depend on how informed the researcher is about the natural history and biology of the plant. If the research is ultimately motivated by profit, then some measure of economic value of the individual should be included in the analysis. Like any other trait, the aspects of phenotype associated with economic value are likely to be correlated with the plant's response to atmospheric changes, since these responses are expected to directly affect growth, survivorship, and reproduction.

The Measurement of Natural Selection

It is clear from the foregoing discussion that how the population is sampled will be crucial in determining what kind of information is obtained.

With regard to which traits and how many ought to be measured, the time scale of the prediction needs to be considered. If one only wishes to know if there is selection acting on a trait at the present moment, then only that trait need by measured. However, if predictions about response to selection are the goal, then it is important to know how much of the overall selection acting on a trait is direct and indirect. Similarly, Gillespie and Winner (1990) found that artificial selection for O_3 resistance in *Raphanus sativus* resulted in a correlated response to SO_2. Plants resistant to O_3 were also resistant to SO_2. Continuing the example used in the last section, if direct selection acts only on conductance, then the distribution of assimilation phenotypes will change as well. This is due to indirect selection on assimilation as a function of its correlation with conductance. Directly and indirectly selected traits may behave in entirely different ways evolutionarily, since their genetic bases may be entirely different. For example, even though direct selection may be strongly positive and indirect selection weaker and negative in a particular instance, if the correlated trait causing the indirect selection has the higher heritability, evolution may be driven primarily by the weaker correlated selection (see discussion in Roose,

Chapter 5, this volume). One should therefore attempt to measure traits suspected of being both correlated with the trait of primary interest and having direct effects on fitness. Determining which traits are directly affected by pollutants and what traits they interact with may not be a simple matter. Especially when the interest is in chronic rather than acute pollutant effects, extensive experimental work may be necessary before a reasonably complete but manageable set of traits can be chosen.

The second sampling question which must be addressed is what scheme should be used for choosing the individuals to be measured in the population. Define precisely what kind of information is desired. If detection of selection is the only goal, two extreme phenotypic classes can be compared for their fitnesses. This is essentially what Gregorius (1990) describes as the paired sampling technique; see Muller-Starck (1985) and Bergman and Scholz (1987), for an application of this method. When measures are difficult or extremely expensive, this may maximize the ability to detect selection. It will only detect directional selection, however, since any intermediate fitness maximum or minimum will go unmeasured. If instead one desires information about how selection is acting on the study population, to predict that population's evolutionary response to the pollutant, then an unbiased random sampling scheme is appropriate. If an understanding of how selection will act on the species in general is the goal, then the fitness function should be estimated. To do this, a stratified sampling scheme might be used to increase accuracy in the fitness estimates at the extremes of the population's distribution. A stratified sample is one in which all phenotypes are equally represented, regardless of their proportional representation in the population. These three approaches are illustrated in Fig. 3.

For most species, the best estimate of the strength of selection comes from a lifetime study of cohorts in a population (see Kalisz, 1986). The drawback is that it can take a long time. Instead the phenotypic distribution in all extant cohorts during one season can be measured, keeping track of fecundity by phenotype and age/stage/size class as well. The shift in mean phenotype from one age/stage/size to the next is assumed to be due to the action of natural selection, and the entire shift between the first class and the reproductive class (weighted by fecundity in the reproductive individuals) is the measure of the strength of selection. An implicit assumption of this method is that the present day seedlings are drawn from the same phenotypic distribution as were the seedlings of the present day adults. When this assumption is violated, estimates of the strength of selection can be inaccurate. For example, if rapid evolution in response to the selective agent is occurring, this age/stage/size class-comparative method will underestimate selection, since the mean of the extant seedlings will be shifted toward the fitness optimum from the mean of the previous generation's seedlings. The difference in mean phenotype between the postselection adults and the putatively preselection seedlings will be smaller than would the comparison between individuals of one generation meas-

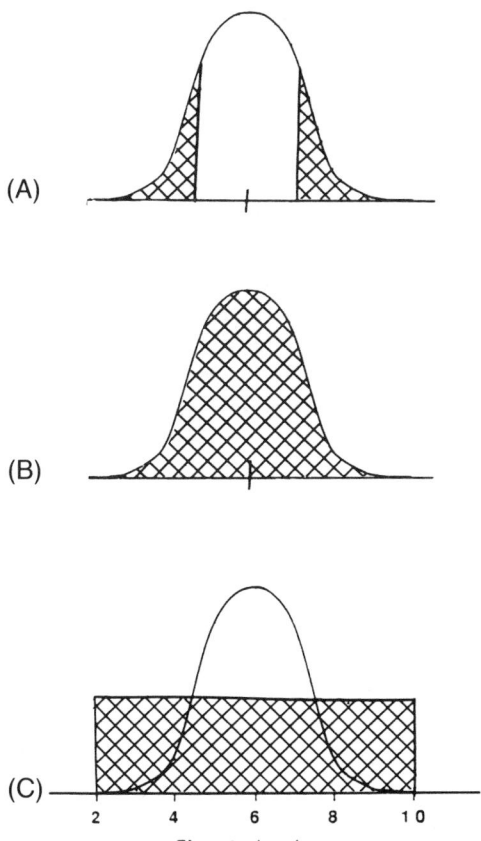

FIGURE 3. Sampling scheme for the population should be a function of the question to be addressed. The normal distributions symbolize the actual distribution of phenotypes in the population. *Crosshatched areas* signify the frequency of each phenotype in the sample which is measured. (A) Fitness differences between extremes of the distribution are of interest. (B) Quantification of the magnitude and direction of selection in a particular population. (C) Quantification of the fitness function for any phenotypic distribution in a particular environment.

ured at the onset and end of their lives. Except in the case of both high heritability and a large magnitude of selection, this difference will not appreciably affect the estimate of selection.

Effects on Population Dynamics and Age Structure

Air pollution can produce one of several outcomes in a population, ranging from extinction to total adaptation. With adaptation mean fitness may be greater than or equal to fitness in the prepollutant population. In the case of population extinction, the detrimental effects of air pollution may be indiscriminate, such that the whole population does poorly in the presence of pollution. An extreme example of this can be found in certain forests of Europe, where large tracts of adult trees are dying and local extinction of species may be taking place (see Scholz et al. 1990). In general, the effects of air pollution are expected to be more subtle. Rather than mortality of all individuals within a population, a range in damage to plant tissue, such as leaf necroses or mottling in the presence of O_3 and SO_2 are expected (reviewed in Roose et al. 1982). There is also clear evidence that air pollution is not just altering the appearances of the adult individuals or adult survivorship. The viability of pollen, ovules, seeds, seedlings, or saplings can be decreased as well (e.g., Ernst et al. 1985; Houston and Dochinger 1977). In the case of the eastern red pine populations in the Indiana Dunes area in the United States, no effects of air pollution are seen in the adult pines. However, none of the cones produced by these adults have viable seed and no seedlings were found in the study area (D. Karnosky, personal communication). In this situation, air pollution is postulated to disrupt the natural age-structure of the population by blocking seed production. Without recruitment of new trees from the seed crop this population could go extinct. Changes in the number of individuals in each stage class (e.g., seeds, seedlings, saplings, small trees, large trees) will alter the population growth rate and the stability of the population. The extent to which the population dynamics are affected will be a function of the magnitude and duration of the perturbation and the ability of the population to respond to the stress. How foliar damage, or other types of damage translate into changes in the individual plant's growth rate (Bell and Clough 1983), stage-specific survivorship probabilities, and reproductive output that in turn effect population growth and stability are virtually unstudied and best understood through the application of demographic analyses.

In the following section, general demographic techniques and the application of matrix models useful for monitoring the effects of air pollution on population stability and growth rate will be discussed. Again, the utility and interpretation of these measures to applied science will be stressed rather than the theory which underlies the models. The basic theory of demography can be found in any basic ecology text (e.g., Begon et al. 1986). Specific applications of demography to plant populations can be found in Harper and White (1974) and Harper (1974). The reader is referred to van Groenendal et al. (1988) for an introduction to the use of projection matrices in population biology, to Hartshorn (1975) and Enright and Ogden (1979) for their application to forest systems, and to Caswell (1989, and references therein) for a thorough and comprehensible treatment of the theory of matrix population models.

Demographic Analyses

Demographic analyses require that survivorship (l_x) and fecundity (m_x) be measured throughout the life cycle, where x denotes the stage or age of an individual (Leslie 1945; Deevey 1947; Caswell and Werner 1978). In matrix notation, stage/age-specific probabilities of survival and fecundity are elements, a_{ij}, in the matrix. These elements determine the dynamics of the population, the stable stage/age distribution, and the reproductive value of each stage/age (Fisher 1930; Charlesworth 1980). The geometric growth rate of the population, λ, describes whether the population is increasing, static or decreasing in size. If $\lambda > 1$, then the population increases exponentially, if $\lambda = 1$ then the population is stable and if $\lambda < 1$ then the population declines exponentially. The reproductive value of a stage/age describes the expected contribution of individuals in that stage/age to population growth. The stable stage/age distribution at equilibrium is the proportional representation of each stage/age-class once the population has reached a stable growth rate. The geometric growth rate, the reproductive value, and the stable stage/age distribution are primary descriptors of the future population dynamics if conditions remain as they were when l_x and m_x were measured in the population. Therefore, they can be used to evaluate the present status of a population and project its status into the future. There are a number of software packages available that will calculate demographic parameters from life tables, such as Eispack (Eigensystem Subroutine Package, Release 2; Argonne National Laboratory). Other software designed for matrix manipulation (SAS, IML package, and APL) allow for short programs to be easily written for calculating the demographic parameters. Although greater initial effort is required with the latter options, in the long run they may be more efficient, since bootstrapping or jackknifing variance estimates can be built into the program.

Whenever possible, the results from demographic analyses of polluted populations should be compared with the results of demographic analyses performed on data from the same population *prior* to the onset of pollution. When these data are not available, data from plants from the same provenances or similar habitat may be used, especially if multiple polluted and unpolluted sites are available. These comparisons will determine if the affected population is actually adapting to air pollution stress (λ affected population $\geq \lambda$ original population) or has reduced in fitness (λ affected population $\geq \lambda$ original population). Significance tests of the estimates of these parameters can be made using resampling techniques. Caswell (1989, pp 185–195) provides two methods for calculating confidence intervals for λ. Calculation of these confidence intervals for both the polluted and nonpolluted populations will establish if the ranges of λ estimated for the two populations overlap.

If the affected population is stable or is growing relative to the original population, two possible interpretations can be made: either the population is resistant (preadapted) or it has evolved resistance. If the affected population is in decline relative to the unaffected population, the population is not capable of coping with the pollution stress or has not yet evolved pollution resistance. Attempting to predict evolution of resistance to pollution is similar to the problem a doctor faces in diagnosing a patient with a potentially deadly disease. There are two processes going on a once. First, the disease is affecting the patient. Second, the body's immune system is developing an immune response to the infectious agent. The evolution of resistance is analogous to the latter. The development of resistance may be quite likely, given infinite time, but the immune system is in a race with the first process, the declining health of the body and the possibility of death should the disease continue. This is analogous to the declining demographic health of the population during exposure to pollution. To issue an accurate prognosis, the physician must both assess the likelihood and rate of progress of the development of resistance and determine the likelihood that the body will remain functioning long enough for resistance to develop.

Assessment of the likelihood of evolving resistance to the pollutant (i.e., adapting to the pollutant) therefore involves: (1) estimating the likelihood and rate of evolution of resistance, and (2) estimating the likelihood of population extirpation and the time until extirpation. The likelihood of evolving resistance can be assessed by looking at the most fit phenotypes in the population described in the previous section. If it can be shown that there is a substantial heritable basis to their superior fitness, then one can reasonably assume that such a phenotype can come to be the mode for the population. One can then ask whether a population composed of such individuals would be capable of replacing itself. A projection matrix based on their life history traits can be constructed and used to calculate λ for the imaginary future population composed of these superior phenotypes. Based on heritability estimates for the life history traits in the population, one can calculate likely rates of the evolution of higher fitness (i.e., increasing λ).

There are a number of possible outcomes of such an exercise. First, suppose that the fittest phenotypes do replace themselves, and their life history traits are highly heritable. Suppose also that the population is declining but only very slowly. The potential for the evolution of resistance is quite high in this case. Second, imagine again that the population is declining very slowly (i.e. λ close to 1.0), but the most-fit genotypes do not replace themselves even if they were the mode of the population. If there is heritable variation in life history attributes within the population, resistance may eventually be achieved, but the rate at which that occurs depends largely on the rate at which new variation is introduced to the population. A prognosis in this case is unpredictable due to a lack of knowledge about processes at the level of the gene (e.g., mutation rates, recombination rates, and rates of origin of variation in quantitative tracts), although a more informed guess can be made after reading Roose's section on the generation of genetic variation in this book.

Given that a populations is not heading toward rapid extinction, examination of the transition probabilities between life history stages in the polluted population can indicate stages in the life cycle which are experiencing high mortality. These bottlenecks, through which few individuals pass, are common features of plant population dynamics even in unpolluted populations and are due to the natural biotic and abiotic sources of stress and/or mortality in the population. The extent to which this natural stage-specific mortality is exaggerated by the populations' sensitivity to pollution and its effect on the population dynamics is the concern of plant breeders and foresters. Comparisons between polluted and nonpolluted populations' transition probabilities are necessary to determine if fitness is decremented over naturally occurring mortality. In addition, if phenotypic variation in resistance is seen at those stages in the life cycle which correspond to bottlenecks, breeders may wish to concentrate their efforts on selecting those individuals which have highest fitnesses during these sensitive stage/age-classes for inclusion in the breeding program.

Perturbation Analyses

In a perturbation analysis, the results of a demographic analysis are re-examined. The question asked is: holding all other transition probabilities constant, how sensitive is λ to a small change in each of the transition probabilities? By performing perturbation analyses, the role of each element in the matrix in determining population growth rate can be examined individually. If stage/age-specific decrements in survival or fecundity as a result of air pollution are known, perturbation analyses can be used to examine the magnitude of these effects directly on population growth rates.

There are two types of perturbation analyses. Sensitivity analyses permit the comparison of the importance of particular elements in the matrix influencing λ for the single population under consideration (see Caswell 1986, for methods; Caswell and Werner 1978, for an application). Sensitivity values, s_{ij}, are the partial derivative of λ taken with respect to each element, a_{ij}. Elasticity analyses, a second type of perturbation analysis, examine the proportional effects of proportional changes in the element on λ (see Caswell 1978, 1989; deKroon et al. 1986 for methods; Crouse et al. 1987; Kalisz and McPeek 1991 for an application). Elasticity values, e_{ij}, are the partial derivative of the $\ln(\lambda)$ taken with respect to the $\ln(a_{ij})$. A particularly useful aspect of elasticity analyses is the fact that the e_{ij} values are expressed as a proportion and the all e_{ij}s sum to one. This means that the elasticity values calculated for several populations can be used in comparative studies. The among-population differences in the effect of individual elements among population within a species or even the differences between species with similar stage-structure can be compared.

Conclusions

The utility of selection, response to selection, and demographic analyses for understanding responses of plant populations to anthropogenic changes in the atmosphere is great for both natural and managed systems. Studies of natural selection in the field can be employed to identify traits which are important for resistance. From the methods presented, one can choose an approach that optimizes the balance between cost, time budget, and information gained. Studies of selection in the field coupled with experimental determination of inheritance will provide fundamental insights into the likelihood of the evolution of resistance in plant populations. Demographic analyses coupled with the above approaches can provide the link between predictions of short-term response and predictions of long-term evolutionary health. The ability to interpret these kinds of studies depends on a thorough understanding of the ecology of the population and the biology of the plant. Multidisciplinary, collaborative approaches are advocated.

Acknowledgments. We thank JA Teeri for many enlightening discussions. This work was supported in part by NSF grants BSR85-06401 and BSR89-06283 to SJT and BSR87-13967 to SK. This is Kellogg Biological Station Contribution No. 661.

References

Arnold SJ, Wade MJ (1984a) On the measurement of natural and sexual selection: theory. Evolution 38:709–719

Arnold SJ, Wade MJ (1984b) On the measurement of natural and sexual selection: applications. Evolution 38:720–734

Aycock MK, Jr (1972) Combining ability estimates for weather fleck in *Nicotiana tabacum* L. Crop Science 12:672–674

Becker WA (1984) Manual of quantitative genetics, 4th ed. Academic Enterprises, Pullman

Begon J, Harper JL, Townsend CR (1986) Ecology. Blackwell Scientific, London,

Bell JNB, Clough WS (1973) Depression of yield in ryegrass exposed to sulphur dioxide. Nature 241:47–49

Bergmann F, Scholz F (1987) The impact of air pollution on the genetic structure of Norway spruce. Silvae Genetica 36:80–83

Caswell H (1978) A general formula for the sensitivity of population growth rate to changes in life history parameters. Theoretical Population Biology 14:215–230

Caswell H (1986) Matrix models and the analysis of complex plant life cycles. Lectures on Mathematics in Life Sciences 18:171–234

Caswell H (1989) Matrix population models. Sinauer Press, Sunderland, MA

Caswell H, Werner PA (1978) Transient behavior and life history analysis of Teasel (*Dipsacus sylvestris* Huds.). Ecology 59:53–66

Charlesworth B (1980) Evolution in age-structured populations. Cambridge University Press, Cambridge

Crespi BJ, Bookstein FL (1989) A path-analytic model for the measurement of selection on morphology. Evolution 43:18–28

Crouse DT, Crowder LB, Caswell H (1987) A stage-based population model for loggerhead sea turtles and implications for conservation. Ecology 68:1412–1423

Deevey ES (1947) Life tables for natural populations of animals. Quarterly Review of Biology 22:277–285

DeVos NE, Hill RR Jr, Pell EJ, Cole RH (1982) Quantitative inheritance of ozone resistance in potato. Crop Science 22:992–995

de Kroon H, Plaisier A, van Groenendael J, Caswell H (1986) Elasticity as a measure of the relative contribution of demographic parameters to population growth rate. Ecology 67:1427–1431

Endler JA (1986) Natural selection in the wild. Princeton University Press, Princeton, NJ

Engle RL, Gabelman WH (1989) Inheritance and mechanism for resistance to ozone damage in onion, *Allium cepa* L. American Society for Horticultural Science 5:423–430

Enright N, Ogden J (1979) Applications of transition matrix models in forest dynamics: *Araucaria* in Papua, New Guinea and *Nothofagus* in New Zealand. Australian Journal of Ecology 4:3–23

Ernst WHO, Tonneijck AEC, Pasman FJM (1985) Ecotypic response of *Silene cucubalus* to air pollutants (SO_2, O_3). Journal of Plant Physiology 118:439–450

Falconer DS (1981) Introduction to quantitative genetics, 2nd edn, Longman Inc, New York.

Fisher RA (1930) The genetical theory of natural selection. Claredon Press, Oxford

Gillespie CT, Winner WE (1990) Development of radish lines differing in resistance to O_3 and SO_2. New Phytologist (in press)

Gregorious HR (1990) The attribution of phenotypic variation to genetic or environmental variation in ecological studies. In: Scholz F, Gregorious HR, Rudin D (eds) Genetic effects of air pollution in forest tree populations. Springer-Verlag, Berlin

Harper JL (1977) Population biology of plants. Academic Press, New York

Harper JL, White J (1974) The demography of plants. Annual Review of Ecology and Systematics 5:419–463

Hartshorn GS (1975) A matrix model of tree population dynamics. In: Golley FB, Medin E (eds) Tropical ecological systems. Springer-Verlag, New York, pp 41–51

Houston DB, Dochinger LS (1977) Effects of ambient air pollution on cone, seed and pollen characteristics in Eastern white and red pines. Environmental Pollution 12:1–5

Kalisz S (1986) Variable selection on the timing of germination in *Collinsia verna* (Scrophulariaceae). Evolution 40:479–491

Kalisz S, McPeek MA (1991) The demography of an age-structured annual: resampled projection matrices, elasticity analyses and seed bank effects. Ecology (in review)

Karnosky DF (1977) Evidence for genetic control of response to sulfur dioxide and ozone in *Populus tremuloides*. Canadian Journal of Forest Research 7:437–440

Krause GHM, Arnt U, Brandt CJ, Bucher J, Kenk G, Matzner E (1986) Forest decline in Europe: development and possible causes. Water, Air and Soil Pollution 31:647–668

Lande R, Arnold SJ (1983) The measurement of selection on correlated characters. Evolution 37:1210–1226

Leslie PH (1945) On the use of matrices in certain population mathematics. Biometrika 33:183–212

Manly BFJ (1985) The statistics of natural selection on animal populations. Chapman and Hall, London

Mitchell-Olds T, Shaw RB (1987) Regression analysis of natural selection: statistical interference and biological interpretation. Evolution 41:1149–1161

Muller-Starck G (1985) Genetic differences between "tolerant" and "sensitive" beeches (*Fagus sylvatica* L.) in an environmentally stressed adult forest stand. Silvae Genetica 34:241–247

Pell EJ (1987) Ozone toxicity—Is there more than one mechanism of action? In: Hutchinson TC, Meema KM (eds) Effects of atmospheric pollutants on forests, wetlands and agricultural ecosystems. Springer-Verlag, Berlin, pp 229–240

Phillips PC and Arnold SJ (1989) Visualizing multivariate selection. Evolution 43:1209–1222

Pitelka LF (1988) Evolutionary responses of plants to anthropogenic pollutants. Trends in Ecology and Evolution 3(9):233–236

Roose ML, Bradshaw AD, Roberts TM (1982) Evolution of resistance to gaseous air pollutants. In: Unsworth MH, Ormrod DP (eds) Effects of gaseous air pollution in agriculture and horticulture, Butterworth Scientific, London, pp 379–409

Scheiner SM (1989) Variable selection along a successional gradient. Evolution 43:548–562

Schluter D (1988) Estimating the form of natural selection on a quantitative trait. Evolution 42:849–861

Scholz F, Gregorious HR, Rudin D (1990) Genetic effects of air pollution in forest tree populations. Springer-Verlag, Berlin

Sokal RR, Rohlf FJ (1981) Biometry, second edition. Freeman, San Francisco

Taylor GE, Jr (1978) Genetic analysis of ecotypic differentiation within an annual plant species, *Geranium carolinianum* L., in response to sulfur dioxide. Botanical Gazette 139:326–368

Taylor GS (1968) Ozone injury on Bel W-3 tobacco controlled by at least two genes. Phytopathology 58:1069

Tonsor SJ, Kalisz S, Fisher J (1991) The role of the seed bank in population genetic structure evidence from *Plantago lanceolata* L. Evolution (in review)

Treshow M (ed) (1984) Air pollution and plant life. Wiley, New York

Troyanowsky C (1985) Air pollution and plants. Weinheim. Deerfield Beach, FL

van Groenendael J, de Kroon H, Caswell H (1989) Projection matrices in population biology. Trends in Ecology and Evolution 3:264–269

Wade MJ, Kalisz S (1989) The additive partitioning of selection gradients. Evolution 43:1567–1569

Wade MJ, Kalisz S (1990) The causes of natural selection. Evolution 44:1947–1955

Winner WE, Mooney HA (1980a) Ecology of SO_2 resistance: II. Photosynthetic changes of shrubs in relation to SO_2 absorption and stomatal behavior. Oecologia 44:296–302

Winner WE, Mooney HA (1980b) Ecology of SO_2 Resistance: I. Effects of fumigations on gas exchange of deciduous and evergreen shrubs. Oecologia 44:290–295

Winner WE, Mooney HA, Goldstein RA (1985) Sulfur dioxide and vegetation. Physiology, ecology and policy issues, Stanford University Press, Stanford, CA

Commentary to Chapter 10

Ecological Genetics of Plant Populations in Polluted Environment

A. Jonathan Shaw

A wide range of techniques have been described in this book for analyzing the responses of plant populations to air pollution. The technique that any researcher employs is dependent upon the questions being asked, practical considerations such as limitations on time, space, or monetary resources, and biological characteristics of the organisms. Obviously, techniques appropriate for long-lived trees may not be ideal for annual species. Likewise, techniques for analyzing populations of diploid seed plants are unlikely to be appropriate for haploid organisms such as bryophytes or algae. This commentary will serve as an overview of the analytical methods discussed in this book for studying processes involved in the responses of plant populations to air pollution stress.

A fundamental problem in population and evolutionary biology is to distinguish environmental from genetic sources of phenotypic variation. Moreover, understanding the sources of phenotypic variation in resistance to air pollutants is prerequisite to making predictions about evolutionary responses to present and future environmental changes. Several relatively simple techniques can be employed to partition phenotypic variation into genetic and environmental components (Table 1).

In common garden experiments a compromise is often made between maximizing control over experimental conditions and providing a realistic setting from which extrapolations to the natural environment are likely to be valid. There is considerable discussion in this book especially among those studying physiological responses of plants to pollution stress, of the advantages and disadvantages of field versus growth chamber measurements of physiological parameters. One sacrifices a certain amount of environmental control while the other compromises realism. There seems to be general agreement that a place exists for both approaches: there is a need for both realism and rigorously controlled studies, and the contrasting approaches are complementary.

Agriculturalists are quite familiar with the problem of trying to predict plant performance in new environments from experimental studies completed at a particular site (e.g., Sprague and Federer 1951; Findlay and

TABLE 1. Summary of population-level techniques for ecological genetic studies of pollution stress and plant populations.

Techniques for distinguishing environmental and genetic sources of phenotypic variation
    Common garden experiments
    Reciprocal transplants
    Paired comparisons
    Open-pollinated sibships
Techniques for determining the genetical architecture of phenotypic variation within
    populations
    Covariances among relatives
    Diallel analysis
    Haploid sib analysis
    Artificial selection experiments
Techniques for demonstrating past or present natural selection
    Correlations between phenotypic traits and environmental characteristics
    Temporal changes in the frequency distribution of phenotypic traits within populations
    Regression analysis

Wilkinson 1964; Breese 1969). Genotype-environment interactions complicate the interpretation of any common garden experiment, no matter how realistic the experimental conditions. The component of phenotypic variation in a population attributable to interactions between genotypes and microenvironmental factors can be estimated directly from common garden experiments with the appropriate design and statistical analysis. Reciprocal transplant experiments, however, provide another level of environmental variation that can be directly analyzed. Phenotypic variation can be partitioned into components attributable to genotypic and environmental sources, as well as to the interactions between genotypes and both macroenvironmental (between sites) and microenvironmental (within sites, between experimental blocks, for example) variation. To the evolutionist, genotype × environment variation is at least as relevant as genotypic differences per se, since it is the variation among genotypes in their responses to nonrandom environmental variation that provides the potential for natural selection and adaptive evolution.

Reciprocal transplant and common garden experiments are often impossible with forest trees that cannot be reproduced clonally and grown rapidly. When trees differing in the degree of visible pollution damage are observed in a natural population, it is difficult to determine if this variation is due to microenvironmental differences or genetic variability in resistance. In order to partition this phenotypic variation into genotypic and environmental components, the method of "paired comparisons" (Table 1) has been employed by European forest ecologists (Gregorius 1989). Generally, a tree is randomly selected from the population, and its nearest neighbor is sampled for a paired comparison in which both members of the pair experience the same or very similar environments. Isozymes can be used as genetic markers in order to determine if there are consistent allelic differences between apparently resistant versus tolerant individuals.

This technique provides evidence that can be suggestive of either genetic or environmental causes for the observed variation but is likely to be inconclusive in many cases. Using paired comparisons to distinguish genetic from environmental sources of variability assumes that microenvironmental factors which could affect the apparent resistance of trees (e.g., small scale differences in actual pollutant concentrations) occur independently of factors that might affect the distribution of marker alleles. If there is no correlation between the particular isozyme markers and the genes that confer resistance (due either to pleiotropic effects, linkage, or a causal relationship between isozyme characteristics and resistance), this approach could miss a genetic component to resistance when in fact it exists. The opposite problem i.e., that microenvironmental causes of variation are misinterpreted as genetic could result if small-scale environmental variation affects phenotypic responses to pollution and the genetic markers in a correlated manner. The assumption that differences in morphological characters among closely adjacent plants within populations are due to genetic factors because the environment must be uniform over such short distances has been sharply criticized in the plant systematics literature (e.g., Wyatt et al. 1982). Environmental uniformity over short distances relative to the sizes of forest trees may be valid in the case of air pollutants that vary over regional scales, and it may be that correlated microenvironmental effects on pollution responses and the distribution of enzyme markers are rare. Nevertheless, studies utilizing paired comparisons should be undertaken and interpreted with caution.

More complicated methods of analysis are necessary when specific information about the genetical architecture of within-population variation is needed. Variation among seed collections made from individual maternal parents in the field (i.e., so-called open-pollinated sibships) provides solid evidence for genetic variation when differences run in family lines. The partitioning of genetic variation into additive versus nonadditive components, however, requires further analysis. Open-pollinated sibships are likely, on average, to be intermediate between full and half sibs (e.g., Ellstrand 1984), and a precise genetic interpretation of family differences is therefore impossible. Nevertheless, if distinguishing environmental from genetic components of variation is sufficient for the particular research objective, data from field-collected, open-pollinated sibships are less likely to include nongenetic carry-over effects than data based on clonally propagated samples.

More precise genetic information can be obtained if seed from natural populations can be collected to serve as parents in experimentally controlled crosses. Observations on phenotypic resemblances among relatives such as parents and their offspring or half and full sibs, allow for additive and nonadditive components of genetic variation to be distinguished under certain conditions (Cockerham 1954; Falconer 1981). These methods allow estimation of narrow-sense heritabilities; i.e., the ratio of the additive component of genetic variation to the total phenotypic variation. One of the outstanding problems in pollution ecology is to predict future evolutionary responses of plant populations to continuing, and in many geographic regions, increasing levels of atmospheric pollutants. Since the response to natural selection is proportional to the amount of additive genetic variation for that trait, reliable predictions about future evolutionary changes require a knowledge of the genetic architecture of variation in natural populations.

Mitchell-Olds and Rutledge (1986) summarized the application of quantitative genetic theory to the study of natural plant populations. Improperly or incompletely executed quantitative genetic analyses can provide misleading estimates of genetic parameters and erroneous predictions of future evolutionary change. It is worthwhile here to briefly summarize the most important sources of ambiguity in such studies. This summary is based to a large extent on the discussions of Falconer (1981) and Mitchell-Olds and Rutledge (1986).

The estimation of such genetic parameters as heritabilities and genetic correlations requires a series of biological and methodological assumptions. Some of these are probably violated in many studies of natural populations, while violation of others can be avoided with an appropriate experimental design. It is assumed, for example, that individuals for genetic analyses are selected at random from the population, and that there is no genotype-environment correlation under the subsequent experimental conditions. These are relatively minor problems if attention is paid to random sampling from the source population, and a randomized experimental design is employed.

Other problems arise, however, that result from the biological characteristics of the study organism and are more difficult to control. The estimation of unbiased genetic parameters assumes disomic inheritance and no linkage or linkage disequilibrium. It also assumes that (field-collected) parents are equally inbred and that the progeny are noninbred. Although linkage disequilibrium is apparently not common in predominantly outbreeding plants, it may be more important in species that exhibit a high incidence of inbreeding (Allard et al. 1972). Few studies of quantitative genetic variation have also assessed the degree of linkage disequilibrium in the same population. Likewise, it difficult or impossible to know for most populations whether individuals selected for genetic analyses are equally inbred. In fact, it is likely that this assumption is frequently violated to a greater or lesser extent. The degree to which this will bias estimates of genetic parameters in natural populations is difficult to judge.

Most standard techniques for estimating heritabilities assume that inheritance is disomic, a large number of genes affect the trait, there are no maternal effects, and epistatic variance is absent or minimal. Diallel analyses can be used to estimate maternal and nonadditive genetic effects (Falconer 1981). It has become apparent from discussions in this book that little is known about the role of major genes, as opposed to classical polygenic inheritance, in determining differences in the level of resistance to pollutants among plants within natural populations. Macnair (Commentary to Chapter 5, this volume) has shown how difficult it is in practice to determine if major genes are involved, owing to inevitable imprecision in the experimental measurement of resistance. If resistance is affected by few genes, genetic parameters based on the assumption of many genes each with small effect can be strongly biased.

The assumption of disomic inheritance may be problematic in polyploid species. Moreover, techniques for estimating quantitative genetic parameters were developed for studies of diploid higher plants and animals and are not necessarily appropriate for cryptogamic plants. If general predictions about the responses of plants to air pollutants are to be developed, it is important to investigate the responses of a diverse range of organisms. Algae, ferns, fern allies, and bryophytes have life cycles which make some of the standard techniques for studying population-level processes inappropriate. On the other hand, these organisms sometimes offer opportunities to address issues that are much more difficult to get at with diploid seed plants.

One of the assumptions of methods utilizing covariances among relatives is that the epistatic component of variation is minimal or absent. Covariance among half sibs, for example, includes an epistatic component which is generally ignored when calculating heritabilities (Falconer 1981). Although the assumption of minimal or absent epistasis is almost invariably made in theoretical discussions of polygenic evolution (e.g., Lande 1981), there is very little empirical support from studies of natural populations. Genetic analyses of variation in domesticated plants and animals frequently suggest a substantial epistatic component (Jinks et al. 1969; Geiger 1988). The main problem is that epistasis is very difficult to detect in diploids and involves a relatively complicated and time-consuming breeding program that is rarely practical in studies of natural populations.

Organisms with a free-living haploid stage in the life cycle, such as bryophytes or ferns, provide an opportunity to estimate the epistatic component of variation by measuring variance within and among so-called haploid sib families (Shaw et al., 1989). Haploid sibs are meiotic progeny derived from a single diploid sporophytic individual and are genetically comparable to the population of gametes derived from a diploid plant or animal.

Multiple sporophytes can be sampled from natural populations and the haploid meiotic progeny grown as single spore isolates. Traits such as resistance to pollutants can be readily measured, and the total variation in the population partitioned into components attributable to differences among families (sporophytes) and haploid sibs within families. From these observational components of variation, the additive and epistatic components of genetic variation in the population can be estimated because dominance is absent in such haploid organisms.

A valuable approach with any group of organisms, but so far infrequently employed in pollution studies, is to practice artificial selection for resistance (see, however, Winner et al. Chapter 7, this volume). For practical purposes of predicting the responses of plant populations to natural selection, realized heritabilities derived from observed responses to artificial selection provide the most direct possible evidence (Roose, Chapter 5, this volume).

The past or present occurrence of natural selection for resistance to pollutants (or other traits) can be inferred, or more directly demonstrated, in a variety of ways (Endler 1986). Natural selection itself can be defined in more than one way. Endler (1986) defines selection as a process in which there is phenotypic variation, some variants have a higher reproductive output than others (i.e., the phenotypic variation is correlated with fitness), and at least some of the phenotypic variation is heritable. In contrast, Tonsor and Kalisz (Chapter 10, this volume) use a within-generation definition of natural selection following the approach of Arnold and Wade (1984a, 1984b): selection is the process in which some phenotypic variants produce more offspring than others. The response to selection, a function of both the intensity of selection and the degree of heritability for the trait, is a between-generation process that can be conceptually distinguished from selection itself and can be measured independently.

The response to selection for resistance to atmospheric pollutants has been inferred by indirect methods such as correlations between trait distributions and environmental factors, such as the concentration of a particular pollutant. Habitat-correlated variation, or ecotypic differentiation, provides retrospective and indirect, though often convincing evidence of past selection. In this book, such evidence for the evolution of increased levels of pollution resistance has been presented for several plant species (Bell et al., Chapter 3, this volume), including *Populus tremuloides*, *Silene cucubalus*, and *Lupinus* sp. The occurrence of population differentiation provides indirect evidence of past selection but reveals nothing about the present. Indirect evidence of ongoing selection is possible at sites that can be repeatedly visited, such as Philips Park in Manchester, UK, (Bradshaw and McNeilly, Chapter 2, and Bell et al., Chapter 3, this volume). This approach is still indirect, since it assumes that temporal changes in phenotypic distribution result from selection.

Evidence for the occurrence of stabilizing, disruptive, or directional natural selection can be obtained by regression analyses of phenotypic traits on relative fitness (Arnold and Wade 1984a, 1984b; Kalisz 1986; Tonsor and Kalisz, Chapter 10, this volume). This technique provides a powerful yet simple tool for assessing which of several possible physiological or other phenotypic traits may have large effects on fitness of plants in polluted environments. It should be kept in mind, however, that regression analyses suffer from the shortcomings of any descriptive approach; thus, Mitchell-Olds and Shaw (1987) emphasize that it should be used primarily to suggest hypotheses about the effects of selection on phenotypic traits. These hypotheses can then be tested experimentally.

Ambiguities in biological interpretation arise because observed correlations between phenotypic traits and relative fitness do not necessarily mean that the correlated traits directly affect fitness. In addition, Mitchell-Olds and Shaw (1987) point out potential problems such as the existence of unmeasured but selectively important traits and difficulties with significance testing, that can make interpretations difficult or even misleading. The choice of traits to measure in such studies is of critical importance and requires previous knowledge of pollution effects on the organism under study. Nevertheless, the regression method is the most direct method available for investigating the present occurrence of natural selection and is likely to provide valuable insights into the evolution of populations in habitats subjected to high levels of atmospheric pollution.

Conclusions

The subject of microevolutionary processes in natural plant populations has been an active and exciting area of research in recent years. Studies on the evolution of heavy metal resistance have contributed a great deal to evolutionary biology in general, in part because when present in excessive concentrations, toxic metals result in strong selection for resistance. Selection for resistance to regional pollutants such as O_3 is considerably more subtle and therefore more difficult to detect and study. Experimental studies of population responses to higher, acute levels of pollutants are easier, but it has been emphasized repeatedly in this book that the physiological, genetic, and evolutionary consequences of acute exposure are likely to differ from those related to lower and more chronic exposure. The latter constitute the major focus of this book. It appears that artificial selection experiments represent an underutilized but extremely valuable approach to fulfilling the need for predictions about future changes in natural populations.

The regression technique discussed by Tonsor and Kalisz (Chapter 10, this volume) for evaluating the processes of natural selection is an exciting approach that has not yet been applied to natural populations in polluted environments. With due regard to the potential biological and statistical complications, this approach is likely to provide new insights into ecological genetics and air pollution stress in natural plant populations.

References

Allard RW, Babbel GR, Clegg MT, Kahler AL (1972) Evidence for coadaptation in *Avena barbata*. Proceedings of the National Academy of Science USA 69:3043–3048

Arnold SJ, Wade MJ (1984a) On the measurement of natural and sexual selection: Theory. Evolution 38:709–719

Arnold SJ, Wade MJ (1984b) On the measurement of natural and sexual selection: applications. Evolution 38:720–734

Breese EL (1969) The measurement and significance of genotype-environment interactions in grasses. Heredity 24:27–44

Cockerham CC (1954) An extension of the concept of partitioning hereditary variance for analysis among relatives when epistasis is present. Genetics 39: 859–882

Ellstrand NC (1984) Multiple paternity within fruits of the wild radish, *Rhaphanus sativus*. American Naturalist 123:819–828

Endler JA (1986) Natural selection in the wild. Princeton University Press, Princeton, NJ

Falconer DS (1981) Introduction to quantitative genetics, 2nd ed. Longmans, London

Findlay KW, Wilkinson GN (1963) The analysis of adaptation in a plant-breeding programme. Australian Journal of Agricultural Research 14:742–754

Geiger HH (1988) Epistasis and heterosis. In: Weir BS, Eisen EJ, Goodman MM, Namkoong G (eds) Proceedings of the Second International Conference Quantitative Genetics, Sinauer, Sunderland, MA, pp 395–399

Gregorius HR (1989) The attribution of phenotypic variation to genetic or environmental variation in ecological studies. In: Scholz F, Gregorius HR, Rudin D (eds) Genetic effects of air pollutants in forest tree populations. Proceedings of the Joint Meeting of the IUFRO Working Parties, Springer-Verlag, Berlin, pp 1–13

Jinks JL, Perkins JM, Breese EL (1969) A general method for the detection of additive, dominance, and epistatic variation for metrical traits. II. Application to inbred lines. Heredity 24:419–429

Kalisz S (1986) Variable selection on the timing of germination in *Collinsia verna* (Scrophulariaceae). Evolution 40:479–491

Lande R (1981) A quantitative genetic theory of life history evolution. Ecology 63:607–615

Mitchell-Olds T, Rutledge JJ (1986) Quantitative genetics in natural plant populations: a review of the theory. American Naturalist 127:379–402

Mitchell-Olds T, Shaw RG (1987) Regression analysis of natural selection: statistical inference and biological interpretation. Evolution 41:1149–1161

Shaw J, Beer SC, Lutz J (1989) Potential for the evolution of heavy metal tolerance in *Bryum argenteum*, a moss. I. Variation within and among populations. Bryologist 92:73–80

Sprague GF, Federer WT (1951) A comparison of variance components in corn yield trials: II. Error, year × variety, location × variety, and variety components. Agronomy Journal 43:535–541

Wyatt R, Lane DM, Stoneburner A (1982) The misuse of mixed collections in bryophyte taxonomy. *Taxonomy* 1:698–704

11
Ecological Genetics and Changes in Atmospheric Chemistry: The Application of Knowledge

DAVID F. KARNOSKY

Introduction

Ample evidence indicates that plants vary markedly in their responses to air pollutants. Furthermore, it is known that air pollution can exert a selective force on plants at both the sporophytic and gametic levels. The application of knowledge regarding air pollution and ecological genetics focuses strongly on two central concepts. First, since plants vary from extremely sensitive to strongly resistant to air pollution, it is possible to select and breed for plants with either high or low sensitivity for use in managed ecosystems. Second, the maintenance of genetic diversity in plant populations in both unmanaged and managed ecosystems should be given more serious attention. This chapter reviews research activities in these two areas and suggests where additional research is needed. It focuses on trees and forest ecosystems, as others (Bell et al. Chapter 3, this volume) discuss similar aspects with other plants.

For purposes of discussion, the term resistance is used to indicate those plants that are not injured by air pollution. In most cases with trees, we do not know enough about the mechanisms to distinguish between tolerance and avoidance.

Managed Ecosystems

Developing Pollution-Resistant Plants

Numerous researchers have demonstrated that it is possible to select and breed plants with improved air pollution resistances through traditional genetic approaches (Bell et al. 1980; Garsed and Rutter 1982; Rohmeder and Schonborn 1965). However, this approach to resolving air pollution problems must never be viewed as a permanent solution. Rather, it should be viewed as a temporary action to complement continued efforts at reducing air pollution at the source.

Pollution-resistant plants could be useful in revegetating local areas affected by point-source pollutants, particularly when the pollution abatement costs are prohibitive or if abatement technology is undeveloped. Pollution-resistant plants could also be beneficial in planting where non-point-source pollutant problems are of a transboundary nature, especially if the political or economic situation in the polluting country prevents cleanup at the source and in urban areas where accumulated effects of multiple pollutants create very complex abatement scenarios.

POINT SOURCES

Although many of North America's major point-source pollution problems have been corrected through installation of tall stacks or emission-control devices or the use of lower sulfur fuel and other technologies, inumerable, smaller, localized, point-source pollution situations still exist where vegetation is experiencing acute exposure to SO_2 and other pollutants. The proliferation of small facilities generating energy from refuse burning is an example of an industrial process which has created many localized air pollution problems in the United States.

Large-scale point-source pollution by SO_2 and heavy metals remains a major problem in some countries, particularly in Eastern Block and developing countries. In Poland, for example, SO_2 pollution is having a significant negative economic impact on the forest industry (Kabala 1989). Field selection and subsequent breeding of Scot's pine (*Pinus sylvestris*) for resistance to point-source pollution is being used in Poland (Bialobok et al. 1980) and Czechoslovakia (Kanak 1986) to identify resistant trees for reforesting heavily impacted areas where the natural forests have been destroyed by SO_2 and heavy metals.

HEAVILY-IMPACTED, NON-POINT-SOURCE REGIONS

Much of Great Britain and Europe continues to be impacted by elevated air-pollution levels. For example, Fowler and Cape (1982) estimated that there were about 1 million hectares in Western Europe where the annual mean SO_2 concentration exceeded 26 ppb. The Association of German Engineers recommends that 13 ppb SO_2 (as a growing season mean) should not be exceeded if sensitive plants are to be protected (Knabe 1976). The mountainous regions along the borders between East and West Germany and between Poland and Czechoslovakia illustrate the destructive nature of region-wide air pollution problems in Europe (Mazurski 1986; Paces 1985; Prinz et al. 1987). Here, widespread forest effects threaten a forest industry that has been a stable economic force for centuries.

The transboundary and Eastern-versus-Western-Block nature of these pollution problems illustrates the tasks yet to be accomplished in attacking pollution at its sources. These are certainly high priority regions for the development and deployment of pollution resistant planting stock. It is little wonder that the Czechoslovakian (Kanak 1986) and Polish efforts (Bialobok 1980) to select and breed pollution-resistant trees are among the most advanced in the world.

URBAN AREAS

The characterization of the relative air pollution resistances of commonly used urban trees has been under way for the past 2 decades in the United States. The relative resistances to common urban air pollutants such as ozone (O_3) and SO_2 have been determined for numerous tree species (Davis and Coppolino 1974; Davis et al. 1981), seed sources (Karnosky and Steiner 1981; Townsend and Dochinger 1974), and cultivars (Karnosky 1981). This work has been aimed at identifying plant material that can withstand the relatively high air pollution doses common to major urban areas. No breeding work has been done to try to further increase resistance levels.

Developing Pollution-Sensitive Bioindicator Plants

Monitoring multiple pollutants with equipment designed to accurately characterize ambient air pollution loading is a costly and rigorous exercise. Consequently, monitoring stations are sparsely scattered across the landscape. Some states still only have a handful of certified monitoring stations. Because of the large capital expense and the intensive and skilled labor requirement to calibrate, maintain, and operate air-monitoring equipment, alternative methods for detecting air pollution are desirable for augmenting the mechanical monitoring networks. One alternative is biological-sensitive indicator plants or biomonitors (Manning and Feder 1980). Bel W-3 tobacco is probably the best-known bioindicator plant, having been extensively used to monitor O_3 levels in the United States (Feder et al. 1975; Heck and Heagle 1970) and other countries (Ashmore et al. 1980; Bell and Cox 1975; Goren and Donagi 1980; Horsman 1981).

Plants that respond predictably to air pollutants can be used independently or in conjunction with mechanical monitors to biomonitor the presence and amount of air pollution. The rapid response of plants to air pollution, the relatively inexpensive nature of biomonitoring versus monitoring with instruments, and the fact that biomonitors can be used in remote sites are advantages of using bioindicators.

There are also some disadvantages. First, the effects of air pollution on plants are under the influence of many factors such as climate, soil, water, pathogens, plant age, and the genetic variability of the plants (Posthumus 1976). Thus, the response of a given plant to a pollutant dose may not be consistent. Second, different pollutants can induce similar symptoms on some plants, as occurs with tipburn injury on *Pinus*. Third, the responses of plants are often difficult to standardize and may look very different from day to day as symptoms develop. Generally, bioindicators only allow us to determine that a given pollutant was present but not the dose. Finally, bioindicators may require considerable attention to maintain or evaluate, limiting their usefulness in remote areas.

There is certainly a need for additional development of long-lived, easy-to-maintain bioindicator plants that are sensitive to specific single pollutants. In addition, developing bioindicators with a range of pollutant sensitivities would help in determination of pollutant doses. Manning and Feder (1980) describe the types of plants that can be used as starting plants for the development of bioindicators for a range of air pollutants.

Reciprocal or side-by-side grafting of sensitive and resistant genotypes offers a simple, conventional method to develop more sophisticated plant biomonitors. In much the way horticulturists have established five-in-one apple varieties, single bioindicator plants could be developed with a range of sensitivities to one or more pollutants. Reciprocal grafting has been used to demonstrate the genetic control of air pollution responses and to show the independence of shoot and root responses for eastern white pine (Dochinger and Seliskar 1965) and trembling aspen (Karnosky unpublished data).

Germplasm Preservation and Maintenance of Diversity

The evidence that air pollution is impacting the gene pool of forest trees has been reviewed by Pitelka (1988), Sinclair (1969), and Treshow (1980). Although early studies showed ecosystem impacts near point sources of pollution (Gordon and Gorham 1963; Hedgcock 1914; Scheffer and Hedgcock 1955), this discussion will focus on more subtle regional pollution. Kriebel and Leben (1981) examined eastern white pine in three field experiments in northern Ohio, a region with high SO_2 levels in recent decades. The plantations were forest genetic research sites with a range of provenances and families. Trees ranged in age from 8 to 13 years and were scored for needle yellowing, retention, and length. Seed-lot differences in sensitivity to air pollution were statistically significant. There was a significant correlation with symptoms and the estimated atmospheric sulfur loading of the seed source. More trees of the western, low ambient SO_2 seed sources were affected by Ohio's relatively high SO_2 levels than were trees of east-central, high SO_2-level provenances. Since two of the plantations contained similar seed sources collected 11 years apart (1955 and 1966), it was possible to compare the available gene pool at two points in time. The finding that the second collection had virtually no sensitive genotypes while the earlier collection had many present suggests that sensitive genotypes had been lost. Based on their study, the authors speculated that $\geq 40\%$ of the original eastern white pine population in northern Ohio had been lost.

Sensitive eastern white pine genotypes are slower growing than neighboring resistant trees, resulting in increased mortality rate (Karnosky 1980,
1989). Natural mortality was 10 times higher for sensitive trees than for the
resistant trees in a mildly polluted area of southern Wisconsin over the 15-
year period from 1971 to 1986. Pollutant monitoring in this area documented O_3 to be the air pollutant most commonly found in phytotoxic
concentrations.

Results similar to those of Kriebel and Leben (1981) have been reported
for trembling aspen (*Populus tremuloides*) and red maple (*Acer rubrum*)
collected from areas with different pollution histories and then exposed to
O_3 in fumigation chambers (Berrang et al. 1986a, 1986b). Field studies
with trembling aspen suggested that the chamber and field responses to O_3
were similar (Berrang et al. 1989). As with the eastern white pine research
of Kriebel and Leben (1981), the northern Ohio trembling aspen trees
were the most tolerant. This region lies in the industrial Midwest where
pollution loading has been high for the past century. Ozone has been
shown to decrease the growth, change the root/shoot ratio, and cause
premature leaf drop in trembling aspen trees (Keller 1988; Wang et al.
1986). Clonal differences in sensitivity were found in these studies. Keller
(1988) concluded that the physiological stresses caused by O_3 might determine survival in O_3-polluted areas with severe competition.

Changes in the diversity of managed stands is causing concern in Europe
today. The situation is particularly significant because of the long history of
forest cultivation and the limited natural areas where germplasm has been
preserved. Several European scientists have cautioned about the dangers
of germplasm loss (Anonymous 1985; Gregorius 1986; Gregorius et al.
1985). Numerous population studies have been conducted to demonstrate
that gene frequencies can be changed by pollution-induced natural selection. For example, enzymatic studies by Mejnartowicz (1983) showed that
the F_1 generation was impoverished in some genes and genotypes for Scot's
pine trees growing in a Polish forest impacted by fluoride and SO_2. Scholz
and his coworkers have presented evidence for isozyme differences in
Norway spruce (*Picea abies*) trees that can be interpreted as pollution-
caused change (Scholz and Bergmann 1984; Bergmann and Scholz 1985,
1987, 1989). Since Norway spruce (*Picea abies*) is one of the most sensitive
tree species in Central Europe as well as one of the most economically
important, continued efforts to further understand the population structure of this species are needed.

Two additional species of concern to German foresters are European
beech (*Fagus sylvatica*) and European silver fir (*Abies alba*). Muller-Stark
(1985, 1989) presented evidence for pollution-induced loss of multiplicity
in European beech as determined by isozyme studies. The lack of natural
variability in European silver fir has been theorized as a cause of this
species' generally high degree of susceptibility to pollution-related forest
decline or other stresses (Larsen 1986). More recent work by Larsen et al.

(1988) suggests that the more southern provenances of European silver fir are generally more SO_2 resistant. These same provenances are generally thought to be more variable than the northern provenances, which are more seriously impacted by forest decline.

Elevated SO_2 levels are also suspected to be affecting tree populations in several other locations in Europe. The scarcity of Scot's pines in the industrial Pennines has been correlated with elevated SO_2 levels (Farrar et al. 1977). The forests of the mountainous regions along the borders of Poland and Czechoslovakia, the Tatra and Sudetan regions, harbor two regions of European larch (*Larix decidua*) that are severely impacted by pollution (Mazurski 1986). This particular problem has taken on significance beyond these mountains as these two seed sources are in demand from larch growers in other parts of Western Europe, Great Britain, and eastern North America. Seedlings from this region have consistently performed better than those from other sources. Since European larch is endemic to such a limited area of Europe, this problem generates a concern over air-pollution impact on the extinction of two of the five regional populations of the species.

Unmanaged Ecosystems

Biodiversity

The question of biological diversity being threatened by air pollution assumes major importance for natural, unmanaged ecosystems. In North America, these areas include state and federal preserves, parks, and wilderness areas. These areas often contain rare and endangered vegetation types or untapped gene pools for commercial or noncommercial plant species. Woodwell (1989) proposed that there is a mandate to ensure that air pollution does not seriously impact these areas. Some ecologists and environmentalists would argue than any pollution-related genetic change is undesirable in wilderness ecosystems. However, economists quickly point out that there is a very high cost to protect wilderness ecosystems from air pollution. A recent conference to address this issue for Class I wilderness areas of National Forests provided initial estimates of susceptibility to critical loadings for sulfur, nitrogen, and ozone (Fox et al. 1989). This conference highlighted the paucity of information on air-pollution sensitivity of noncommercial plant species in wilderness areas and focused attention on the difficult task of a land manager in determining the potential impacts of new emission sources on natural ecosystems.

Although there are numerous examples to illustrate this problem of air-pollution impacts in natural areas, three will be discussed: decline of high elevation red spruce (*Picea rubens*) in the northeastern United States, deterioration of birch (*Betula* sp.) stands along the coast of New England

and eastern Canada, and mortality in rare pine (*Pinus* sp.) stands in the mountains surrounding Mexico City, Mexico.

Red spruce decline has been reported at high elevations in Vermont (Siccama et al. 1982; Vogelmann et al. 1985), New York (Scott et al. 1984), and North Carolina (Bruck 1984). At all three locations, the sequence of symptoms appears to be the same. There is abscission of early-age leaves preceding loss of older foliage, and progression of defoliation from branch tips and crown, culminating in death of the trees. Large acreages at some higher elevation sites are showing spruce-decline symptoms, suggesting that spruce will no longer be a major component of these areas. Although the cause has not yet been determined climatic stress, acid rain, acidic fog, and ozone have all been suggested, operating either by direct foliar effects or through soil-mediated changes (Johnson 1983; McLaughlin and Norby, Chapter 4, this volume.).

The deterioration of birch stands along the eastern North American coast was first reported by the Canadian Forest Insect and Disease Survey in 1979 (Magasi 1985). The first region impacted was along the north shore of the Bay of Fundy in New Brunswick. Subsequent observations of similar symptoms were made along the coast of Maine. Although a positive causal agent has not yet been identified, the highly acidic nature of the fog in this area is suspected of being a major contributing factor to this problem (Cox et al. 1989). The fact that O_3 affects the type and amount of leaf waxes present has also been speculated to contribute to this problem because of the likelihood that affected leaves would be more susceptible to acidic precipitation (K. Percy, Canadian Forestry Service, personal communication).

The mountainous regions of Mexico contain many unique pine species and ecotypes. These trees occur in limited natural ranges and are threatened by degradation of air quality. The mountains surrounding Mexico City are experiencing losses of pines at an accelevating rate (Tovar 1989) in a situation similar to the smog impact in the San Bernardino Mountains of southern California (Miller 1973; Miller et al. 1963).

Although it is difficult to justify concern for any of these three situations based on economic values, it is proposed that alterations in community structure and the loss of biological diversity are serious consequences, particularly in wilderness areas. The loss of diversity resulting in simplification of ecosystems is expected to affect stability (Woodwell 1970; Smith 1974; Klein and Perkins 1988), although this hypothesis is difficult to prove.

Implications of Germplasm Loss

The effect of air pollution in reducing genetic variability has been raised as a concern (Gregorious et al. 1985; Treshow 1980). It is believed that the loss of variability will result in reduced natural adaptability of forest tree populations. In isolated situations, this loss of adaptability could result in local extinction of some populations. Furthermore, the predicted global climate changes (Houghton and Woodwell 1989) could affect the survival of populations with reduced adaptability.

The lack of genetic variability in parts of the range of silver fir in Europe is thought to be a contributing factor in the species' decline. The geographic distribution of silver fir dieback is closely related to that part of the natural range (Central Europe) where the species is characterized by almost no variation among provenances and a small ecological adaptability as measured by a number of ecophysiological measurements following experiments of SO_2 exposure and/or drought (Larsen 1986; Larsen et al. 1988). In contrast, no dieback symptoms are reported from southern regions where silver fir shows pronounced variation between seed sources and a high adaptability. Larsen suggests that the limited genetic variability has predisposed large portions of the silver fir population to air pollution or other stresses. Red spruce, which is suffering from forest decline in the United States (Scott et al. 1984; Siccama et al. 1982; Vogelmann et al. 1985), is another species with a very low amount of natural genetic variability (Eckert 1989). The relationship of this lack of genetic variability and the limited potential to adapt to environmental stress and change for silver fir and red spruce needs to be explored further.

Besides the reduced adaptability, a major concern is the loss of the genes or alleles themselves as sensitive genotypes are lost. This concept is a highly theoretical one and very difficult to prove. Intuitively, it would seem that the loss of a small number of highly sensitive genotypes should not be as important as losing a high proportion of the population, as has been suggested to have occurred in northern Ohio with eastern white pine (Kriebel and Leben 1981). However, if the sensitive genotypes have unique genes or unique alleles for air-pollution sensitivity, then these genotypes may also have other unique alleles that have some role in the growth, development, or reproduction of the population.

To better understand this problem, it is useful to examine the ecological methods for studying germplasm loss; isozyme analysis is the most frequently used technique. The relative frequency of occurrence of various alleles in a population, the relative proportion of heterozygosity, the numbers of alleles per loci and calculated genetic distances are useful in determining the genetic diversity in populations. Scholz and Bergmann (1984) compared the allelic composition at four enzyme loci in a number of Norway spruce clones selected for differences in SO_2 sensitivity; they developed a model of loss of allelic variation by directed selection (Bergmann and Scholz 1985), which was subsequently tested (Bermann and Scholz 1989). The results suggest that a certain but unknown amount of Norway spruce genetic information runs a risk of being lost due to air pollution stress. Mejnartowicz (1983) demonstrated selection pressure caused by SO_2 and fluorides in Scot's pine trees in Poland using isozyme analysis. Resistant trees had different allelic frequencies for the acid phosphatose isoenzyme than did sensitive individuals, and lower heterozygosity was noted in the F_1 generation, suggesting a loss of diveristy.

The complex nature of genetic changes that occur in populations impacted by air pollution is illustrated by the study of European beech in central Europe in which greater heterogeneity was measured in resistant beech trees (Muller-Stark 1985). The resistant cohorts also contained about one-third more different alleles and genotypes. The author interpreted the observations to be a response to the increasingly complex stressed environment in which adaptability was favored by increased heterozygosity of individual trees. Similar results were reported for Scot's pine families in a field trial stressed by air pollution in northern Germany (Geburek et al. 1987) in which genic and genotype multilocus diversities were about 2.5 times greater in tolerant versus sensitive trees. These two situations support the hypothesis of heterozygotes being superior in a heterogeneous environment as proposed by Lerner (1954). The study by Geburek et al. (1987) provided evidence for selection against a specific glutamate dehydrogenase allele, and he speculated that it may be lost if the selection pressure continues.

There is additional need for study in this area of air pollution-induced germplasm loss. The conflicting results which suggest loss of heterozygosity in Scot's pine (Mejnartowicz 1983) and Norway spruce (Bergmann and Scholz 1985; Scholz and Bergmann 1984) but increased heterozygosity in European beech (Muller-Stark 1985) and Scot's pine (Geburek et al. 1987) illustrate the complexity of genetic questions related to environmental stresses and forest trees. Species, environment, study and stress type, and interaction of multiple stresses all contribute to the difficulty in making comparisons between studies.

Methods for Germplasm Preservation

The task of maintaining diversity for future generations is an extremely complex one, especially when dealing with a widespread selection force such as air pollution. The most desirable approach would be to eliminate air pollution problems at their sources. It is apparent that alternate measures must be taken in heavily impacted areas such as central Europe. For example, it has been estimated that forest decline in Germany extends to $\geq 50\%$ of the trees of the major forest species and to $\geq 80\%$ of the silver fir (Muhs 1989). For this reason, the Federal Republic of Germany has initiated a large-scale program to conserve forest gene resources, developing protocols for conservation programs for the priority species. The proposed directions include stand preservation, promotion of natural stand regeneration, seeding and planting in or out of impacted areas, orchard establishment in nonimpacted areas, and preservation of pollen, seed or plant parts in cold storage (Muhs 1989). While many of the concepts and techniques for forest gene resource preservation have been developed through traditional forest genetics and tree improvement programs, the task of preserving diversity is far more difficult than simply preserving a few selected genotypes as occurs in most tree improvement programs (Ledig 1988).

Other European countries have also initiated conservation programs to protect forest resources from air pollution. Lithuanian scientists have been practising the preservation of forest genetic resources in polluted areas since the early 1970s (Vaicys and Armolaitus 1986), having established nearly 200 hectares of plantations of threatened trees in clean-air locations. In Czechoslovakia, researchers have been taking similar measures to preserve Norway spruce that is under severe air pollution stress in the Ore Mountains (Kanak 1986). They have established protected seed stands in impacted areas and developed clonal seed orchards in clean-air sites. They have also resorted to germinating seed from arboretum seed archives to reestablish some populations previously lost.

Future Research

Conventional Research Needs and Applications

Needs for research on the genetic effects of air pollution on forest tree populations were discussed at a joint meeting of the IUFRO Working Parties "Population and Ecological Genetics," "Biochemical Genetics," and "Genetic Aspects of Air Pollution" (Scholz et al. 1989). The main concern addressed at the meeting was the need for more information on methods to detect and quantify genetic change caused by air pollution (Karnosky et al. 1989). Additional priorities included the need to (1) quantify air pollution-induced genetic losses, (2) determine changes in adaptability in plant populations, (3) determine consequences of air pollution-induced loss of adaptability, and (4) develop strategies for the conservation of forest genetic resources.

Although the majority of the forests in Europe have been managed for hundreds of years so that there are few natural populations remaining, the opportunity exists in North America to study evolutionary effects of air pollution on natural populations for acreages have been preserved as wilderness or unmanaged areas. To this end, research involving isozymes to study variability in populations and to examine heterozygosity of populations from polluted and nonpolluted areas is needed. Little is known about the narrow-sense heritability of pollution resistance in any tree species (Roose, Chapter 5, this volume). Also needed are studies to characterize the physiological costs of resistance, effects of air pollution on competition between sensitive and resistance genotypes, the mechanisms and gene action controlling pollution resistance and air pollution effects on in situ reproduction (sexual or asexual).

Biotechnological Applications

The newly developing biotechnological methods for cell and tissue culture, gene isolation, and gene transfer have potential for use in the study of genetic effects of air pollution. In vitro culture of cells, tissues, or protoplasts offer opportunity for screening pollution resistance, examining biochemical responses, and explaining mechanisms of action (Grimes et al. 1983; Illman and Pell 1985; Sigal et al. 1988). Tanaka et al. (1988) found increased SO_2 tolerance of tobacco plants regenerated from paraquat-resistant callus, suggesting in vitro screening for traits related to air pollution could result in improved resistance.

Cell and tissue culture may also be used to mass propagate selected resistant or sensitive genotypes and cryopreservation of germplasm of endangered genotypes. Similarly, it may be a useful technique for studying gene action at the cellular level and isolating genes controlling air pollution resistance. Ideal candidate genes would be those involved in stomatal closure in the presence of O_3 that has been detected in onion (*Allium cepa*) (Engle and Gabelman 1966) or the few major genes involved in O_3 resistance in beans (*Phaseolus vulgaris*) (Butler et al. 1979).

Developing techniques for gene isolation and transfer will soon permit engineering of bioindicator plants with specific responses to given doses of air pollution. Similar, advances will permit resistance genes to be transferred from one species to another to improve the level of resistance in important species that are particularly sensitive to air pollution.

However, one must consider that this field is only in its infancy and that a considerable amount of research is necessary before gene isolation, transfer, and in vitro manipulation are very routinely used in problem solving applications.

Conclusions

There are many applications of ecological genetics related to air pollution stress. It is possible to select individual plants to serve as biomonitors for pollution; these are particularly helpful in monitoring in remote areas. It is also possible to select and/or breed pollution-resistant plants with the understanding, that priority be given to eliminating pollution rather than developing pollution-resistant plants. While additional studies are needed to document the impacts of air pollution on germplasm loss and diversity, studies cited in this paper indicate that air pollution stress, alone or in combination with other stresses, has eliminated sensitive genotypes from forest-tree populations in some areas.

References

Anonymous (1985) Forest genetic material endangered by aerial pollution. Silvae Genetica 34:49–50

Ashmore MR, Bell JNB, Reily CL (1980) The distribution of phytotoxic ozone in the British Isles. Environmental Pollution (Series B) 1:195–216

Bell JNB, Cox RA (1975) Atmospheric ozone and plant damage in the United Kingdom. Environmental Pollution 8:163–170

Bell JNB, Ayazloo M, Wilson GB (1982) Selection for sulphur dioxide tolerance in grass populations in polluted areas. In: Bornkamm R, Lee JA, Seaward MRD (eds) Urban Ecology. Blackwell Scientific, Oxford, pp 171–180

Bergmann F, Scholz F (1985) Effects of selection pressure by SO₂ pollution on genetic structures of Norway spruce (Picea abies). In: Gregorius HR (ed) Population genetics in forestry; Lecture notes in biomathematics, Springer-Verlag, Berlin, pp 267–275

Bergmann F, Scholz F (1987) The impact of air pollution on the genetic structure of Norway spruce. Silvae Genetica 36:80–83

Bergmann F, Scholz F (1989) Selection effects of air pollution in Norway spruce (Picea abies) populations. In: Scholz F, Gregorius HR, Rudin D (eds) Genetic aspects of air pollutants in forest tree populations. Springer-Verlag, Berlin, pp 143–160

Berrang P, Karnosky DF, Mickler RA, Bennett JP (1986a) Natural selection for ozone tolerance in Populus tremuloides. Canadian Journal of Forest Research 16:1214–1216

Berrang P, Karnosky DF, Mickler RA, Bennett JP (1986b) Population changes in eastern hardwoods caused by air pollution. Proceedings of the North American Forest Biology Workshop, pp 3–10

Berrang P, Karnosky DF, Bennett JP (1989) Natural selection for ozone tolerance in Populus tremuloides. II. Field verification. Canadian Journal of Forest Research 19:519–522

Bialobok S (1980) Studies on the effect of sulphur dioxide and ozone on the respiration and assimilation of trees and shrubs in order to select individuals resistant to action of these gases. Polish Academy of Sciences, Report P1-FS 74

Bialobok S, Karolewski P, Oleksyn J (1980) Sensitivity of Scot's pine needles from mother trees and their progenies to the action of SO₂, O₃, a mixture of these gases, NO₂ and HF. Aboretum Kornickie 25:289–303

Bruck RI (1984) Decline of montane boreal ecosystems in central Europe and the southern Appalachian mountains. Proceedings of the Technical Association of the Pulp and Paper Industry 159–163

Butler LK, Tibbitts TW, Bliss FA (1979) Inheritance of resistance to ozone in Phaseolus vulgaris L. Journal of the American Society of Horticultural Science 104:211–213

Cox RM, Spavold-Tims J, Hughes RN (1989) Coastal white birch deterioration in areas receiving acid fog and other pollutants around the Bay of Fundy, Canada. In: Bucher JB and Bucher-Wallin I (eds) Air pollution and forest decline, vol 2. EAFV, Birmensdorf, Switzerland. pp 393–396

Davis DD, Coppolino JB (1974) Relative ozone susceptibility of selected woody ornamentals. HortScience 9:537–539

Davis DD, Umbach DM, Coppolino JB (1981) Susceptibility of tree and shrub species and response of black cherry foliage to ozone. Plant Disease 65:904–907

Dochinger LS, Seliskar CE (1965) Results from grafting chlorotic dwarf and healthy eastern white pine. Phytopathology 55:404–407

Eckert RT (1989) Genetic variation in red spruce and its relation to forest decline in the Northeastern United States. In: Bucher JB and Bucher-Wallin I (eds) Air pollution and forest decline, vol 1. EAFV, Birmonsdorf, Switzerland pp 319–324

Engle RL, Gabelman WH (1966) Inheritance and mechanism for resistance to ozone damage in onion (*Allium cepa* L.) American Society of Horticultural Science 89:423–430

Farrar JF, Relton J, Rutter AJ (1977) Sulphur dioxide and the scarcity of *Pinus sylvestris* in the industrial pennines. Environmental Pollution 14:63–68

Feder WA, Kelleher TJ, Riley WD, Perkins I (1975) Ozone injury on tobacco plants on Nantucket Island is caused by long-range transport of ozone from the mainland. Proceedings of the American Phytopathologists Society 2:97

Fowler D, JN Cape (1982) Air pollutants in agriculture and horticulture. In: Unsworth MH, DP Ormrod (eds) Effects of gaseous air pollution in agriculture and horticulture. Butterworth, London, pp 3–26

Fox DG, Bartuska AM, Byrne JG, Cowling E, Fisher R, Likens GE, Lindberg SE, Linthurst RA, Messer J, Nichols DS (1989) A screening procedure to evaluate air pollution effects on Class I wilderness areas. USDA Forest Service Gen Tech Rep RM-168

Garsed SG, Rutter AJ (1982) Relative performance of conifer populations in various tests for sensitivity to SO_2 and the implications for selecting trees for planting in polluted areas. New Phytologist 92:349–367

Geburek T, Scholz F, Knabe W, Vornweg A (1987) Genetic studies by isozyme gene loci on tolerance and sensitivity in an air polluted *Pinus sylvestris* field trial. Silvae Genetica 36:49–53

Gordon AG, Gorham E (1963) Ecological aspects of air pollution from an iron-sintering plant at Wawa, Ontario. Canadian Journal of Botany 41:1063–1078

Goren AI, Donagi AE (1980) Assessment of atmospheric ozone levels in Israel through foliar injury to Bel-W3 tobacco plants. Oecologia 44:418–421

Gregorius HR (1986) The importance of genetic multiplicity for tolerance of atmospheric pollution. Proceedings of the 18th IUFRO World Congress, Division 2, vol 1. Ljubljana, Yugoslavia, pp 295–305

Gregorious HR, Hattemer HH, Bergmann F, Muller-Starck G (1985) Umweltbelastung und Anpassungsfähigkeit von Baumpopulationen. Silvae Genetica 34:230–241

Grimes HD, Perkins KK, Boss WF (1983) Ozone degrades into hydroxyl radical under physiological conditions: a spin trapping method. Plant Physiology 72:1016–1020

Heck WW, Heagle AS (1970) Measurement of photochemical air pollution with a sensitive monitoring plant. Journal of the Air Pollution Control Association 20:97–99

Hedgcock, GG (1914) Injury by smelter smoke in southeastern Tennessee. Journal of the Washington Academy of Science 4:70–71

Horsman DC (1981) A survey of ozone in Melbourne using tobacco as an indicator plant. Environmental Pollution (Series B) 2:69–77

Houghton RA, Woodwell GM (1989) Global climatic change. Scientific American 260:36–44

Illman BL, Pell EJ (1985) Characterization of ozone response of potato leaf protoplasts. Canadian Journal of Botany 63:1936–1941

Johnson AH (1983) Red spruce decline in the Northeastern US: Hypotheses regarding the role of acid rain. Journal of the Air Pollution Control Association 33:1049–1054

Kabala SJ (1989) The economic effects of sulfur dioxide pollution in Poland. Ambio 18:250–251

Kanak K (1986) Possibilities for using species of *Pinus* for afforestation at emission areas. Proceedings of the 18th IUFRO World Forestry Congress, Division 2, vol 1. Ljubljana, Yugoslavia, pp 39–47

Karnosky DF (1980) Changes in southern Wisconsin white pine stands related to air pollution sensitivity. In: Miller PR (ed) Proceedings of the Symposium on the Effects of Air Pollutants on Mediterranean and Temperate Forest Ecosystems. USDA Gen Tech Rep PSW-49, p 238

Karnosky DF (1981) Chamber and field evaluations of air pollution tolerances of urban trees. Journal of Arboriculture 7:99–105

Karnosky DF, Steiner KC (1981) Provenance and family variation in response of *Fraxinus americana* and *F. pennsylvanica* to ozone and sulfur dioxide. Phytopathology 71:804–807

Karnosky DF (1989) Air pollution induced population changes in North American forests. In: Bucher JB, Bucher-Wallin I (eds) Air pollution and forest decline, vol 1. EAFV, Birmonsdorf, Switzerland, pp 315–317

Karnosky DF, Scholz F, Geburek T, Rudin D (1989) Implications of genetic effects of air pollution on forest ecosystems—knowledge gaps. In: Scholz F, Gregorius HR, Rudin D (eds) Genetic aspects of air pollutants in forest tree populations. Springer-Verlag, Berlin, pp 199–201

Keller T (1988) Growth and premature leaf fall in American aspen as bioindications for ozone. Environmental Pollution 52:183–192

Klein RM, Perkins TD (1988) Primary and secondary causes and consequences of contemporary forest decline. Botany Review 54:1–43

Knabe W (1976) Effects of sulfur dioxide on terrestrial vegetation. Ambio 5:213–218

Kriebel HB, Leben C (1981) The impact of air pollution on the gene pool of eastern white pine. Proceedings of the 17th IUFRO World Congress, Division 2. Kyoto, Japan, pp 185–189

Larsen JB (1986) Silver fir decline: a new hypothesis concerning this complex decline syndrome in *Abies alba* (Mill.) Forstwirtschaft Centralblatt 105:381–396

Larsen JB, Qian XM, Scholz F, Wagner I (1988) Ecophysiological reactions of different provenances of European silver fir (*Abies alba* Mill.) to SO_2 exposure during winter. European Journal of Forest Pathology 18:44–50

Ledig FT (1988) The conservation of diversity in forest trees. BioScience 38:471–479

Lerner IM (1954) Genetic homeostatis. Oliver and Boyd, Edinburgh

Magasi LP (1985) Forest pest conditions in the Maritimes in 1985. Information Report M-X-159. Canadian Forestry Service—Maritimes, pp 29–32

Manning WJ, Feder WA (1980) Biomonitoring air pollutants with plants. Applied Science Publishers, London

Mazurski KR (1986) The destruction of forests in the Polish Sudetes Mountains by industrial emissions. Forest Ecology Management 17:303–315

Mejnartowicz LE (1983) Changes in genetic structure of Scots pine (*Pinus silvestris* L.) population affected by industrial emission of fluoride and sulphur dioxide. Genetica Pollonica 24:41–47

Miller PR (1973) Oxidant-induced community change in a mixed conifer forest. Advances in Chemistry 122:101–117

Miller PR, Parmeter JR, Taylor OC, Cardiff EA (1963) Ozone injury to the foliage of ponderosa pine. Phytopathology 53:1072–1076

Muhs HJ (1989) Measures for the conservation of forest gene resources in the Federal Republic of Germany. In: Scholz F, Gregorious HR, Rudin D (eds) Genetic aspects of air pollutants in forest tree Populations. Springer-Verlag, Berlin, pp 187–198

Muller-Starck G (1985) Genetic differences between "tolerant" and "sensitive" beeches (*Fagus sylvatica* L.) in an environmentally stressed adult forest stand. Silvae Genetica 34:241–247

Muller-Starck LG (1989) Genetic implications of environmental stress in adult forest stands of *Fagus sylvatica* L. In: Scholz F, Gregorius HR, Rudin D (eds) Genetic aspects of air pollutants in forest tree Populations. Springer-Verlag, Berlin, pp 127–142

Paces T (1985) Sources of acidification in Central Europe estimated from elemental budgets in small basins. Nature 315:31–36

Pitelka LF (1988) Evolutionary response of plants to anthropogenic pollutants. Trends in Ecology and Evolution 3:233–236

Posthumus AC (1976) The use of higher plants as indicators for air pollution in the Netherlands. In: Karenlampi L (ed) Proceedings of the Kuopio Meeting on Plant Damage Caused by Air Pollution. Kuopio, Finland, pp 115–120

Prinz B, Krause GHM, Jung KD (1987) Development and causes of novel forest decline in Germany. In: Hutchinson TC, Meema KM (eds) Effects of atmospheric pollutants on forests, wetlands and agricultural ecosystems. Springer-Verlag, Berlin, pp 1–24

Rohmeder E, Schonborn UA (1965) Der Einfluss von Umwelt und Erbgut auf die Widerstandsfähigkeit der Waldbäume gegenüber Luftverunreinigung durch Industrieabgase. Forstwirtschaft Centralblatt 84:1–13

Scheffer TC, Hedgcock GG (1955) Injury to northwestern forest trees by sulfur dioxide from smelters. USDA Technical Bulletin 1117

Scholz F, Gregorius HR, Rudin D (eds) (1989) Genetic effects of air pollutants in forest-tree populations. Springer-Verlag, Berlin

Scholz F, Bergmann F (1984) Selection pressure by air pollution as studied by isozyme-gene-systems in Norway spruce exposed to sulphur dioxide. Silvae Genetica 33:238–241

Scott JT, Siccama TG, Johnson AH, Breisch AR (1984) Decline of red spruce in the Adirondacks. New York, Bulletin of the Torrey Botany Club 111:438–444

Siccama TG, Bliss M, Vogelmann H (1982) Decline of red spruce in the Green Mountains of Vermont. Bulletin Torrey Botany Club 109:162–168

Sigal LL, Eversman S, Berglund DL (1988) Isolation of protoplasts from loblolly pine needles and their flow-cytometric analysis for air pollution effects. Environmental and Experimental Botany 28:151–161

Sinclair WA (1969) Polluted air: potent new selective force in forests. Journal of Forestry 69:305–309

Smith WH (1974) Air pollution. Effects on the structure and function of the temperate forest ecosystem. Environmental Pollution 6:111–129

Tanaka K, Furusawa I, Kondo N, Tanaka K (1988) SO$_2$ tolerance of tobacco plants regenerated from paraquat-tolerant callus. Plant Cell Physiology 29:743–746

Tovar, DC (1989) Air pollution and forest decline near Mexico City. Environmental Monitoring Assessment 12:49–66

Townsend AM, Dochinger LS (1974) Relationship of seed source and development stage to the ozone tolerance of *Acer rubrum* seedlings. Atmospheric Environment 8:957–964

Treshow M (1980) Pollution effects on plant distribution. Environmental Conservation 7:279–286

Vaicys M, Armolaitis K (1986) Gas resistance and regeneration of forests damaged by industrial emission. Proceedings of the 18th IUFRO World Forestry Congress, Division 2, vol 1. Ljubljana, Yugoslavia, pp 360–367

Vogelmann HW, Badger GJ, Bliss M, Klein RM (1985) Forest decline on Camels Hump, Vermont. Bulletin Torrey Botany Club 112:274–287

Wang D, Karnosky DF, Bormann FH (1986) Effects of ambient ozone on the productivity of *Populus tremuloides* Michx. grown under field conditions. Canadian Journal of Forestry Research 16:47–55

Woodwell GM (1970) Effects of pollution on the structure and physiology of ecosystems. Science 168:429–433

Woodwell GM (1989) On causes of biotic impoverishment. Ecology 70:14–15

Commentary to Chapter 11(A)

Plant Ecological Genetics and Air Pollution Stress: A Commentary on Implications for Natural Populations

DAVID J. PARSONS and LOUIS F. PITELKA

Introduction

The concept of preserving examples of natural ecosystems has traditionally focused on protecting native biotic communities from the direct impacts of local human activities (Leopold et al. 1963; Graber 1983). Increasing levels of atmospheric pollutants and potential global scale changes in climate, radiation, and carbon dioxide (Mooney et al. 1987; McLaughlin and Norby Chapter 4, this volume) focus attention on the inadequacy of our existing system of ecological reserves (Graham 1988). Set aside to preserve examples of native ecosystem elements and processes for the benefit of future generations (Hendee et al. 1978), most wilderness areas, national parks, and other nature preserves are not large enough to escape the impacts of regional pollution and global climate change caused by activities outside their boundaries. Managers of such areas must be cognizant of a new set of intrusions that threaten the integrity of the areas they manage.

The possibility that plant populations may evolve in response to selection pressures exerted by changes in atmospheric chemistry poses a major challenge to managers of natural areas. Important questions to be answered include:

1. What are the potential long-term consequences of evolutionary change in response to air pollutants in natural populations?
2. How does evolutionary change in response to air pollutants compare with other changes caused by human activities in terms of the seriousness of the threat to natural populations and ecosystems?
3. What options are open to land managers to deal with evolutionary changes?

It is critical to evaluate the consequences of genetic changes, and yet the scientific understanding of the phenomena is not yet adequate to make such assessments possible. This commentary reviews some of the major issues concerning implications of evolutionary changes for natural populations.

Genetic and Evolutionary Costs of Resistance

The evolution of resistance, be it tolerance or avoidance, is a primary mechanism in permitting organisms to continue inhabiting an area in the face of environmental change. In the case of pollution stress, there are now numerous documented examples of evolution of resistance in plant populations (Bradshaw and McNeilly, Chapter 2, this volume; Bell et al., Chapter 3, this volume). The evolutionary costs and consequences of such resistance are largely unknown. For example, it is not known to what extent the loss of sensitive individuals (genotypes) of Ponderosa pine in the mountains of southern California (Miller 1973) has compromised the evolutionary potential of those populations by constraining further adaptive change. The possibility that natural selection resulting from air pollution stress will lead to a significant loss of genetic diversity in sensitive species has received considerable attention in the literature (Gregorius 1986; Scholz et al. 1990) and is a major topic of discussion (Barrett and Bush, Chapter 6, this volume; Karnosky, Chapter 11, this volume). Because the potential loss of significant genetic diversity must be regarded as a major threat to the integrity of individual species and entire ecosystems, it is important to determine the likelihood of such losses.

Numerous factors will influence the extent to which selection imposed by air pollution will lead to loss of genetic diversity. Of primary importance is the strength of selection. If air pollution stress is extreme and very few individuals carry genes for resistance, then the resulting heavy mortality will cause a population to go through a bottleneck similar to what happens when a population is founded by a few colonizing individuals. In such cases significant genetic diversity may be lost (including genes whose function is unrelated to the cause of mortality) through random loss of alleles owing to genetic drift. Populations must decline to extemely small sizes to result in significant losses of alleles by drift. This will most likely happen only in areas severely affected by pollution. Thus, land managers can use the two criteria of severe pollution and high mortality linked to the pollution to identify those situations where attempts to actively preserve genetic diversity will be necessary. It is important to emphasize that any stress causing high mortality could lead to losses of genetic diversity through drift.

Another factor that will determine the likelihood of losses of genetic diversity is the number of genes responsible for resistance to the pollutant. If resistance is determined by one or a few genes, strong selection could result in the loss of sensitive genotypes (alleles) from the population. However, if this means that only one or a few alleles are lost and these do not carry linked but unrelated alleles, then the future evolutionary potential of the population is not likely to be compromised significantly. In contrast, if resistance is a quantitative trait involving many genes that are also important for traits not related to pollution stress (Roose, Chapter 5, this volume), strong selection conceivably could lead to the loss of many alleles with more serious implications. An important mitigating factor in this case is that nonadditive effects and the likelihood of negative correlations with other fitness components (Winner et al., Chapter 7, this volume) will make it extremely unlikely that many alleles would disappear from a population. If selection is weak enough so that reproduction and recombination occur before large numbers of sensitive genotypes die, the chance that large numbers of alleles will be lost is reduced further. Thus, the threat to genetic diversity might at first seem greater when a quantitative trait is involved, but the inherent complexities of such systems reduce the speed and likelihood of significant losses. Unfortunately, we do not often know how many genes are involved in most forms of resistance to anthropogenic pollutants.

Even if selection for resistance to changes in atmospheric chemistry does not result in a significant loss of genetic diversity, there are other reasons that managers of natural ecosystems might be concerned about such evolutionary changes. In the case of national parks, some have argued that any changes caused by human activities are undesirable and, where possible, should be prevented (Graber 1983). This viewpoint must be reconciled with the realization that evolutionary changes may make populations better able to survive the presence of inescapable pollution.

Another more fundamental issue concerns the potential physiological and, ultimately, the fitness costs of resistance. Adaptations that confer resistance may have costs associated with them that make individuals less able to survive other stresses or less able to compete when the pollutant stress is absent (Winner et al., Chapter 7, this volume). Costs of adaptation to pollutants probably vary depending on the mechanism and likely will vary among co-occurring resistant species all exposed to the same pollutant. These differences may alter competitive balances and species composition.

Since the detailed ramifications for natural ecosystems of the loss of sensitive genotypes and the evolution of pollutant resistance are largely unknown, the prudent approach appears to be one of caution coupled with expanded research and monitoring. An increased emphasis on studying the effects of the loss of alleles or genotypes on the vigor, competitive interactions, and future adaptability of sensitive species is needed. Since such studies can seldom be adequately carried out solely under greenhouse or fumigation chamber conditions, consideration should be given to utilizing our few remaining natural areas for long-term observation (Parsons 1989). Geographic areas of limited genetic diversity, as identified through surveys of isozyme polymorphism, should be given special emphasis. Such information might prove especially valuable in predicting areas where loss of species might occur.

Evolution of Tolerance vs Ecological Replacement

Another important consideration is that the greatest threats to both the genetic and ecological integrity of natural ecosystems are imposed by ecological rather than evolutionary processes. Rates of genetic change within a population are largely irrelevant if the population is rapidly going extinct and being replaced by other more resistant species. Thus, differences in sensitivity to pollution among species and consequent shifts in competitive balances within a community may result in rapid population decline or extinction of the most sensitive species. In the face of predicted rapid global scale changes in climate and other environmental factors, the resulting ecological changes may overwhelm more subtle shifts in genetic makeup. Ecological changes due to a changing global environment can be expected to include effects on physiology, competition, plant-disease and plant-animal interactions, and nutrient cycling. Although the ultimate effects on species distributions and ecosystem processes are unknown, it seems clear that the greatest threats to the genetic diversity and integrity of natural systems will result from the ecological extinction of entire populations or species rather than from evolutionary changes within populations.

From a management perspective, one must evaluate to what extent managers need to be concerned with the costs of the evolution of resistance to pollutants when the affected ecosystems are faced with potentially overwhelming shifts in species distributions and competitive interactions. The lack of basic information both on projected atmospheric changes and on predicted ecological and genetic consequences of those changes inhibits the development of accurate predictive models. Yet, even should such models exist, the paucity of long-term inventory, monitoring and research programs in undisturbed ecosystems would limit the ability to test them.

Hands-on Mitigation

A number of hands-on management actions can be used to mitigate the effects of changing atmospheric chemistry on terrestrial vegetation. These include the breeding or selection of pollution resistant plants, planting of resistant genotypes or species, artificial seeding or promotion of natural regeneration, and preservation of seeds and pollen. Whereas such methods are common in managed ecosystems (Karnosky, Chapter 11, this volume), they are of more limited application to natural areas. The concept of intensive management, including the introduction of non-native genetic material counters the policy mandate of the National Park Service and other wilderness management agencies (Christensen 1989). Nevertheless, on a limited basis, such actions may well be appropriate. Should it be shown that selection for pollution-resistant genotypes results in significant losses of genetic diversity or other undesirable changes in the gene pools of native populations, the preservation of seeds and pollen would assure that genetic material is preserved for possible future use. Should a sufficient number of individuals be lost due to selection against sensitive genotypes, land managers may be faced with deciding whether to introduce pollution-resistant genotypes (or in some cases even new, resistant species) as opposed to allowing nature to determine what individuals or species will colonize vacated niches. In addition to questions of genetic integrity, such decisions may need to consider the loss of visual resource quality from allowing dead and dying trees to persist and the impact on the public's perception and appreciation of natural ecosystems if alien species have been intentionally planted. The most acceptable compromise would likely be the use of locally collected resistant genotypes of the same species that is being impacted.

Other Issues

A number of other issues regarding the effects and implications of changes in atmospheric chemistry on natural plant communities deserve attention. Resolution of many of these depend in large part on scientific unknowns as well as questions of policy and practicality.

One of the principal questions that must be addressed is how to determine the acceptability of different types and levels of change in natural ecosystems. This is primarily a question of policy. It must reflect societal value judgments as well as practicality. Operational definitions of "natural" and "acceptability" must reflect compromises between ideals and what is practical and possible. In a world of burgeoning population and increasing global pollution, idealism seems to take the back seat with increasing frequency.

In evaluating potential impacts of atmospheric pollution on vegetation, we must be cautious of generalizations and extrapolations. Since most research has focused on a limited number of species, it is virtually impossible to predict the full ecological ramifications of increasing pollution on any wilderness ecosystem. In general, no more than a couple of species have been studied from any given ecosystem. The temptation to extrapolate a limited knowledge to a wide array of species and ecosystems must be controlled. Similarly, caution must be used in generalizing among growth forms. Annual plants, for example, can be expected to have a much more rapid rate of evolution, and thus an ability to adapt to changing conditions, than do long-lived trees (Barrett and Bush, Chapter 6, this volume). Caution must also be used in generalizing between pollutants or geographic areas. Responses to SO_2, CO_2, and heavy metals, for example, often vary markedly for the same species. Similarly, differences in land use history between Europe and North America create very different situations. Many intensively managed European forests have unnaturally low genetic diversity, having been planted from selected seed source, and thus may be more susceptible to environmental stresses.

The concept of using pollution-sensitive bioindicator plants to detect specific air pollutants in remote areas has some potential for alerting land managers of imminent problems. Karnosky (Chapter 11, this volume) reviews the potential and limitations of biomonitors.

Conclusions

Despite considerable uncertainty as to the details of future changes, it is apparent that changing atmospheric chemistry will continue to place considerable stress on many natural ecosystems. Although the implications of this changing environment for the future of wilderness, parks, and other natural areas are largely speculative, there is little doubt that it poses a serious problem. If predictions of global warming and consequent shifting species distributions come true, management agencies will be faced with new requirements for understanding ecosystem dynamics. As some species, or perhaps entire communities migrate in response to changing environmental and competitive stresses, while others flourish or suffer, the challenge of ecosystem management (Agee and Johnson 1988) will be severely tested. Expanded research knowledge, increasingly active and sophisticated management actions, and more effective public education will be needed. In addition, land management agencies charged with managing natural areas will need to become more knowledgeable of and responsive to new advances in knowledge of genetics, physiology, plant growth, community interactions, and ecosystem dynamics.

This discussion indicates that land managers probably need not be overly concerned about effects of selection for resistance to pollution, especially in situations where pollution is not extreme and mortality rates are low. However, our understanding of the evolutionary responses of plant populations to pollution and other stresses is still meager. It is critical to carefully follow those situations where air pollution is significantly affecting natural ecosystems and to undertake the necessary genetic studies to determine whether and how rapidly selection for resistance is occurring. In addition, land managers should be aware that the evolution of resistance to pollutants may involve undesirable effects other than loss of genetic diversity.

References

Agee JK, Johnson DR (1988) Ecosystem management for parks and wilderness. University of Washington Press, Seattle, Washington

Christensen NL (1989) Wilderness and natural disturbance. Forum for Applied Research and Public Policy 4(2):46–49

Graber DM (1983) Rationalizing management of natural areas in national parks. The George Wright Forum 3(4):48–56

Graham RW (1988) The role of climatic change in the design of biological preserves: the paleoecological perspective for conservation biology. Conservation Biology 2:391–394

Gregorius HR (1986) The importance of genetic multiplicity for tolerance of atmospheric pollution. Proceedings of the 18th IUFRO World Congress, Ljubljana, Yugoslavia, Div. 2, vol. 1:295–305

Hendee JC, Stankey GH, Lucas RC (1978) Wilderness management. USDA Forest Service misc. Public. No. 1365

Leopold AS, Cain SA, Cottam C, Gabrielson IN, Kimball TL (1963) Wildlife management in the national parks. American Forests 69(4):32–35, 61–63

Miller PC (1973) Oxidant-induced community change in a mixed conifer forest. In: Naegle JA (ed) Air pollution damage to vegetation. Advances in Chemistry Series 122. American Chemical Society, Washington DC, pp 101–117

Mooney HA, Vitousek PM, Matson PA (1987) Exchange of materials between terrestrial ecosystems and the atmosphere. Science 238:925–932

Parsons DJ (1989) Evaluating national parks as sites for long-term studies. In: Likens GE (ed) Long-term studies in ecology. Springer-Verlag, New York, pp 171–173

Scholz F, Gregorius H-R, Rudin D (eds) (1990) Genetic aspects of air pollutants in forest tree populations, Springer-Verlag, Berlin, p 201

Ecological Genetics and Changes in Atmospheric Chemistry: A Commentary on Application of Knowledge in Managed Forest Systems

Patricia A. Layton and Alan A. Lucier

Introduction

The term "managed forest systems" encompasses many different forest types, management objectives, and silvicultural practices. Closed forests occupy only 2.8 billion hectares of the world and of that number only 1 billion hectares are managed. This includes all closed forest in the Soviet Union, which represents about three-fourths of total managed forests (World Resources 1987). In the United States there are 102 million hectares of managed forests (53% of closed forests) (U.S. Forest Service 1980).

Forest management objectives range from biomass or energy production to watershed protection, with most forests being managed for multiple purposes. Management intensities are equally diverse and range from very high on short-rotation tree farms to minimal on lands reserved for non-timber uses. The majority of forests have generally low to moderate management, consisting primarily of fire protection, timber harvest, and either natural regeneration or planting. Parsons and Pitelka (Commentary to Chapter 11B, this volume) discuss the influences of changes in atmospheric chemistry on forests with little management.

Forest products companies and other landowners with a strong interest in timber production may use genetically improved seedlings, weed control, and fertilization on selected sites to increase yields. Foresters have a range of control over the genetic background of trees in a managed forest. In a clonal plantation, the forester may know the identity and growth characteristics of every tree, while in a naturally regenerated forest the processes of natural selection are allowed to determine the genetic composition. Most forests in the world are reproduced through natural regeneration practices. In the southeastern United States, only one-third of the pine forests are planted (Bones et al. 1989), representing almost 70% of U.S. tree planting, and in Germany, with its long history of forest management, Muhs (1989) reports that only about 30–50% of the forests are planted, mostly conifers. Frequency of weed control or fertilization on any one site is usually no greater than once or twice in a rotation of 25 to 100 years. Long rotations and limited stand management opportunities place a premium on long-range planning.

The amount and type of impact that can be achieved through applying our knowledge of ecological genetics and breeding to introduce resistance to changing atmospheric chemistry may be limited. The following discussion will highlight what options the forest manager has in managing for changing atmospheric chemistry as well as identifying barriers to management changes.

Long-Term Implications of Ozone and Acidic Deposition for Forest Genetic Resources

Managed forests in many parts of the world have suffered greatly under various pressures from human populations. Combinations of exploitive use, neglect and conversion to other land uses have seriously reduced the health and extent of managed forests in many parts of the world. Air pollutants impose additional stresses that vary greatly in intensity over space and time (McLaughlin and Norby, Chapter 4, this volume). Selection pressure from air pollutants and possible global atmospheric changes could slowly alter the genetics of forests.

Foresters have traditionally considered variability as an inherent feature of tree populations, which can be managed through selective cutting and utilized in breeding programs. From this perspective variability in tree resistance to pollution stress may be viewed as providing a measure of protection from forest damage. In some cases, variability has been reduced in plantation forestry. Reforestation in Germany after World War II is an example where limited provenances of seed were used to reforest vast acreages of forest. Forest managers must insure that as the need for yield improvement or seedlings grows, selection does not limit the gene pool for resistance or adaptability over time.

One focus of recent research on air pollutants and forests has been on the potential for reductions in current tree health and stand productivity (Barnard et al. 1990; McLaughlin and Norby, Chapter 4, this volume). Current reductions might occur through direct pollutant effects on tree tissues, winter hardiness, or drought sensitivity, or through indirect effects on site quality (e.g., soil fertility, climate). Ozone and sulfur dioxide injuries to tree foliage are the most common direct effects. Reductions in soil fertility due to acidic deposition or trace metal accumulation are the most common indirect effects. Effects from ultraviolet-B radiation and increasing carbon dioxide levels may have more far-reaching impacts than traditional pollutants. Global climate change may indeed eliminate forests from some regions and change the composition in others in a manner as drastic as the last period of glaciation.

Forests in the vicinity of point sources and large cities have been damaged through both direct and indirect effects of pollutants. Air pollutants at toxic levels can rapidly change a forest's population genetics. But what about the effect of elevated levels of an air pollutant—levels that can reduce growth but will not kill sensitive trees and eliminate them from the breeding population? The lower levels of selection pressure that may result from this type of situation may slowly over many generations change the genetic composition of forests by eliminating alleles. Trees, however, are very long-lived and may reproduce for hundreds of years whereas air pollution stress may change considerably over the same time span. Therefore, the amount of genetic change that occurs in a forest may be difficult to quantify.

Drawbacks to Implementation of Management Alternatives to Reduce the Impact of Atmospheric Change

Forest managers are frequently confronted with serious and immediate threats to forest health and productivity including: (i) pests, drought, fire, flood; (ii) soil damage by careless logging contractors and recreational vehicles; and (iii) shortsighted tax and land use policies which discourage wise management. Scientific theories about the effects of atmospheric pollutants are much less tangible and in some cases have been greeted with some degree of skepticism.

Possible effects from point-source and regional air pollutants have been described in earlier chapters (McLaughlin and Norby, Chapter 4, this volume). The most serious effect, however, may be to reduce the tree's capacity for dealing with other stresses such as drought or insects (Winner et al., Chapter 7, this volume). This type of effect as well as reduced growth is hard to measure currently. Forest managers may therefore perceive such effects as largely beyond their control.

A first step in getting forest managers and forest tree breeders to deal with the potential problems is to demonstrate that they exist and that there is potential through genetic change to deal with changes in atmospheric chemistry. Forest management is a business and therefore demonstration of the problem must present economic analyses of growth losses or other damage and the return from investing in preventative breeding programs.

Management Opportunities

The management opportunities in forest systems can be viewed along two basic lines of thought. The first is to assess injury and its impact on forest health and productivity and the second is to manage forests and breeding programs to deal with pollution problems.

Assessment of Injury

The development of test procedures for understanding both injury and growth decline is critical. If seedlings or small clonal plants are used, correlations with older, larger trees must be developed. It is important to understand the impact of pollution at various growth stages so that tests accurately reflect potential impact. Studies should also include understanding the method and impact of damage during reproductive stages. Initial studies of this type have been described by Cox (1989) and Venne et al. (1989).

A second issue is the prediction of what pollutants will be important during the anticipated lifetime of forest trees. This means anticipating pollution conditions 10 to 100 years in the future. The suite of future pollutants is expansive and includes ozone, ultraviolet-B radiation, nitrogen oxides, carbon dioxide, heavy metals, and aspects of global change. Based on point source and regional effects, Roberts (Commentary to Chapter 4, this volume) pointed out that central European countries anticipate some pollutants to decline in the next 30 years as the region uses soft coal reserves. Sulfur dioxide will likely be a minor pollutant in the future because of government regulations on stack gas emissions. Therefore selection of future important pollutants is critical.

Management Alternatives

Management alternatives are primarily breeding for resistance to pollutants. Horizontal resistance is identified as the only logical choice for trees due to their long life cycle and the various stresses that they normally are exposed to during their life span. Karnosky (Chapter 11, this volume) discussed several programs in Europe that are currently developing trees resistant to large-scale point sources of pollution.

Forests should be composed of trees from the best adapted seed sources for a site. Matching both species and seed source to particular forest sites enhances the general vigor and health of a forest and thereby reduces stresses that may compound the potential effects of air pollution. It is important to recognize that air quality is now a factor of site quality for a forest just as climate and soils have been the traditional factors. The use of bioindicator plants to aid in monitoring pollution on a site to give some indication of site quality is discussed by Karnosky (Chapter 11, this volume).

Progeny and provenance tests of trees should be replicated in air pollution high hazard areas in order to test genetic responses to pollutants. These tests, similar to ones that have been done for disease (Sohn and Goddard 1979), provide assessments of performance but also field validation for short-term or laboratory testing programs. They may be especially helpful where pollution levels are at a high chronic state rather than at an acute level. Comparisons among sites, both low and high hazard areas, can provide good measures of the impact of chronic levels of pollutants.

In today's world, urbanization, pollution, and loss of diversity are reducing the large genetic resource base that has existed in forest ecosystems. This resource is the basis for future breeding and genetic change in trees and must be conserved before dramatic gene pool losses occur. Muhs (1989) described a model gene conservation program ongoing in the Federal Republic of Germany. Genes with low frequencies must be preserved as well as those genes that have higher frequencies. The genetic extremes of the population are at an even greater risk of destruction and programs should begin soon to protect these resources. Conservation of genetic resources should not be limited to species that are currently planted and/or bred but should extend to commercially important nonplanted species as well. Saving these genetic resources may be critical for biodiversity in future generations. Large-scale efforts of this type may require governmental action such as that in Germany.

An option for the future is biotechnology. There are several barriers to its use in the near term. These include: (i) finding acceptable genes and promoters to cause resistance to air pollution; (ii) developing techniques for inserting the new genes into a species; (iii) developing clonal techniques for commercial vegetative propagation; and (iv) overcoming the public outcry that may occur if genetically engineered trees are planted in the forest. To achieve public acceptance, trees will probably have to be sterile, forcing vegetative propagation of all trees to be released. Biotechnology is therefore more of a long-term option than one which can be used today. This should not, however, discourage research to identify genes for resistance and investigation of their mode of action for future use.

Finally, in addition to breeding, foresters must learn to deal with such indirect effects as reduction in soil fertility. Management options such as liming or fertilization must be examined as an option in maintaining site quality for existing or future stands. Management options are extremely important when one considers that the option to impact the genetic makeup of a stand occurs only once during a rotation. After the initial selection of what to plant, foresters must then use other silvicultural options to maintain a viable, healthy forest.

Summary

Intensively managed forests are a small part of the world's forest resource. Therefore, solutions to atmospheric changes such as planting resistant trees may have a relatively small impact on the overall growth and existence of the world's trees. Although the challenge may seem overwhelming it is important to begin management practices immediately to protect forests. These must include a mix of practices. First, the site must be matched to the species and seed sources. Second, the best management practices possible to reduce nonpollution stresses must be employed thereby reducing the possibility of interactions. Healthy, vigorous trees may have a higher probability of success in dealing with potential anthropogenic stresses. Third, test methods must be developed that predict a tree's response to pollutants throughout its life stages, including growth and reproductive fitness. Fourth, genetic tests must be utilized to incorporate horizontal resistance to pollution into the selected populations that will form the breeding and production populations for future managed forests; vertical resistance strategies are not on option. Fifth, the genetic resource in forest populations that exist today must be protected to avoid air pollution and other anthropogenic stresses from eliminating valuable resources. Finally, as Karnosky (Chapter 11, this volume) concluded, the development of pollution resistant plants should not be viewed as a permanent solution, but rather as a temporary action to bridge forests to a time when air pollution has been reduced at the source.

References

Barnard JE, Lucier AA, Johnson AH, Brooks RT, Karnosky DF, Dunn PH (1990) Changes in forest health and productivity in the United States and Canada. State of Science and Technology Report No. 16, National Acid Precipitation Assessment Program, Washington DC

Bones JT, Carey AE, Joyce LA, Schlatterer EF (1989) The national overview. In: An analysis of the land base situation in the United States: 1989–2040, Chapter 1. USDA Forest Service General Technical Report RM-181

Cox RM (1989) Natural variation in sensitivity of reproductive processes in some boreal forest trees to acidity. In: Scholz F, Gregorius HR, Rudin D (eds) Genetic effects of air Pollutants in forest tree populations. Springer-Verlag, Berlin, pp 77–88

Muhs HJ (1989) Measures for the conservation of forest gene resources in the Federal Republic of Germany. In: Scholz F, Gregorius HR, Rudin D (eds) Genetic effects of air pollutants in forest tree populations. Springer-Verlag, Berlin, pp 187–198

Sohn S, Goddard RE (1979) Influence of infection present on improvement of fusiform rust resistance in slash pine. Silvae Genetica 28:173–180

U.S. Forest Service (1980) An assessment of the forest and rangeland situation in the U.S. U.S. Department of Agriculture, Washington, DC

Venne H, Scholz F, Vornweg A (1989) Effects of air pollutants on reproductive processes of poplar (*Populus* spp.) and Scots pine (*Pinus sylvestris* L.). In: Scholz F, Gregorius HR, Rudin D (eds) Genetic effects of air pollutants in forest tree populations. Springer-Verlag, Berlin, pp 89–106

World Resources Institute (1987) An assessment of the resource base that supports the global economy. Basic Books, New York

Subject Index

Species Index